M. Wada, K. Shimazaki, M. Iino (Eds.)

Light Sensing in Plants

M. Wada · K. Shimazaki · M. Iino (Eds.)

Light Sensing in Plants

With 46 figures, including 4 in color

Springer The Botanical Society of Japan

Masamitsu Wada, Dr.
Department of Biology, Graduate School of Science, Tokyo Metropolitan University
1-1 Minami Osawa, Hachioji, Tokyo 192-0397, Japan

Ken-ichiro Shimazaki, Dr.
Department of Biology, Faculty of Science, Kyushu University
Ropponmatsu, Fukuoka 810-8560, Japan

Moritoshi Iino, Ph.D.
Botanical Gardens, Graduate School of Science, Osaka City University
2000 Kisaichi, Katano, Osaka 576-0004, Japan

Library of Congress Control Number: 2004117723

ISBN 4-431-24002-0 Springer-Verlag Tokyo Berlin Heidelberg New York

Springer is a part of Springer Science+Business Media
springeronline.com

Printed in Japan

Typesetting: SNP Best-set Typesetter Ltd., Hong Kong
Printing and binding: Hicom, Japan

Printed on acid-free paper

Preface

Plants utilize light not only for photosynthesis but also for monitoring changes in environmental conditions essential to their survival. Wavelength, intensity, direction, duration, and other attributes of light are used by plants to predict imminent seasonal change and to determine when to initiate physiological and developmental alterations. Most plants sense red/far-red light and blue light through photoreceptors: phytochromes detect red/far-red light, while there are several kinds of blue-light receptors, including cryptochromes, phototropins, and ZLP/FKF/LKP/ADO. The typical phytochrome responses known as red/far-red photoreversible phenomena were discovered in 1952 by Borthwick et al. and the phytochrome was characterized as a chromoprotein in 1959 by Butler et al. However, blue-light receptors were not identified until cryptochrome was found in 1993 by Cashmore's group. Now we are in an exceptional period of discovery of blue-light receptors such as phototropins, ZLP/FKF/LKP/ADO, and PAC in *Euglena*. Thus, it is very timely to publish this book on light sensing and signal transduction in plant photomorphogenesis written by leading scientists gathered at Okazaki from all over the world in June 2004. It was a great opportunity to discuss new discoveries in the field. It also marked the retirement of Prof. Masaki Furuya, who has contributed substantially to this field for many years.

This volume, published as part of the special-issue series of The Botanical Society of Japan, presents the advances made over the last 5 to 10 years in many of the related fields. Included are Prof. Furuya's "History and Insights" of plant photomorphogenesis, three overviews of the main photoreceptors, and Prof. Briggs' epilogue comparing the status of research in 1986 and 2004, when the XVI and the LVIII Yamada Conferences on plant photomorphogenesis were held at Okazaki. I believe that this book will prove indispensable and will contribute to the advancement of the study of photomorphogenesis.

I express my sincere gratitude to Yamada Science Foundation and to the executive members of the Foundation for their generosity, which made it possible for us to publish this book.

Masamitsu Wada

Contents

Part III Cryptochrome

Part IV Phototropin

Part V Other Photoreceptors

Part VI Signal Transduction

Part VII Photoperiodism and Circadian Rhythm

Part VIII Epilogue

Contributors

Ahmad, Margaret (p. 131)
Université de Paris VI, 4 Place Jussieu, Casier 156, 75005 Paris Cedex, France

Alvey, Richard M. (p. 299)
Department of Biology, Indiana University, 1001 East 3rd Street, Bloomington, IN 47405, USA

Balabas, Barbara E. (p. 299)
Department of Biology, Indiana University, 1001 East 3rd Street, Bloomington, IN 47405, USA

Balashov, Sergei P. (p. 213)
Department of Physiology and Biophysics, University of California Irvine, Irvine, CA 92697, USA

Bogomolni, Roberto A. (p. 147, 163)
Department of Chemistry and Biochemistry, University of California Santa Cruz, Santa Cruz, CA 95060, USA

Bouly, Jean-Pierre (p. 131)
Université de Paris VI, 4 Place Jussieu, Casier 156, 75005 Paris Cedex, France

Briggs, Winslow R. (p. 139, 147, 163, 357)
Department of Plant Biology, Carnegie Institution of Washington, 260 Panama Street, Stanford, CA 94305, USA

Brücker, Gerhard (p. 103)
Plant Physiology, Free University, Königin-Luise-Strasse 12–16, D-14195 Berlin, Germany

Casal, Jorge José (p. 285)
Faculty of Agronomy, IFEVA, University of Buenos Aires, Av. San Martín 4453, 1417 Buenos Aires, Argentina

Cashmore, Anthony R. (p. 121)
Plant Science Institute, Department of Biology, University of Pennsylvania, Philadelphia, PA 19104, USA

DENG, XING WANG (p. 253)
Department of Molecular, Cellular, and Developmental Biology, Yale University, New Haven, CT 06520-8104, USA

EBREY, THOMAS G. (p. 213)
Department of Biology, University of Washington, Seattle, WA 98195, USA

FANKHAUSER, CHRISTIAN (p. 277)
Department of Molecular Biology, University of Geneva, Sciences III, 30 quai E. Ansermet, CH-1211 Geneva 4, Switzerland

FARRENS, DAVID L. (p. 203)
Biochemistry and Molecular Biology, Oregon Health and Science University, Portland, OR 97239-3098, USA

FENG, SUHUA (p. 253)
Department of Molecular, Cellular, and Developmental Biology, Yale University, New Haven, CT 06520-8104, USA

FOLTA, KEVIN M. (p. 239)
Plant Molecular and Cellular Biology Program and Horticultural Sciences Department, University of Florida, Gainesville, FL 32611, USA

FURUYA, MASAKI (p. 3)
Retired, the University of Tokyo in 1987, Riken Frontier Research Program in 1992, and Hitachi Advanced Research Laboratory in 2001
Permanent address: 6-2-10 Kugahara, Ota-ku, Tokyo 146-0085, Japan

GÄRTNER, WOLFGANG (p. 307)
Max-Planck-Institute for Bioinorganic Chemistry, Stiftstr. 34–36, D-45470 Mülheim, Germany; and Department of Organic Chemistry, University of Duisburg Essen, Duisburg, Germany

GIOVANI, BALDISSERA (p. 131)
Université de Paris VI, 4 Place Jussieu, Casier 156, 75005 Paris Cedex, France

HÄDER, DONAT-PETER (p. 223)
Plant Ecophysiology, Friedrich-Alexander University, Staudtstrasse 5, D-91058 Erlangen, Germany

HAGA, KEN (p. 269)
Botanical Gardens, Graduate School of Science, Osaka City University, Kisaichi, Katano, Osaka 576-0004, Japan

HASUNUMA, KOHJI (p. 315)
Kihara Institute for Biological Research, Graduate School of Integrated Science, Yokohama City University, Maioka-cho, Totsuka-ku, Yokohama 244-0813, Japan

HEBERLE, JOACHIM (p. 155)
Forschungszentrum Jülich, IBI-2: Structural Biology, D-52425 Jülich, Germany

Hertel, Rainer (p. 231)
Institut Biologie III, Albert-Ludwigs-Universitaet, Schaenzlestrasse 1, D-79104 Freiburg, Germany

Hughes, Jon (p. 103)
Plant Physiology, Justus Liebig University, Zeughaus, Senckenbergstrasse 3, D-35390 Giessen, Germany

Iino, Moritoshi (p. 269)
Botanical Gardens, Graduate School of Science, Osaka City University, Kisaichi, Katano, Osaka 576-0004, Japan

Inagaki, Noritoshi (p. 111)
Department of Plant Physiology, National Institute of Agrobiological Sciences, Tsukuba, Ibaraki 305-8602, Japan

Iseki, Mineo (p. 223)
National Institute for Basic Biology, Okazaki, Aichi 444-8585, Japan; and PRESTO, Japan Science and Technology Agency, Kawaguchi, Saitama 332-0012, Japan

Izawa, Takeshi (p. 333)
National Institute of Agrobiological Sciences, Plant Molecular Genetics Group, Tsukuba, Ibaraki 305-8602, Japan

Kehoe, David M. (p. 299)
Department of Biology, Indiana University, 1001 East 3rd Street, Bloomington, IN 47405, USA

Kim, Jeong-Il (p. 57)
Kumho Life and Environmental Science Laboratory, Gwangju 500-712, South Korea; and Environmental Biotechnology National Core Research Center, Gyeongsang National University, Jinju 660-701, South Korea

Kohchi, Takayuki (p. 37)
Graduate School of Biostudies, Kyoto University, Kyoto 606-8502, Japan

Lariguet, Patricia (p. 277)
Department of Molecular Biology, University of Geneva, Sciences III, 30 quai E. Ansermet, CH-1211 Geneva 4, Switzerland

Lee, Bumkyu (p. 315)
Kihara Institute for Biological Research, Graduate School of Integrated Science, Yokohama City University, Maioka-cho, Totsuka-ku, Yokohama 244-0813, Japan

Li, Lina (p. 299)
Department of Biology, Indiana University, 1001 East 3rd Street, Bloomington, IN 47405, USA

Lin, Chentao (p. 325)
Department of Molecular, Cell, and Developmental Biology, University of California, Los Angeles, CA 90095, USA

Maeda, Akio (p. 213)
Department of Biochemistry, University of Illinois at Urbana/Champaign, Urbana, IL 61801, USA; and Department of Biology, University of Washington, Seattle, WA 98195, USA

Mittmann, Franz (p. 103)
Plant Physiology, Justus Liebig University, Zeughaus, Senckenbergstrasse 3, D-35390 Giessen, Germany

Nagatani, Akira (p. 69)
Department of Botany, Graduate School of Science, Kyoto University, Sakyo-ku, Kyoto 606-8502, Japan

Nagy, Ferenc (p. 79, 261)
Plant Biology Institute, Biological Research Center, Temesvari krt. 62, H-6726 Szeged, Hungary

Nakasako, Masayoshi (p. 43)
Department of Physics, Faculty of Science and Technology, Keio University, 3-14-1 Hiyoshi, Kohoku-ku, Yokohama 223-8522, Japan; and RIKEN Harima Institute, Sayo, Hyogo 679-5148, Japan

Ni, Min (p. 293)
Department of Plant Biology, University of Minnesota, St. Paul, MN 55108, USA

Ntefidou, Maria (p. 223)
Plant Ecophysiology, Friedrich-Alexander University, Staudtstrasse 5, D-91058 Erlangen, Germany

Quail, Peter H. (p. 21)
Plant Gene Expression Center, 800 Buchanan Street, Albany, CA 94710, USA; and Department of Plant and Microbial Biology, University of California, Berkeley, CA 94720, USA

Repp, Alexander (p. 103)
Plant Physiology, Free University, Königin-Luise-Strasse 12–16, D-14195 Berlin, Germany

Ritz, Thorsten (p. 243)
Department of Physics and Astronomy, Frederick Reines Hall, University of California, Irvine, CA 92697-4575, USA

Rüdiger, Wolfhart (p. 171)
Department Biologie I (Botanik), Ludwig-Maximilians-Universität, Menzinger Strasse 67, D-80638 Munich, Germany

SAKAI, TATSUYA (p. 179)
RIKEN Plant Science Center, Yokohama, Kanagawa 230-0045, Japan

SCHÄFER, EBERHARD (p. 79, 261)
Institute of Botany, University of Freiburg, Schaenzlestrasse 1, D-79104 Freiburg, Germany

SEIB, LAURA (p. 299)
Department of Biology, Indiana University, 1001 East 3rd Street, Bloomington, IN 47405, USA

SHARROCK, ROBERT (p. 51)
Department of Plant Sciences and Plant Pathology, Montana State University, Bozeman, MT 59717, USA

SHIMAMOTO, KO (p. 339)
Laboratory of Plant Molecular Genetics, Nara Institute of Science and Technology, 8916-5 Takayama, Ikoma 630-0101, Japan

SHIMAZAKI, KEN-ICHIRO (p. 185)
Department of Biology, Kyushu University, Ropponmatsu 4-2-1, Fukuoka 810-8560, Japan

SHINOMURA, TOMOKO (p. 111)
Hitachi Central Research Laboratory, Hatoyama, Saitama 350-0395, Japan

SINESHCHEKOV, VITALY A. (p. 95)
Physicochemical Biology, M.V. Lomonosov Moscow State University, Moscow 119899, Russia

SOMERS, DAVID E. (p. 347)
Department of Plant Cellular and Molecular Biology/Plant Biotechnology Center, Ohio State University, Columbus, OH 43210, USA

SONG, PILL-SOON (p. 57)
Kumho Life and Environmental Science Laboratory, Gwangju 500-712, South Korea; Environmental Biotechnology National Core Research Center, Gyeongsang National University, Jinju 660-701, South Korea; and Department of Chemistry, University of Nebraska, Lincoln, NE 68588-0304, USA

STOWE-EVANS, EMILY L. (p. 299)
Department of Biology, Indiana University, 1001 East 3rd Street, Bloomington, IN 47405, USA
Current address: Biology Department, Bucknell University, Lewisburg, PA 17837, USA

SWARTZ, TREVOR E. (p. 147, 163)
Department of Chemistry and Biochemistry, University of California Santa Cruz, Santa Cruz, CA 95060, USA; and Department of Plant Biology, Carnegie Institution of Washington, Stanford, CA 94305, USA

TAKAGI, SHINGO (p. 87)
Department of Biology, Graduate School of Science, Osaka University, Toyonaka, Osaka 560-0043, Japan

TAKANO, MAKOTO (p. 111)
Department of Plant Physiology, National Institute of Agrobiological Sciences, Tsukuba, Ibaraki 305-8602, Japan

TOKUTOMI, SATORU (p. 43)
Research Institute for Advanced Science and Technology, Osaka Prefecture University, 1-2 Gakuen-cho, Sakai, Osaka 599-8570, Japan

WADA, MASAMITSU (p. 193)
Department of Biological Science, Graduate School of Sciences, Tokyo Metropolitan University, Hachioji, Tokyo 192-0397, Japan; and Division of Photobiology, National Institute for Basic Biology, Okazaki, Aichi 444-8585, Japan

WATANABE, MASAKATSU (p. 223)
National Institute for Basic Biology, Okazaki, Aichi 444-8585, Japan; and Department of Photoscience, School of Advanced Sciences, Graduate University for Advanced Studies, Shonan Village, Hayama, Kanagawa 240-0193, Japan

XIE, XIANZHI (p. 111)
Department of Plant Physiology, National Institute of Agrobiological Sciences, Tsukuba, Ibaraki 305-8602, Japan

YANAGAWA, YUKI (p. 253)
Department of Molecular, Cellular, and Developmental Biology, Yale University, New Haven, CT 06520-8104, USA

YOKOI, SHUJI (p. 339)
Laboratory of Plant Molecular Genetics, Nara Institute of Science and Technology, 8916-5 Takayama, Ikoma 630-0101, Japan

YOSHIDA, YUSUKE (p. 315)
Kihara Institute for Biological Research, Graduate School of Integrated Science, Yokohama City University, Maioka-cho, Totsuka-ku, Yokohama 244-0813, Japan

YU, XUHONG (p. 325)
Department of Molecular, Cell, and Developmental Biology, University of California, Los Angeles, CA 90095, USA

Zeidler, Mathias (p. 103)
Plant Physiology, Justus Liebig University, Zeughaus, Senckenbergstrasse 3, D-35390 Giessen, Germany

Part I
Prologue

Chapter 1
History and Insights

Masaki Furuya[1,2]

Genesis (Legend to 1950s)

Human beings have always relied on plants to provide their staple foods and raw materials for diverse tools, and since prehistoric times must have known that sunlight greatly influences plant development and reproduction. From the Renaissance onwards, careful observations of nature led to a growing awareness that both higher and lower plants respond variously to light in terms of irradiation dosage for photosynthesis, direction for phototropism, timing and duration for photoperiodism, and wavelengths for photomorphogenesis. Joseph Priestley (1772) discovered that green plants utilize light as their source of energy for the production of complex organic substances. Julius Sachs (1864) demonstrated that only the blue region of visible light resulted in phototropic bending of plants. Charles Darwin and his son (1881) carried out a pioneering experiment on light-signal transduction of phototropism, in which they separated the photoreceptive site from the responding growth region in monocot seedlings. In 1910, Georg Klebs gathered a lot of evidence that the environmental light greatly influences growth and development of seed plants and ferns. However, the molecular basis of light perception and signal transduction in plants was not elucidated until quite recently.

The physiological capacity of plants to adjust processes throughout their life cycle to the seasonal change of environment is crucial for their survival. Julien Tournois (1914), a graduate student of the École Normale Supérieure in Paris, discovered that night length rather than day length was the determining factor for flowering time of his experimental material, Japanese hop. Wightman Garner and Harry Allard (1920) at the Arlington Farm of USDA carried out comprehensive experiments on flowering time in several plants by changing the night length using three dark houses. They discovered that most of the plants tested

[1] *Retired*, the University of Tokyo in 1987, Riken Frontier Research Program in 1992, and Hitachi Advanced Research Laboratory in 2001 (see Furuya 2004)
[2] Permanent address: 6-2-10 Kugahara, Ota-ku, Tokyo 146-0085, Japan

could be classified as "short-day" or "long-day" plants, and established the concept of photoperiodism. Karl Hamner and James Bonner (1938) made a decisive contribution to photoperiodism research by finding that a brief exposure of light in mid-night, given under normally inductive conditions for flowering, caused cocklebur, a short-day plant, to remain completely vegetative.

Recognition that many responses of plants to light have a common underlying cause came from the measurement of action spectra using a custom-built spectrograph (Parker et al 1949). The year 1952 was a momentous year in the history of plant photomorphogenesis, because Harry Borthwick and his colleagues of USDA in Beltsville discovered the red (R) and far-red (FR) photoreversible effect on seed germination in lettuce and night-break of photoperiodic floral induction in cocklebur (Borthwick et al 1952). They soon formulated the unique idea that reversible changes in the optical density of appropriate tissues might result from irradiating the sample alternately with actinic R and FR light. This hypothesis was proved by Warren Butler, Karl Norris, Bill Siegelman and Sterling Hendricks (1959), who showed repeatedly photoreversible absorption changes at 660 and 730 nm regions upon alternately given R and FR actinic light in etiolated maize tissues and a crude extract of the relevant proteinaceous pigment. Shortly after this discovery, the term "phytochrome" was half-jokingly used by Butler in their laboratory, then published by Borthwick and Hendricks (1960). It is remarkable that the members of the same institution discovered all key phenomena such as the photoperiodism, the R/FR reversible effect and the photoreceptor phytochrome (Sage 1992).

The Era of Spectrophotometry, Physiology, and Biochemistry (1960s–1980s)

Photoreversible Regulation and Molecular Properties of Phytochrome

The discoverers of phytochrome had proposed a simple hypothesis that phytochrome in its red light absorbing form (P_r) is physiologically inactive, and is only active in its far-red absorbing form (P_{fr}). In the following few years, they attempted to prove this hypothesis photometrically and biochemically (Siegelman and Butler 1965), but the puzzle did not prove to be simple. P_{fr} was found to undergo non-photochemical transformations in vivo such that both P_{fr} decay and P_{fr} reversion to P_r took place in the dark (Butler et al 1963). However, in crude extracts, P_{fr} showed neither decay nor reversion, and P_r and P_{fr} appeared quite stable in vitro (Furuya et al 1965). After a dual-wavelength difference spectrophotometer, Ratiospect R2, became commercially available in 1963, several laboratories began to measure phytochrome in vivo to examine the correlation of photoreversible responses of plants to R and FR light with photometrically measured phytochrome content, initial P_{fr} state and dark transformation of P_{fr} in vivo. However, most of these attempts failed to find any correlation (Hillman 1967). This presented an obstacle, which persisted for some time, and is reflected in the fact that the number of publications on spectrophotometric measurements of phytochrome in vivo reached a plateau of ca. 20 papers/year by 1966.

In an alternative approach, workers were attempting to clarify the structure and molecular properties of phytochrome. The Beltsville group initially developed a procedure for the isolation and purification of phytochrome, finding its average molecular weight as a monomer to be ca. 40 kilodaltons (kDa). However, other larger forms of phytochrome, including degraded "small" (<60 kDa) and undegraded "large" (114–118 kDa) phytochromes were subsequently discovered (Briggs and Rice 1972, Pratt 1982), culminating in the isolation of full-length "native" (124 kDa) phytochrome by Vierstra and Quail (1982). In parallel with these efforts, Wolfhart Rüdiger and his colleagues spent two decades engaged in determining the nature of the phytochrome chromophore, and were finally able to describe the chemical structure of phytochromobilin in both P_r and P_{fr} forms (Rüdiger et al 1983). Lagarias and Rapoport (1980) discovered the structure of the A ring of phytochromobilin and demonstrated the manner of its linkage to the phytochrome peptide. Since the 1970s, Pill-Soon Song has developed his model of phytochrome molecules in terms of photoreversible change of the chromophore topography between P_r and P_{fr} and inter-domain crosstalk between the chromophore and the apoprotein (Park et al 2000, Chapter 6).

Although phytochrome was long believed to be easily extractable from plant tissues using a simple buffered solution, Rubinstein et al (1969) provided an evidence for bound phytochrome fraction in oat cells. Quail et al (1973) found that the pelletability of phytochrome from crude extracts was enhanced by a brief irradiation of etiolated tissues with R light. Using immunocytochemistry, Mackenzie et al (1975) observed a photoreversible redistribution of P_{fr} sequestering in the cytoplasm, but were not able to demonstrate the physiological significance of this process. In contrast, Wolfgang Haupt (1970) clearly demonstrated a role for membrane-bound phytochrome in chloroplast movement in *Mougeotia* using a microbeam irradiation technique.

During this era, evidence accumulated in support of the existence of two physiologically, photometrically, and immunochemically distinct phytochrome pools controlling R/FR reversible reactions in higher plants. Namely, "labile" type I phytochrome (phyI) is synthesized as P_r in the dark and P_{fr} is destroyed rapidly in the light, whereas "stable" type II phytochrome (phyII) is produced constitutively and stays in cells for longer time irrespective of the light conditions (Furuya 1993). In fact, the hottest issue in the Yamada Conference held at Okazaki in 1986 (Furuya 1987) was "green" phytochrome. At the enthusiastic request of the participants, an extra session was organized, in which Yukio Shimazaki from the Pratt laboratory, Jim Tokuhisa from the Quail laboratory, and Hiroshi Abe from my laboratory told their latest stories on biochemically and immunochemically distinguishable phytochromes from etiolated and "green" tissues. Further, during the past several decades, researchers in Wageningen genetically isolated many photomorphogenic mutants, including a cucumber long hypocotyl mutant (*lh*) that was immunochemically determined to be a phyII-deficient mutant (López-Juez et al 1992).

A Period of Groping in Studies on Photomorphogenesis

From the 1960s to the 1980s, only phytochrome was the known photoreceptor for photomorphogenesis, and its action could only be recognized in R/FR photoreversible, low fluence (LF) responses. During this period, researchers had become aware that plants respond to light in a variety of other ways, but the corresponding photoreceptor pigments were not known.

In many early studies, we suffered from a significant effect on photomorphogenesis of the extremely dim "green safe light" used in dark rooms, which did not cause significant change of spectrophotometrically measured phytochrome in vivo. To avoid this effect and to prepare totally etiolated samples, we had to grow plants in lightproof aeration boxes. Blaauw et al (1968) found that red light of very low fluence (VLF) inhibited growth in Avena seedlings, and that this effect was not reversed by far-red light. Similar reports about VLF effects in etiolated plants increased time being, but further analysis was technically very difficult in those days.

Hans Mohr and his colleagues in Freiburg had extensively investigated the effect of blue and far-red light on photomorphogenesis in terms of sensor pigments, signal amplification, and gene expression, and established the concept of the High Energy Reaction (Mohr and Schäfer 1983), which was later renamed the high irradiance reaction (HIR). In a crucial experiment using bichromatic light, Karl Hartmann (1966) was able to show that although the HIR does not show R/FR reversibility and does not obey the reciprocity law, it is indubitably mediated by phytochrome.

Since the early report of Sachs (1864), blue and near-UV light effects on development and metabolism were widely documented in the plant kingdom and microbes (Senger 1980), but at this time we understood little about the photoreceptor pigments for these phenomena. One of the reasons for this frustrating situation was that plant cells contain a number of natural compounds that absorb light in the blue and/or near-UV spectral regions. Using only the conventional spectrophotometric, biological, and biochemical methods of the day, it was very difficult to identify any of them as photoreceptors for specific phenomena.

It is our good fortune that we can look back at the early history of phytochrome studies in the book by Linda Sage (1992) and of photomorphogenesis in the proceedings of symposia (Mitrakos and Shropshire 1972, Smith 1976 1983, De Greef 1980, Furuya 1987, Thomas and Johnson 1990), an encyclopedia (Shropshire and Mohr 1983), and other more advanced treatises (Kendrick and Kronenberg 1994).

The Era of Molecular Genetic Approaches (1990s)

Differential Photoperception by Phytochromes

The year 1989 was another turning point for phytochrome research, because of the discoveries of the phytochrome gene family by Bob Sharrock and Peter Quail (1989) and of the *det* mutant, which caused morphogenesis to follow the photomorphogenic path in complete darkness, by Joanne Chory and her collaborators (1989). These findings caused a great sensation among us and provided new questions about whether individual phytochrome family members have discrete physiological or photosensory functions, and whether each has a discrete primary mechanism of action and a unique signal transduction pathway.

To answer these questions, molecular genetic approaches using Arabidopsis mutants soon became a main highway in this field during the 1990s, while transgenic overexpression of each phytochrome gene (*PHY*) proved to be less fruitful. Individual phytochrome photoreceptor mutants were reported in 1993, and phytochrome A (phyA) null mutant (*phyA*) and phytochrome B (phyB) null mutant (*phyB*) were soon being extensively used. One of Maarteen Koornneef's Arabidopsis mutants, *hy3* (Koornneef et al 1980), was found to have mutations in the *PHYB* gene by Reed et al (1993), whereas *hy1* and *hy2* were shown to be chromophore-deficient mutants. Several different groups screened mutant seedlings under continuous FR light and identified *phyA* mutants, finding that *phyA*-null mutants of Arabidopsis display a WT phenotype in white light (Whitelam et al 1993), and that phyA and phyB showed overlapping functions in Arabidopsis development (Reed et al 1994). Despite the apparently unique photoperception of phyA and phyB under continuous irradiation with FR light (cFR) and R light (cR) respectively, evidence soon accumulated for redundancy between phyA and phyB effects and for mutual antagonism between the actions of these phytochromes (Whitelam and Devlin 1997). Fifteen *PHYA*-regulated genes identified by fluorescent differential display screen were expressed photoreversibly by R/FR exposures, suggesting redundancy among phyA, phyB, and other phyII type phytochromes (Kuno et al 2000).

Using the relevant Arabidopsis *phyA* and *phyB* mutants, Shinomura et al (1996) determined separate action spectra for phyA- and phyB-specific induction of seed germination at Okazaki large spectrograph. We discovered that phyA induces seed germination photo-irreversibly in response to VLF light in the range 300–780 nm, while phyB regulates germination in R/FR reversible manner of LF light, identical with the result by Borthwick et al (1952). The classic HIR is now known to include phyA-, phyB-, and blue-UV photoreceptor-mediated HIRs. Shinomura et al (2000) found that the phyA-HIR can in fact be replaced by intermittent irradiation with FR pulses if given at intervals of 3 min for 24 h, and that the action spectra for phyA-HIR determined by such intermittent treatment of 300–800 nm lights using Arabidopsis WT, *phyB*-, and *phyAphyB*-mutants had peaks at blue and FR regions and was very similar to the action spectra constructed for the HIR in *Sinapis* (Mohr and Schäfer 1983). Very similar differential photoperception by phytochromes was recently shown in rice using *phyA*-, *phyB*- and *phyC*-mutants (Chapter 12).

In addition to photoreceptor mutants, putative mutants for early steps in light signal transduction were isolated in Arabidopsis in the laboratories of Peter Quail (Chapter 2), Nam-Hai Chua (Bolle et al 2000) and several others. These mutants were characterized for their epistasis with *phyA* and *phyB* mutations, allowing some (FHY1, FHY3, FAR1, and PAT1) to be assigned to phyA signaling and others (PEF2, PEF3, and RED1) to phyB, while a third group (PIF3, PSI2, and PEF1) could be assigned to both (see review by Hudson 2000). However, it seems too early to assemble the entire phytochrome signaling pathway upon these mutant studies. The constitutively de-etiolate mutants, *cop/det/fus*, mimic the phenotype of light-grown seedlings when grown in the dark and appear to act at later stages of light signal transduction in association with the COP1/COP9 signalosome (Chapter 29).

Apart from the mutant analyses described above, a new field of phytochrome signaling studies was born in this era, based on the growing recognition that light-induced nuclear import of cytosolic phytochromes is a multi-step signaling process. The first evidence came from the demonstration by immunocytochemistry and *PHYB::GUS* transgenic techniques that phyB was translocated into the nucleus under cR (Sakamoto and Nagatani 1996). This observation has subsequently been extended to all five Arabidopsis phytochromes, using *PHYA-E::GFP* fusion proteins in transgenic plants (Nagy and Schäfer 2002), and indicates the importance of phytochromes in the control of gene expression. The intracellular distribution of native phytochromes has also been observed using cryosectioning and immunochemical staining techniques at the optical (Hisada et al 2000) and electron microscope (Hisada et al, 2001) levels. Another victory in this era was the successful chemical synthesis of phytochromobilin and its diverse derivatives by the group of Katsuhiko Inomata in Kanazawa, enabling us at long last to analyze the relationship between chromophore structure and phytochrome function in vitro and in vivo (Hanzawa et al 2001, 2002).

Thanks to recent genome projects, phytochrome-related proteins have been discovered in cyanobacteria and eubacteria, and this has opened new avenues for investigating biliprotein photosensory function and the evolution of phytochromes in the entire plant kingdom (Montgomery and Lagarias 2002, Chapter 3). The diversity of phytochrome gene families reflects the diverse evolutionary histories of plants, and it would be of interest to investigate a possible relationship between the most functionally advanced phytochrome, phyA, and the evolutionary emergence of seed plants.

Discovery of Blue Light Photoreceptor Pigments

Every meeting on plant photomorphogenesis during the 1970s and 1980s consisted of two major sessions, respectively dealing with phytochrome and blue-UV absorbing pigments. In the latter of these sessions we had long been frustrated with our inability to identify photoreceptor pigments. However, in 1993 Margaret

Ahmad and Tony Cashmore have opened this heavy door using one of Koornneef's Arabidopsis mutants, *hy4*, which was defective in blue light-dependent photomorphogenesis. They isolated a T-DNA tagged *hy4* allele, which allowed the cloning of the *HY4* gene (Ahmad and Cashmore 1993). The protein encoded by *HY4* was a member of the photolyase family and was named cryptochrome (cry). Chentao Lin and colleagues (1996) cloned and characterized a second member of the cry family containing a distinct C-terminal sequence, which named cry2, and the *HY4*-encoded cry renamed as cry1. Since that time, Arabidopsis cryptochromes have been shown to be nuclear proteins that mediate light control of stem elongation, leaf expansion, photoperiodic flowering, and the circadian clock (Chapters 13, 14, 38).

Jiten Khurana and Ken Poff (1989) isolated several Arabidopsis mutants specifically defective in phototropic responses. Winslow Briggs and his colleagues cloned and characterized genes (*NPH1–4*) of these mutants, and showed that the gene product of *NPH1* was a blue light receptor, which was renamed phototropin 1 (Chapter 15). Phototropin research is the most rapidly moving area of photomorphogenesis research at the moment (Chapters 15–22).

The most recently discovered blue photoreceptor, FKF1, is essential for photoperiodic-specific light signaling in Arabidopsis (Imaizumi et al 2003). Looking through the literature of blue light effects and pertinent pigments (Table 1), it is quite likely that we will find other blue light receptors in future.

Problems and Dreams

A Working Hypothesis of Phytochrome Actions

The recent rapid progress of molecular genetic approaches to the study of phytochrome has increased our knowledge enormously, but I feel that we are still sailing on a boat cast adrift on a dark ocean (Furuya 2004). To get out of this situation, we need to provide a marine chart for further sailing. Let us try to draw a chart using the accumulated evidence about the different modes of photoperception by phytochromes. Here I present a tentative chart (Figure 1) as a model for discussion, assuming that: (1) all phytochromes are synthesized as P_r in cytoplasm; (2) upon light irradiation, all phytochromes produce "functionally indistinguishable P_{fr}" as the active form, and differential functional activities among their gene family members arise from different kinetics of intracellular P_{fr} translocation; (3) phytochrome degradation occurs mainly in nucleus; (4) VLF light is sufficient for photoconversion of $phyI_r$ to $phyI_{fr}$, kI_1, whereas that of kII_1 requires LF; (5) $kI_3 >>>> kI_4$; most $phyI_{fr}$ binds to a hypothetical carrier protein(s) very soon after its photoconversion to P_{fr}, so that only a minimal amount of $phyI_{fr}$ remains in cytoplasm; (6) in contrast, $kII_3 <<< kII_4$; the binding affinity of $phyII_{fr}$ to the carrier is significantly low, so the majority of $phyII_{fr}$ stays in cytoplasm for a long time, and results in a slow escape reaction; (7) the affinity of the carrier proteins to phytochromes is speculated from physiological

TABLE 1. Identification of photoreceptors for blue and UV-A light-dependent phenomena in plants (after Table 2 in Wada and Kadota 1989, with additions)

Organism Phenomena	Photoreceptors	References
Anthophyta		
Stem elongation	cry1, cry2	Ahmad and Cashmore 1993, Folta and Spalding 2001, Chapter 13
	phyA	Shinomura et al 2000
	phot1	Folta and Spalding 2001
Leaf expansion	cry2	Lin et al 1998
	phot1, phot2	Sakai et al 2001, Sakamoto and Briggs 2002, Chapter 15
Phototropism	phot1, phot2	Huala et al 1997, Sakai et al 2001, Chapter 15
Chloroplast relocation		
Accumulation response	phot1, phot2	Sakai et al 2001, Chapter 22
Avoidance response	phot2	Kagawa et al 2001, Chapter 22
Stomata opening	phot1, phot2	Kinoshita et al 2001, Chapter 21
Circadian clock	cry1, cry2	Somers et al 1998, Devlin and Kay 2000, Chapters 38–41
Photoperiodic flowering	cry2	Guo et al 1998, Chapters 38–41
	FKF1	Imaizumi et al 2003
Cytosolic Ca^{2+} increase	phot1, phot2	Baum et al 2001, Harada et al 2003
Ca^{2+} current	phot1, phot2	Stoelzle et al 2003
Anthocyanin synthesis	cry1, cry2	Jackson and Jenkins 1995
Pteridophyta		
Spore germination		
Protonema elongation		
Phototropism	phot?, phy3	Kawai et al 2003
Polarotropism	phot?, phy3	Kawai et al 2003
Apical swelling		
Cell cycle regulation (G1 phase)		
Chloroplast relocation		
Accumulation response	phot?, phy3	Kawai et al 2003, Chapter 22
Avoidance response	phot2, phy3	Kagawa et al 2004, Kawai et al 2003, Chapter 22
Membrane potential		
Bryophyta		
Phototropism		
Polarotropism		
Chloroplast movement	photA, photB	Kasahara et al 2004, Chapter 22
Branching	cry1a, cry1b	Imaizumi et al 2002
Chlorophyta		
Hair whorl formation		
Cap formation		
Chloroplast movement		
Vaucheriophyta		
Growth promotion		
Phototropism		
Apical swelling		
Branching		
Chloroplast movement		
Cortical fiber reticulation		
Electric current		

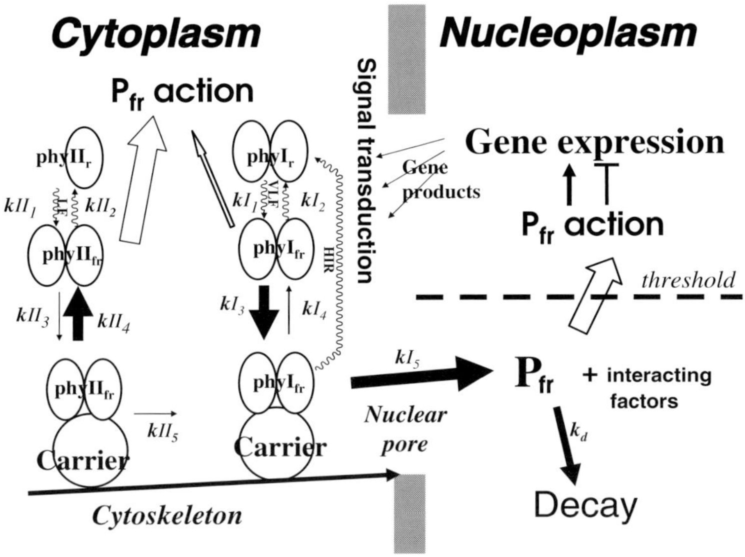

FIG. 1. A model for the P_{fr} action accommodates all differential photoperception modes, VLFR and HIR by type I (labile) and LFR by type II (stable) phytochromes. See text for details. *Wavy lines with arrows*, phototransformation of phytochrome; *Arrows*, Elementary process in cells. The size of arrow indicates speculated amounts of phyI- and phyII-flows, but does not mean faster or slower speed of each process

results to be phyA$_{fr}$ >> phyC$_{fr}$ >>> phyB$_{fr}$, and no P_r of any of the phytochromes binds to the carriers; and (8) $kI_5 = kII_5$, or similar rate; these P_{fr}-carrier complexes would transfer to nucleus at the same or similar speed along the cytoskeleton (Smith and Raikhel 1999).

The evidence that the peaks of action spectra for VLFR and LFR are essentially the same as those of the absorption spectrum of P_r strongly suggests that the both reaction would initiate from the phototransformation of P_r to P_{fr}, and that the difference between VLFR and LFR is the required amount of P_{fr}. The model (Figure 1) explains why only a small amount of phyI$_{fr}$ is enough to exceed the threshold level in the nucleus, while a higher level of phyII$_{fr}$ in the cytoplasm to support the required level of nuclear import. As discussed for a long time by Freiburg workers and others (Kendrick and Kronenberg 1994), real HIR processes would probably be more complicated than the scheme in Figure 1. If, however, phyA-HIR occurs anyway in this cycle, this model not only can account for all three modes of phytochrome photoperception, VLF, LF, and HIR, but also can explain why type I phytochrome is labile while type II stable. This model also explains why no major overall differences have been observed between *PHYA*- and *PHYB*-overproducers in Arabidopsis.

In this model, we expect photoreversible effects of all phytochromes, though VLFR was reported as photo-irreversible (Shinomura et al 1996). However, VLFR could be photoreversible if plants are exposed to extremely short R and FR pulses, and we have indirect evidence to support this idea. The reciprocity law holds when Arabidopsis seed germination is induced by exposure to 760 nm light of 5 μmol m^{-2} for 3 s or longer, but not if exposure times are less than 3 s (Shinomura, unpublished), indicating an involvement of some slow rate-limiting process such as the interaction with carriers.

The NH_2-terminal chromophoric domain (N-domain) of phyA alone is light stable in transgenic Arabidopsis (Wagner et al 1996), probably because it cannot bind as monomer to the carriers. In contrast, the COOH-terminal domain (C-domain) of phyB exists as dimer in vivo and when fused with *GUS* (Sakamoto and Nagatani 1996) or *GFP* (Chapter 7) translocates into nucleus irrespective of the light conditions. Both phyA_{fr} (Wagner et al 1996) and phyB_{fr} (Chapter 7) can only induce their biological effects as dimers.This evidence, together with the fact that the N-domain contains the determinants for the differences in photosensory specificity and photolability between phyA and phyB (Quail 1997) suggests a possibility that differential nuclear import of phytochromes could result from the N-domain dependent change of surface properties of C-domain in terms of hydrophobicity and reactability. In such a case, the C-domains of all phytochromes in P_r form would be so hydrophilic that they stay in cytosol, whereas the C-domain of phyI_{fr} is most hydrophobic and that of phyII_{fr} is less hydrophobic, so they interact with other proteins accordingly. However, we have no idea at present whether only P_{fr}–P_{fr} homodimer can bind with the carrier, or whether P_r–P_{fr} heterodimer is also translocatable to nucleus (Furuya and Schäfer 1996).

Phytochrome effects clearly show a great variation in the lag period between light exposure and the onset of detectable responses in plants, from 2.5 s (Chapter 9) to several hours, and even days (Table 4 in Furuya 1968) and in the escape rate in photoreversible reactions, from a few minutes to many hours (Table 5 in Furuya 1968). From these observations, I assume that there are two essentially different sites of phytochrome primary action; the cytoplasm and the nucleoplasm (Figure 1). Phytochrome action in the cytoplasm rapidly regulates cytoplasmic properties (Chapter 9), while its action in the nucleus occurs more slowly through up- or down-regulation of gene expression (Chapter 2). In this respect, it would be interesting to know where, when and how each phytochrome interacts with PIF3 (Chapter 30), NDPK2 (Im et al 2004), and other interacting factors (Chapter 29).

Despite many attempts since its discovery, none has yet succeeded to develop an in vitro assay system for the primary action of phytochrome molecules. The model (Figure 1) suggests it may not be so easy to find such an assay system, but it could be achieved if it would allow us to identify the hypothetical carrier protein(s) chemically, and to carry out binding assays of the identified carrier protein(s) with phyA_{fr}, phyB_{fr}, or any others by an affinity sensor.

Crosstalk of Light-, Clock-, and Hormone-Dependent Signaling

The overlapping effects among phytochrome family members are widely observed in plants, and the model in Figure 1 will give a hint of candidate sites for their crosstalk. Cryptochromes also may act by interacting with phytochromes, COP1, and clock proteins (Chapters 13, 33 and 38). Interaction between signal transduction pathways from phytochrome and phototropins is evident (Figure 1 in Chapter 22). Besides light, plants respond to other physical stimuli like gravity for which signaling pathways are also likely to involve crosstalk with light signaling (Chapter 32). Light signaling pathways interact widely and diversely with the circadian clock in not only eukaryotes but also prokaryotes. Several models for crosstalk between downstream phytochrome signaling and the clock are proposed in other chapters of this book (Chapters 38–41).

During the last century, plant physiologists spent enormous time and energy to understand the action of plant hormones, starting from auxins in 1920s, gibberellins in 1930s, cytokines in 1950s, abscisic acid and ethylene in 1960s, and more recently expanding to brassinolides and jasmonic acid. They encountered very complicated interactions among these hormones, and could find no clear molecular mechanism for their crosstalk. It seems that it is now our turn as photomorphogenesis researchers to struggle with this old but fundamental problem in plant development, as crosstalk between light- and hormone signaling has now been widely discovered in plants (Chapter 31). Again, it is not yet clear where and how the above-mentioned crosstalk occur in plant cells. The reality of interactions among light-, clock-, and hormonal signaling pathways appears too complicated to allow the analysis of each separate interaction down to its to elementary processes by conventional methods and equipment, so we need a totally new approach to address the extremely complex system of a cell in its entirety.

Application of Photobiology to Plant Industry

We all now know that a wide range of growth and developmental processes in plants are controllable by environmental light, and it follows that the efficiency of productivity in agriculture, horticulture, forestry, and animal husbandry could be improved through manipulation of relevant photoregulatory systems in target plants. However, I know only two examples of applied photobiology in plants; namely, the production of chrysanthemum flowers and a spinach-like vegetable (*Salsola komarori*, Amaranthaceae) become possible throughout the year using the classic night-break of photoperiodism. The fact that our knowledge of photoregulation in plants has been not applied widely to these industries results from the wide gap between the basic photobiology and the industrial application. For example, the shade avoidance syndrome (Smith and Whitelam 1997) may be a good candidate for application in plant industries, but appropriate methods and inexpensive devices for large-scale irradiation with lights of specifically designed wavelength and timing in industrial fields have not been developed. To bridge this gap, new investment to support collaborations between photobiologists and diverse types of engineers will be required. With growing awareness of the need to avoid chemical pollution and other environmental damage, an increased emphasis on applied photobiology and the development of new technology is warranted in the near future.

Acknowledgments. I thank Pill-Soon Song for valuable comments and Jim Weller for careful editing of the manuscript.

References

Ahmad M, Cashmore AR (1993) HY4 gene of *A. thaliana* encodes a protein with characteristic of a blue-light photoreceptor. Nature 366: 162–166

Baum G, Long JC, Jenkins GI, Trewavas AJ (1999) Stimulation of the blue light phototropic receptor NPH1 causes a transient increase in cytosolic Ca^{2+}. Proc Natl Acad Sci USA 96: 13554–13559

Blaauw OH, Blaauw-Jensen G, Van Leeuwen WJ (1968) An irreversible red-light-induced growth response in *Avena*. Planta 82: 87–104

Bolle C, Koncz C, Chua NH (2000) PAT1, a new member of the GRAS family, is involved in phytochrome A signal transduction. Genes Dev 14: 1269–1278

Borthwick HA, Hendricks SB (1960) Photoperiodism in plants. Science 132: 1223–1228

Borthwick HA, Hendricks SB, Parker MW, Toole EH, Toole VK (1952) A reversible photoreaction controlling seed germination. Proc Natl Acad Sci USA 38: 662–666

Briggs WR, Rice HV (1972) Phytochrome: Chemical and physical properties and mechanism of action. Annu Rev Plant Physiol 23: 293–334

Butler WL, Norris KH, Siegelman HW, Hendricks SB (1959) Detection, assay, and preliminary purification of the pigment controlling photoresponsive development of plants. Proc Natl Acad Sci USA 45: 1703–1708

Butler WL, Lane HC, Siegelman HW (1963) Nonphotochemical transformations of phytochrome in vivo. Plant Physiol 38: 514–519

Chory J, Peto C, Feinbaum R, Pratt LH, Ausubel F (1989) *Arabidopsis thaliana* mutant that develops as a light-grown plant in the absence of light. Cell 58: 991–999

Darwin C, Darwin F (1881) The power of movement in plants. Appleton, London

De Greef J (ed) (1980) Photoreceptors and plant development. Antwerpen University Press

Devlin PF, Kay SA (2000) Cryptochromes are required for phytochrome signaling to the circadian clock but not for rhythmicity. Plant Cell 12: 2499–2510

Folta KM, Spalding E (2001) Unexpected roles for cryptochrome 2 and phototropin revealed by high-resolution analysis of blue light-mediated hypocotyl growth inhibition. Plant J 26: 471–478

Furuya M (1968) Biochemistry and physiology of phytochrome. Progr Phytochem 1: 347–405

Furuya M (ed) (1987) Phytochrome and photoregulation in plants. Academic, Tokyo

Furuya M (1993) Phytochromes: their molecular species, gene families, and functions. Annu Rev Plant Physiol Plant Mol Biol 44: 617–645

Furuya M (2004) An unforeseen voyage to the world of phytochromes. Annu Rev Plant Biol 55: 1–21

Furuya M, Schäfer E (1996) Photoperception and signalling of induction reactions by different phytochromes. Trends Plant Sci 1: 301–307
Furuya M, Hopkins WG, Hillman WS (1965) Effects of metal-complexing and sulfhydryl compounds on nonphotochemical phytochrome changes in vivo. Arch Biochem Biophys 112: 180–186
Garner WW, Allard HA (1920) Effect of the relative length of day and night and other factors of the environment on growth and reproduction in plants. J Agric Res 18: 553–606
Guo H, Yang H, Mockler TC, Lin C (1998) Regulation of flowering time by Arabidopsis photoreceptors. Science 279: 1360–1363
Hamner KC, Bonner J (1938) Photoperiodism in relation to hormones as factors in floral initiation and development. Bot Gaz 100: 388–431
Hanzawa H, Inomata K, Kinoshita H, Kakiuchi T, Jayasundera KP, Sawamoto D, Ohta A, Uchida K, Wada K, Furuya M (2001) *In vitro* assembly of phytochrome B apoprotein with synthetic analogs of the phytochrome chromophore. Proc Natl Acad Sci USA 98: 3612–3617
Hanzawa H, Shinomura T, Inomata K, Kakiuchi T, Kinoshita H, Wada K, Furuya M (2002) Structural requirement of bilin chromophore for the photosensory specificity of phytochromes A and B. Proc Natl Acad Sci USA 99: 4725–4729
Harada A, Sakai T, Okada K (2003) phot1 and phot2 mediate blue light-induced transient increase in cytosolic Ca2+ differently in Arabidopsis leaves. Proc Natl Acad Sci USA 100: 8583–8588
Hartmann KM (1966) A general hypothesis to interpret 'high energy phenomena' of photomorphogenesis on the basis of phytochrome. Photochem Photobiol 5: 349–366
Haupt W (1970) Über den Dichroismus von Phytochrom-660 und Phytochrom-730 bei *Mougeotia*. Z Pflanzenphysiol 62: 287–298
Hillman WS (1967) The physiology of phytochrome. Annu Rev Plant Physiol 18: 301–324
Hisada A, Hanzawa H, Weller JL, Nagatani A, Reid JB, Furuya M (2000) Light-induced nuclear translocation of endogenous pea phytochrome A visualized by immunocytochemical procedure. Plant Cell 12: 1063–1078
Hisada A, Yoshida T, Kubata S, Nishizawa NK, Furuya M (2001) Technical advance: An automated device for cryofixation of specimens of electron microscopy using liquid helium. Plant Cell Physiol 42: 885–893
Huala E, Oeller PW, Liscum E, Han IS, Larsen E, Briggs WR (1997) Arabidopsis NPH1: A protein kinase with a putative redox-sensing domain. Science 278: 2120–2123
Hudson ME (2000) The genetics of phytochrome signalling in *Arabidopsis*. Semin Cell Dev Biol 11: 475–483
Im YJ, Kim JI, Shen Y, Na Y, Han YJ, Kim SH, Song PS, Eom SH (2004) Structural analysis of *Arabidopsis thaliana* nucleoside phosphate kinase-2 for phytochrome-mediated light signaling. J Mol Biol 343: 659–670
Imaizumi T, Kadota A, Hasebe M, Wada M (2002) Cryptochrome light signals control development to suppress auxin sensitivity in the moss *Physcomitrella patens*. Plant Cell 14: 373–386
Imaizumi T, Tran HG, Swartz TE, Briggs WR, Kay SA (2003) FKF1 is essential for photoperiodic-specific light signalling in *Arabidopsis*. Nature 426: 302–306
Jackson JA, Jenkins GI (1995) Extension-growth responses and expression of flavonoid biosynthesis genes in the Arabidopsis hy4 mutant. Planta 1997: 233–239
Kagawa T, Sakai T, Suetsugu N, Oikawa K, Ishiguro S, Tabata S, Okada K, Wada M (2001) Arabidopsis NPL1: A phototropin homolog controlling the chloroplast high-light avoidance response. Science 291: 2138–2141

Kagawa T, Kasahara M, Abe T, Yoshida S, Wada M (2004) Function analysis of Acphot2 using mutants deficient in blue light-induced chloroplast avoidance movement of the fern *Adiantum capillus-veneris* L. Plant Cell Physiol 45: 416–426

Kasahara M, Kagawa T, Sato Y, Kiyosue T, Wada M (2004) Phototropins mediate blue and red light-induced chloroplast movements in *Physcomitrella patens*. Plant Physiol 135: 1388–1397

Kawai H, Kanegae T, Christensen S, Kiyosue T, Sato Y, Imaizumi T, Kadota A, Wada M (2003) Responses of ferns to red light are mediated by an unconventional photoreceptor. Nature 421: 287–290

Kendrick RE, Kronenberg GHM (eds) (1994) Photomorphogenesis in plants. Kluwer Academic, Dordrecht

Khurana J, Poff KL (1989) Mutants of *Arabidopsis thaliana* with altered phototropism. Planta 178: 400–406

Kinoshita T, Doi M, Suetsugu N, Kagawa T, Wada M, Shimazaki K (2001) phot1 and phot2 mediate blue light regulation of stomatal opening. Nature 414: 656–660

Klebs G (1910) Alterations in the development and forms of plants as a result of environment. Proc R Soc Lond B 82: 547–558

Koornneef M, Rolf E, Spruit CJP (1980) Genetic control of light-inhibited hypocotyl elongation in *Arabidopsis thaliana* (L.) Heynh. Z Pflanzenphysiol 100: 147–160

Kuno N, Muramatsu T, Hamazato F, Furuya M (2000) Identification by large-scale screening of phytochrome-regulated genes in etiolated seedlings of *Arabidopsis thaliana* using the 047fluorescent differential display technique. Plant Physiol 122: 15–24

Lagarias JC, Rapoport H (1980) Chromopeptides from phytochrome. The structure and linkage of the Pr form of the phytochrome chromophore. J Am Chem Soc 102: 4821–4828

Lin C, Ahmad M, Chan J, Cashmore AR (1996) *CRY2*: A second member of the *Arabidopsis* cryptochrome gene family (accession no. U43397). Plant Physiol 110: 1047

Lin C, Yang H, Guo H, Mockler T, Chen J, Cashmore AR (1998) Enhancement of blue-light sensitivity of Arabidopsis seedlings by a blue light receptor cryptochrome 2. Proc Natl Acad Sci USA 95: 2686–2690

López-Juez E, Nagatani A, Tomizawa K, Deak M, Kern R, Kendrick RE, Furuya M (1992) The cucumber long hypocotyl mutant lacks a light-stable PHYB-like phytochrome. Plant Cell 4: 241–251

Mackenzie JM Jr, Coleman RA, Briggs WR, Pratt LH (1975) Reversible redistribution of phytochrome within the cell upon conversion to its physiologically active form. Proc Natl Acad Sci USA 72: 799–803

Mitrakos K, Shropshire W Jr (eds) (1972) Phytochrome. Academic, London

Mohr H, Schäfer E (1983) Photoperception and de-etiolation. Philos Trans R Soc Lond B 303: 489–501

Montgomery BL, Lagarias JC (2002) Phytochrome ancestry: sensors of bilins and light. Trends Plant Sci 7: 1360–1385

Nagy F, Schäfer E (2002) Phytochromes control photomorphogenesis by differentially regulated, interacting signaling pathways in higher plants. Annu Rev Plant Biol 53: 329–355

Park CM, Bhoo SH, Song PS (2000) Inter-domain crosstalk in the phytochrome molecules. Semin Cell Dev Biol 11: 449–456

Parker MW, Hendricks SB, Borthwick HA, Went FW (1949) Spectral sensitivity for leaf and stem growth of etiolated pea seedlings and their similarity to action spectra for photoperiodism. Am J Bot 36: 194–204

Pratt LH (1982) Phytochrome: the protein moiety. Annu Rev Plant Physiol 33: 557–582

Priestley J (1772) Observations on different kinds of air. Philos Trans R Soc Lond 62: 147–264

Quail PH (1997) An emerging molecular map of the phytochromes. Plant Cell Environ 20: 657–665

Quail PH, Marmé D, Schäfer E (1973) Particle-bound phytochrome from maize and pumpkin. Nat New Biol 245: 189–191

Reed JW, Nagpal P, Poole DS, Furuya M, Chory J (1993) Mutation in the gene for red/far-red light receptor phytochrome B alter cell elongation and physiological responses throughout *Arabidopsis* development. Plant Cell 5: 147–157

Reed JW, Nagatani A, Elich TD, Fagan M, Chory J (1994) Phytochrome A and phytochrome B have overlapping but distinct functions in *Arabidopsis* development. Plant Physiol 104: 1139–1149

Rubinstein B, Drury KS, Park RB (1969) Evidence for bound phytochrome in oat seedlings. Plant Physiol 44: 105–109

Rüdiger W, Thümmler F, Cmiel E, Schneider S (1983) Chromophore structure of the physiologically active form (Pfr) of phytochrome. Proc Natl Acad Sci USA 80: 6244–6248

Sachs J (1864) Wirkungen farbigen Lichts auf Pflanzen. Bot Z 22: 353–358, 361–367, 369–372

Sage LC (1992) Pigment of the imagination, a history of phytochrome research. Academic, San Diego

Sakai T, Kagawa T, Kasahara M, Swartz TE, Christie JM, Briggs WR, Wada M, Okada K (2001) Arabidopsis nph1 and npl1: Blue-light receptors that mediate both phototropism and chloroplast relocation. Proc Natl Acad Sci USA 98: 6969–6974

Sakamoto K, Briggs WR (2002) Cellular and subcellular localization of phototropin1. Plant Cell 14: 1723–1735

Sakamoto K, Nagatani A (1996) Nuclear localization activity of phytochrome B. Plant J 10: 859–868

Senger H (ed) (1980) The blue light syndrome. Springer, Berlin

Sharrock RA, Quail PH (1989) Novel phytochrome sequences in *Arabidopsis thaliana*: structure, evolution and differential expression of a plant regulatory photoreceptor family. Genes Dev 3: 1745–1757

Shinomura T, Nagatani A, Hanzawa H, Kubota M, Watanabe M, Furuya M (1996) Action spectra for phytochrome A- and B-specific photoinduction of seed germination in *Arabidopsis thaliana*. Proc Natl Acad Sci USA 93: 8129–8133

Shinomura T, Uchida K, Furuya M (2000) Elementary processes of photoperception by phytochrome A for high irradiance response of hypocotyl elongations in *Arabidopsis thaliana*. Plant Physiol 122: 147–156

Shropshire W Jr, Mohr H (eds) (1983) Photomorphogenesis. Encyclopedia of plant physiology, New Series vol. 16. Springer, Berlin

Siegelman HW, Butler WL (1965) Properties of phytochrome. Annu Rev Plant Physiol 16: 383–392

Smith H (ed) (1976) Light and plant development. Butterworths, London

Smith H (ed) (1983) Photoperception by plants. The Royal Society, London

Smith HMS, Raikhel NV (1999) Protein targeting to the nuclear pore. What can we learn from plants? Plant Physiol 119: 1157–1163

Smith H, Whitelam GC (1997) The shade avoidance syndrome: multiple responses mediated by multiple phytochromes. Plant Cell Environ 20: 840–844

Somers DE, Devlin PF, Kay SA (1998) Phytochromes and cryptochromes in the entrainment of the Arabidopsis circadian clock. Science 282: 1488–1490

Stoelzle S, Kagawa T, Wada M, Hedirich R, Dietrich P (2003) Blue light activates calcium-permeable channels in Arabidopsis mesophyll cells via the phototropin signaling pathway. Proc Natl Acad Sci USA 100: 1456–1461

Thomas B, Johnson CB (eds) (1990) Phytochrome properties and biological action. Springer, Berlin

Tournois J (1914) Études sur la sexualité du Houblon. Ann Sci Nat Bot Biol Veg Ser IX, 19: 49–191

Vierstra RD, Quail PH (1982) Native phytochrome: Inhibition of proteolysis yields a homogeneous monomer of 124 kilodaltons from *Avena*. Proc Natl Acad Sci USA 79: 5272–5276

Wada M, Kadota A (1989) Photomorphogenesis in lower green plants. Annu Rev Plant Physiol Plant Mol Biol 40: 169–191

Wagner D, Fairchild CD, Kuhn RM, Quail PH (1996) Chromophore-bearing NH_2-terminal domains of phytochromes A and B determine their photosensory specificity and differential light lability. Proc Natl Acad Sci USA 93: 4011–4015

Whitelam GC, Devin PF (1997) Roles of different phytochromes in *Arabidopsis* photomorphogenesis. Plant Cell Environ 20: 752–758

Whitelam GC, Johnson E, Peng J, Carol P, Anderson ML, Cowl JS, Harberd NP (1993) Phytochrome A null mutants of *Arabidopsis* display a wild-type phenotype in white light. Plant Cell 5: 757–768

Part II
Phytochrome

Chapter 2

Phytochrome Overview

Peter H. Quail

Introduction

Masaki Furuya organized the last Yamada conference on phytochrome and photoregulation in 1986. The volume that emerged from that meeting provides both an interesting historical perspective and an informative snapshot of the state of the field at that time (Furuya 1987). In a chapter on the History of Phytochrome (see also Sage 1992), Furuya concisely captures the progression of the field in sequential "eras": GENESIS (The Beltsville Era: 1920–1963), describing the physiological experiments that led to the discovery of phytochrome, its ultimate physical detection by spectroscopic methods, and its initial purification and preliminary biochemical characterization; THE ERA OF DISAPPOINTMENT (1963–1970), describing the lack of quantitative correlation observed between spectrophotometrically measurable phytochrome abundance and photochemical state, and the various photoreversible responses of plants to red and far-red light; THE SEVENTIES: THE ERA OF GROPING, describing further biochemical characterization of the phytochrome molecule, definition of the chemical structure of the chromophore, the introduction of immunochemical approaches and microbeam irradiation technology, with a strong focus on the question of subcellular localization as a means of assessing the "membrane hypothesis" of the primary action of phytochrome; and THE EIGHTIES: THE ERA OF MOLECULAR BIOLOGY, referring to the beginnings of the application of molecular biological approaches in the field.

The articles assembled in this original volume (Furuya 1987) document continued significant focus on the biochemical and spectroscopic properties of the phytochrome molecule itself, as well as its subcellular localization. The impact of molecular biology was by then firmly established and the first *PHY* genes and cDNAs had been cloned and sequenced, revealing the amino acid sequence of the photoreceptor. The first biochemical and immunochemical evidence of the

---

Plant Gene Expression Center, 800 Buchanan Street, Albany, CA 94710, USA; and Department of Plant and Microbial Biology, University of California, Berkeley, CA 94720, USA

possible existence of more than one class of phy was emerging, and the first report of the detection of Ser/Thr-protein kinase activity in purified phy preparations was presented. The use of molecular biological tools to investigate light-induced regulation of the expression of specific cloned genes was increasing. The view was widely held that some form of second messenger system transduced the signals from the photoreceptor to nuclear genes, and the long-debated hypothesis that the phy molecule interacts with cellular membranes as its primary site of action was still under active investigation. Striking, given today's perspective, is the lack of the use of genetic approaches to understanding phy-regulated responses in this volume (only one study mentioning the use of mutants), and the apparent complete absence of even the word "Arabidopsis" from the pages of the text.

Remarkable progress has been made since that first Yamada conference, driven primarily by the revolution in Arabidopsis molecular genetics and genomics. *PHY* genes have been cloned and sequenced in numerous higher and lower plants, revealing that the photoreceptor exists in small families in these organisms, and the discovery of bacterial phys has uncovered the likely evolutionary origins of the molecule, as well as opening a new dimension in efforts to define the activities of the family. The phy molecule has been shown to translocate from the cytoplasm to the nucleus and interact with transcription factors, as a likely signaling pathway, at least in higher plants. Numerous genetic and reverse-genetic studies have provided evidence of the photosensory and physiological functions of the different phy family members, and, together with molecular approaches, have identified a considerable number of candidate intermediates in the phy signaling pathways. These advances have been documented in numerous reviews, a selection of which are referred to here (Kendrick and Kronenberg 1994, Millar et al 1994, Smith 2000, Quail 2000, 2002a,b, Quail et al 1995, Whitelam et al 1998, Fankhauser 2001, Moller et al 2002, Gyula et al 2003, Nagy and Schäfer 2000a,b, 2002, Montgomery and Lagarias 2002, Serino and Deng 2003, Wada et al 1997, 2003, Sullivan et al 2003). This brief overview builds on the earlier advances, with focus primarily on the principal areas of progress over the last 5 years or so.

The phy Family Grows

The discovery and molecular cloning of the five Arabidopsis phys, phyA through phyE, by the early 1990s (Clack et al 1994), has been followed by the steady accumulation of sequence information for phys across all major groups of the plant kingdom (Wada et al 1997, Schmidt and Schneider-Poetsch 2002, Mathews and Sharrock 1997). The complete genome sequences of Arabidopsis and rice, and the near complete sequencing of *Chlamydomonas*, have delimited the absolute size and phylogenetic relationships of the phy families within these representative model organisms. The discovery of prokaryotic phys, provided by the sequencing of the full genome of the cyanobacterium, *Synechocystis*, has been followed by demonstrations of the presence of phy-related sequences across a wide diversity of bacterial groups, establishing the ancient evolutionary origins of the phys (Montgomery and Lagarias 2002, Bhoo et al 2001, Chapter 3).

Phylogenetic comparisons among the eukaryotic phys have provided intriguing insights into the possible evolution and functions of these photoreceptors. A not unexpected early divergence in the phy lineages of the higher and lower plants is apparent (Schmidt and Schneider-Poetsch 2002). However, whereas Arabidopsis has five phys (Clack et al 1994), rice has only three of these (phyA, phyB, and phyC) (Goff et al 2002). Together with survey sequence information from a broad cross section of angiosperm phys, these data suggest that whereas the dicots appear to have evolved a five-membered phy family, the monocots have evolved or retained only three (Mathews and Sharrock 1997). Moreover, interesting diversity of structure among lower plant phy family members has been detected, including the "superchrome, " AcPHY3, of the fern, *Adiantum*, which consists of a fusion of phy and phototropin protein sequences (Wada et al 2003, Chapter 22). Perhaps most intriguingly, the model alga, *Chlamydomonas*, thus far appears to have no *PHY* genes, despite the clear presence of phy sequences in other algae, such as *Mougeotia* and *Mesotaenium* (Schmidt and Schneider-Poetsch 2002). Taken together, these data suggest that the phys have a dynamic evolutionary history, reflecting impressive adaptability to a diversity of biological functions across a wide range of plant and bacterial species.

phy Photosensory and Physiological Functions

The diversity and complexity of light-induced plant responses attributed to the phy photoreceptor system long suggested to physiologists and photobiologists that more than one phy with differential activities was necessary to rationalize all the observed phenomena. The discovery of the five-membered phy family in Arabidopsis, and subsequent studies with null mutants in each, established that individual family members have differential, albeit partially overlapping, photosensory and/or physiological functions at various phases of the life cycle (Franklin et al 2003, Monte et al 2003, Quail 2002a, Quail et al 1995, Smith 2000, Whitelam et al 1998). In some cases, different family members monitor the same light signals but have predominant regulatory roles in different physiological responses. For example, whereas phyB has a predominant role in regulating seedling establishment, phyE appears to function primarily in controlling internode elongation (Smith 2000). Conversely, in other cases, different family members monitor different light signals, but control the same physiological response. For example, both phyA and phyB regulate seedling de-etiolation, but, whereas phyB is activated by continuous monochromatic red light (Rc), phyA is exclusively responsible for monitoring continuous monochromatic far-red light (FRc), in controlling this response (Quail et al 1995, Smith 2000, Whitelam et al 1998). Finally, in yet other cases, multiple phys appear to act additively or partly redundantly in regulating the same physiological response to the same light signal. For example, phyB has long been considered to dominate regulation of seedling de-etiolation in response to Rc. However, the residual, partial responsiveness of *phyB* null mutants to Rc indicates that one or more other phy family members participate in this process (Tepperman et al 2004). The recent isolation of *phyC* mutants (Franklin et al 2003, Monte et al 2003) has shown that phyC can have a role, and there is evidence that other family members, particularly phyA, may also contribute significantly to aspects of Rc-induced de-etiolation (Tepperman et al 2004). Evidence for potentially differential functional roles for the three rice phys is also emerging from recent studies with mutants at these loci (see Chapter 12). Similarly, recent studies on *phy* mutants in the ferns, *Adiantum* and *Physcomitrella*, have begun to define the functional activities of some of these lower plant photoreceptors in such responses as red-light-induced phototropism, polartropism and chloroplast movement (Wada et al 2003, Chapters 11 and 22).

Defining phy Signaling Networks

In higher plants, informational light-signal perception by the phy molecule (Pfr formation) initiates an intracellular transduction process that culminates in the altered expression of target genes responsible for directing the adaptational changes in plant growth and development appropriate for the prevailing environment (Smith 2000, Quail 2002a). Increasingly, over recent years, intense research efforts have been mounted to define the molecular, cellular and biochemical mechanisms involved in this process through identifying molecular components that comprise the signaling and transcriptional networks controlled by the photoreceptor family. The principal general strategies being used to approach this goal include: (a) conventional, forward-genetic screens to identify mutants exhibiting aberrant visible photoresponsiveness phenotypes (morphological phenotype); (b) yeast two-hybrid screens to identify phy-interacting proteins as potential primary signaling partners; (c) molecular phylogeny analyses to identify homologs closely related to previously identified components in multigene families; (d) microarray-based expression profiling, both to define the genome-wide complement of phy-regulated genes (the molecular phenotype), and to identify the most rapidly light-responsive genes in this set as potential direct targets of phy signaling; and (e) bioinformatic analysis of the promoters of coordinately light-responsive genes to identify common DNA sequence elements, and, eventually, their cognate binding-proteins, potentially involved in regulating expression of those genes. Of necessity, the functional relevance to phy signaling or transcriptional regulation of any components identified by strategies (b) through (e) requires subsequent assessment by reverse-genetic methods (such as T-DNA or transposon insertion, antisense or RNAi, Tilling or Delete-a-gene technology; Henikoff and Comai 2003), that provide targeted mutagenesis of the encoding genes, coupled with phenotypic analysis (morphological and/or molecular) for aberrant photoresponsiveness caused by the mutated component. Superimposed on these studies is the powerful cell-biological strategy of tagging phy-signaling-system proteins by fusing them to visible molecular markers, such as GUS and GFP, to permit the subcellular location, and potential colocalization of these components to be monitored. Much of what we have learned thus far has come from studies using the seedling de-etiolation process in Arabidopsis as a model system.

Genetically Identified Signaling-Intermediate Candidates

Conventional forward genetic screens for mutants defective in normal seedling photomorphogenesis have identified numerous non-photoreceptor loci that exhibit aberrant de-etiolation (Moller et al 2002, Quail 2002a,b, Gyula et al 2003). These mutants fall into two broad classes: the *cop/det/fus* class that develop in complete darkness as if they were in the light, and those that develop normally in darkness, but display altered sensitivity to light (photodefective mutants). Most of the *cop/det/fus* class that have been molecularly cloned and characterized are considered to act downstream of the convergence of the phy and blue-light-receptor pathways, and have been shown to function either in a nuclear-localized, ubiquitin-proteosome pathway, by targeted proteolysis of the key bZIP transcription factor, HY5 (Serino and Deng 2003, Seo et al 2003, Saijo et al 2003), or in the brassinosteroid pathway (Nemhauser and Chory 2004). The systematic definition of the COP9 signalosome as a novel component of the proteosome system, and the recent identification of COP1 as a ubiquitin E3 ligase (Seo et al 2003, Saijo et al 2003), are landmark contributions with implications beyond the photomorphogenesis field (see Chapter 29).

Three principal subclasses of mutants exhibiting altered responsiveness upon exposure to light have been identified: those responding aberrantly either to FRc only, to Rc only, or to both FRc and Rc. These different classes of mutants suggest that early steps in the phyA and phyB pathways involve upstream intermediates dedicated to the individual photoreceptors, and that the separate pathways converge downstream in some undefined "signal integration" process that drives later common events in photomorphogenesis and the circadian clock. How the majority of these individual components might function together in signal transduction is still largely unknown. Significantly, however, a major fraction of the components that have been cloned localize to the nucleus (Moller et al 2002, Gyula et al 2003). Taken together, these data suggest, therefore, that early phy signaling events are focused in the nucleus, and involve both synthetic (transcriptional) and degradative (post-translational) regulatory mechanisms (Quail 2002a,b, Serino and Deng 2003, Seo et al 2003, Saijo et al 2003).

Light-Induced Nuclear Translocation of phy Molecules

As alluded to above, for many years, existing cell-fractionation and immunocytochemical evidence was widely accepted in the field as indicating that phy molecules are soluble, cytoplasmically localized proteins, that signal to nuclear genes through some second messenger pathway. In addition, certain types of physiological evidence had led to the more explicit hypothesis that the photoreceptor molecule interacts with the plasmamembrane (or other cellular membranes) as its primary site of action in this signaling process (see Kendrick and Kronenberg 1994, Quail 2000). The results of pharmacological and microinjection experiments with the phy-deficient *aurea* mutant of tomato provided initially exciting evidence consistent with the involvement of the classical G-protein, Ca^{2+}/calmodulin and cGMP second-messenger systems (Millar et al 1994). However, little progress has been reported since the original studies in the mid to late 1990s, and no known components of these pathways have thus far been reported in genetic screens for phy signaling mutants (Quail 2000, 2002a, Gyula et al 2003, Jones et al 2003, Moller et al 2002). Currently, therefore, there is no robust evidence for such a second-messenger pathway.

By contrast, compelling evidence has accumulated since the original report of Sakamoto and Nagatani (1996) that phy molecules are in fact induced to rapidly translocate from the cytoplasm into the nucleus following photoconversion to the Pfr form (Yamaguchi et al 1999, Nagy and Schäfer 2000a,b, Bauer et al 2004), potentially obviating the need for a second messenger system. Data from several groups using phy:GUS and phy:GFP fusion proteins, or immunocytochemical localization procedures, indicate that full-length phys are initially present in the cytoplasm in their Pr form, but translocate to the nucleus within minutes of photoconversion to the Pfr form (Yamaguchi et al 1999, Nagy and Schäfer 2000a,b, 2002, Hisada et al 2000). Evidence from use of a phyB:glucocorticoid-receptor fusion protein in *phyB*-mutant rescue experiments indicates that this nuclear translocation is necessary for phy biological function in the living cell, and combined with photoconversion to the Pfr form, is sufficient for this activity (Huq et al 2003). These data provide strong support for the conclusion that primary phy signaling events occur in the nucleus and, conversely suggests the absence of a cytosolic signaling pathway, in phy-regulated seedling de-etiolation. Because all five Arabidopsis phys have been reported to undergo photoinduced nuclear import (Nagy and Schäfer 2000a,b, 2002), the data support the generality of this mechanism. Interestingly, the phy:GFP fusion proteins form highly dynamic speckles in the nucleus following induced translocation (Bauer et al 2004). The composition and functional significance of these structures is under intense investigation (see Chapter 8).

Phytochrome-Interacting Factors

Several yeast two-hybrid screens of cDNA libraries for phy-interacting proteins that may function as primary phy signaling partners have resulted in identification of three, apparently unrelated, proteins: PIF3 (phytochrome interacting factor 3) (Ni et al 1998), PKS1 (phytochrome kinase substrate 1) (Fankhauser et al 1999), and NDPK2 (nucleoside diphosphate kinase 2) (Choi et al 1999) which are capable of direct binding to phy molecules.

PIF3 is a member of the basic helix-loop-helix (bHLH) superfamily of transcriptional regulators (Toldeo-Ortiz et al 2003). In vitro interaction assays showed that both phyA and phyB bind to PIF3, but only upon light-induced conversion to the biologically-active Pfr form (Ni et al 1999, Martínez-Garcia et al 2000, Quail 2002a). PIF3 localizes constitutively to the nucleus and can bind to a G-box DNA sequence, CACGTG, that is present in a variety of light-regulated promoters (Martínez-Garcia et al 2000, Quail 2000, Bauer et al 2004), and phyB can bind specifically and photoreversibly to PIF3 that is already bound to its cognate DNA binding site. Together with the observed light-induced translocation of phy molecules to the nucleus (Sakamoto and Nagatani 1996, Nagy and Schäfer 2000a,b, 2002), these data suggested that PIF3 could recruit the photoreceptor in its active form to G-box-containing promoters. Recent evidence that the phys colocalize with PIF3 in nuclear speckles suggests that they may interact physically in vivo (Bauer et al 2004). Evidence that the observed phy-PIF3 interactions are relevant to phy signaling in vivo came initially from analysis of antisense-PIF3 seedlings (Ni et al 1998). These seedlings exhibited strongly reduced phenotypic responsiveness to light signals and reduced induction of a subset of rapidly photoresponsive genes, in particular the key genes, *CCA1* and *LHY* (Martínez-Garcia et al 2000). The promoters of both of these genes contain G-box motifs and PIF3 bound to these in sequence-specific fashion, consistent with a direct role in regulating their expression. Significantly, *CCA1* and *LHY* themselves encode MYB-class transcription-factor proteins known to function in regulating the expression of *CAB* genes and/or the circadian clock (Wang and Tobin 1998, Alabadi et al 2001). It was proposed, therefore, that PIF3 may represent the central control point through which the phy system regulates both a major branch of photomorphogenesis and the circadian oscillator, and that this regulation may be executed through a short transcriptional cascade using these MYB-related CCA1 and LHY transcription factors as intermediates (Quail 2002a,b). Because phyB can bind to DNA-bound PIF3 as Pfr, it was proposed further that the phys may function as integral, light-switchable components of transcription-regulator complexes, directly at target promoters (Martínez-Garcia et al 2000).

Despite the attractiveness of this model of PIF3 function and mechanism of action, recent data suggest a more complex, and alternative, picture. Three laboratories have found that *pif3* mutant seedlings exhibit enhanced inhibition of hypocotyl elongation in prolonged Rc (hypersensitivity) (Kim et al 2003, Bauer et al 2004, Monte and Quail unpublished) in direct contrast to the initial hyposensitive phenotype reported for *PIF3*-antisense seedlings (Ni et al 1998). It appears that the PIF3-antisense-line hyposensitivity is due to a T-DNA-induced mutation at another locus, and not to antisense suppression of *PIF3* expression (Monte and Quail unpublished). At face value, these new data suggest that PIF3 functions negatively, rather than positively, as initially reported, in the overall process of seedling de-etiolation. However, *pif3* null mutants are defective in the light-induced changes in expression of a subset of rapidly responsive genes, and other early aspects of seedling de-etiolation, such as greening, upon initial exposure to Rc, suggesting a critical positive function at the initial dark-to-light transition experienced by seedlings (Monte and Quail unpublished).

Of more profound importance to the understanding of the molecular mechanism of phy action is the discovery by Bauer et al (2004) that light induces rapid degradation of the PIF3 protein in the nucleus via a mechanism that may be triggered by direct phy–PIF3 interaction. We have confirmed this basic observation and have preliminary inhibitor-based evidence that the degradation may be mediated via the 26S proteosome (Al-Sady and Quail unpublished). This behavior is consistent with a possible transient function of PIF3 at the initial dark to light transition. The emerging concept, then, is that the phy molecule may function to induce PIF3 degradation via the ubiquitin-proteosome system (UPS) (Sullivan et al 2003) upon light-induced binding, following translocation into the nucleus. This would imply more indirect regulation of gene expression by removal of a transcriptional regulator, as opposed to direct participation in the transcriptional regulatory machinery at target promoters. This represents a clear paradigm shift in the proposed mechanism of action of the photoreceptor. Defining the molecular basis of this newly-discovered activity of phyB toward PIF3 is a central focus of current research efforts.

Following identification of PIF3, a number of other members of the large family of bHLH factors in Arabidopsis have been investigated for potential phy-binding activity and involvement in phy-regulated responses to light. Accumulating evidence indicates that photoactivated phyB molecules can indeed interact with multiple bHLHs. In addition to PIF3, four other members of the family, designated PIF1 (Huq et al 2004), PIF4 (Huq and Quail 2002), PIF5, and PIF6 (Khanna and Quail, unpublished) have been shown to bind to the Pfr form of phyB. These all cluster tightly with PIF3 in subfamily 15 of the Arabidopsis bHLH phylogenetic tree (Toledo-Ortiz et al 2003). All five factors share a conserved sequence motif (designated **A**ctive **P**hytochrome **B**inding (APB) motif), that has been shown to be both necessary and sufficient for binding to phyB, and to be necessary for PIF4 function in light-regulated seedling development (Khanna and Quail, unpublished). PIF1, PIF3, PIF4, HFR1, and PIL1 (Huq et al 2004, Kim et al 2003, Bauer et al 2004, Huq and Quail 2002, Fairchild et al 2000, Soh et al 2000, Spiegelman et al 2000, Salter et al 2003, Monte and Quail unpublished), each belonging to subfamily 15 of the bHLH family, have all been shown to be involved in phy-regulated seedling de-etiolation, but each appears to have a differential role in this process. The emerging evidence suggests, therefore, that these closely related factors may comprise a small network of transcriptional regulators that are common targets of phyB signaling, but which potentially regulate different segments of the transcriptional network.

PKS1 is a novel, constitutively cytoplasmic, protein initially isolated in a yeast two-hybrid screen and later shown to be phosphorylated by purified preparations of oat phyA (Fankhauser et al 1999). A recent study with a *pks1* mutant suggests that PKS1 functions specifically in a phyA-mediated very low fluence response mode, in conjunction with a related protein, PKS2, to provide homeostasis to phyA signaling (Lariguet et al 2003). However, the potential functional role of PKS1 phosphorylation in phy signaling is yet to be directly assessed. The enzyme NDPK2, which appears to localize to both nucleus and cytoplasm, was also isolated in a yeast two-hybrid screen (Choi et al 1999). In vitro assays showed that the enzymatic activity of NDPK2 was enhanced by incubation with the Pfr form of oat phyA. A *ndpk2* mutant displayed reduced sensitivity to both Rc and FRc suggesting a functional role in both phyA and phyB signaling. However, the molecular function of NDPK2 in phy signaling remains unclear. In addition to the non-targeted yeast two-hybrid screens described above, targeted molecular interaction studies with pre-selected proteins have resulted in reports of a variety of proteins that can bind to phyA and/or phyB (Moller et al 2002, Quail 2000, 2002b, Gyula et al 2003). However, the functional relevance of these interactions is presently unclear.

Transcription-Factor Genes Are Early Targets of phy Signaling

A small number of studies aimed at defining the spectrum of photoresponsive genes regulated by phyA and phyB have now used microarrays to examine the changes in expression profiles elicited by Rc- or FRc-irradiation of Arabidopsis seedlings. One set of studies employed glass-slide, spotted-cDNA arrays to measure presumptively end-point, steady-state transcript profiles after prolonged (5- to 6-day) FRc or Rc irradiation (Ma et al 2001, 2002, Wang et al 2002). The other set of studies used Affymetrix oligonucleotide microarrays to follow the time-course of changes in expression over the first 24 hours of irradiation of dark-grown seedlings (Tepperman et al 2001, 2004). Because both types of arrays were based primarily on the EST sequences available at the time of construction, each contained largely the same gene set of 6000 to 8000 genes. Although this gene set represents only about 25% to 30% of the total present in the Arabidopsis genome (about 29 000 genes), highly useful information that is likely representative of the genome-wide pattern has been obtained.

Comparison of the expression patterns in *phyA* and *phyB* null mutants with those of the wild type in FRc and Rc, respectively, have identified the photoresponsive genes regulated by these phy family members. The data verify the exclusive role of phyA in FRc perception, but indicate that other family members share the function of mediating Rc signals with phyB, and suggest organ-specific differences in signaling activity among phy family members (Tepperman et al 2001, 2004).

Analysis of the temporal patterns of light-induced expression has shown that 10% of phy-regulated genes exhibit altered expression within 1 hour of the signal ("early-response" genes) (Tepperman et al 2001, 2004). Of the small group of functionally-classifiable early-response genes, a significant proportion are predicted to encode established or putative transcriptional regulators of multiple classes. Of these, over 70% respond to both Rc and FRc wavelengths. The rapid responsiveness of these genes suggests that they may be integral components of a primary transcriptional network under phyA and phyB control, constituting a master-set, each with a primary role in coordinating the expression of the downstream genes that elaborate one or more major facets of light-induced development (Tepperman et al 2001, 2004, Quail 2002a,b). A recent global, bioinformatic analysis of the promoters of phyA-regulated genes defined by microarray analysis has identified a series of sequence motifs potentially involved in the coordinate transcriptional regulation of these genes (Hudson and Quail 2003). The G-box motif, CACGTG, was prominently enriched in the phyA-responsive promoters compared to the remainder of the genes in the genome. Moreover, significantly, a greater abundance of G-box motifs was found in the most rapidly phyA-responsive genes, and in the promoters of phyA-regulated transcription factors, consistent with the notion that G-box-binding proteins, such as bHLH factors, have key functions early in the phy transcriptional network. Although these correlative data are intriguing, there is a clear need to assess the functional relevance of these early-response genes to phy signaling using genetic/reverse genetic approaches.

The discovery that the phy signaling intermediates, CCA1 and LHY, are integral components of the circadian clock (Alabadi et al 2001) has established that the oscillator is embedded at the apex of the phy-regulated transcriptional network, and is therefore positioned to impose oscillatory behavior on numerous downstream genes in the light-regulated cascade. No oscillations are detectable in the central-oscillator genes *CCA1*, *LHY* and *TOC1* in dark-grown Arabidopsis seedlings (Kikis and Quail unpublished), apparently reflecting a steady-state equilibrium of expression established by the mutual feedback regulation exerted by these components on each other. Exposure to light initiates the oscillations in this loop by rapidly inducing enhanced expression of *CCA1* and *LHY*. The characteristic biphasic profile over the first 24 hours after onset of the light-signal (Tepperman et al 2001, 2004) is consistent with the proposed clock model. The data suggest that the first and second peaks represent the activities of the light-input pathway and clock per se, respectively, as suggested previously (Somers et al 1998) providing the potential to dissect components of each separately.

Biochemical Mechanism of Signal Transfer

The central question of the biochemical mechanism of signal transfer from the photoactivated phy molecule to its primary signaling partner(s) has intrigued researchers for many years. A number of mechanisms have been proposed, but for over the past decade considerable attention has been focused on the attractive possibility that the phy molecule may function as a light-regulated protein kinase, and that transphosphorylation of one or more interacting partners comprises the biochemical mechanism of signal transfer (Fankhauser et al 1999, Fankhauser 2000, Montgomery and Lagarias 2002). Both biochemical and evolutionary evidence have been presented in support of this proposition (see Chapters 1 and 3–12, Fankhauser 2000, Quail 2002a, Montgomery and Lagarias 2002). Several studies have documented that purified native and recombinant preparations of oat phyA exhibit Ser/Thr kinase activity both toward the photoreceptor itself (potentially autophosphorylation), and toward several other purified recombinant proteins in vitro, including PKS1 (Fankhauser et al 1999). The discovery of the bacteriophytochromes, and the compelling evidence that at least some can function as light-regulated histidine kinases, capable of transphosphorylating partner response regulators, in classical two-component fashion (Yeh et al 1997, Bhoo et al 2001), provided evolutionary support for the proposal that the apparent eukaryotic descendants of these molecules had retained the basic activity of the progenitor, but had evolved into an atypical Ser/Thr kinase (Yeh et al 1997, Montgomery and Lagarias 2002, Fankhauser 2000). However, direct evidence that mutagenesis of the eukaryotic phy molecule can eliminate the apparent intrinsic kinase activity, and that such mutants lack signaling activity in the plant cell, is yet to be presented. In fact, to the contrary, recent reports that the photoactive, N-terminal domain of phyB is fully functional in the cell in the complete absence of the putative protein kinase domain, that is located in the C-terminal half of the molecule (Matsushita et al 2003), provide compelling evidence that the normal signal transfer activity of the photoreceptor does not require this postulated kinase activity, and seems unlikely therefore to involve transphosphorylation of signaling partners. The possibility that the biochemical mechanism of phy signal transfer involves a novel intermolecular transaction remains open.

Outlook

The spectacular advances in our understanding of the phy system in the 18 years since the last Yamada Conference are readily apparent from the above discussion. Nevertheless, some of the most central questions currently remain unanswered. What is the molecular mechanism of phy action in the cell? Attention has swung to the possibility that the photoreceptor may induce rapid degradation of phy-interacting transcription factors via the UPS system (Bauer et al 2004), rather than functioning as an integral component of transcriptional regulator complexes directly at target gene promoters, as previously hypothesized (Martínez-Garcia et al 2000). What is the biochemical mechanism of signal transfer to primary reaction partners? Given the evidence that the kinase-related domain of the phy molecule is apparently dispensable for signaling activity in the plant, other possible mechanisms, such as physical nucleation of functionally active multiprotein complexes (for example, those of the ubiquitin E3-ligases and 26S proteosome of the UPS system), or allosteric induction of conformational changes in signaling partner(s) to activate latent function, are likely to be vigorously investigated. What is the mechanism of phy nuclear translocation? What is the basis for the differential activities of phy family members? These questions and other issues, such as the potential for enormous complexity posed by the capacity for combinatorial heterodimerization among phy family members (Chapter 5 by Sharrock), and among phy-interacting bHLH family members

(Quail 2000, Toledo-Ortiz et al 2003), offer exciting challenges. The power of the molecular, genetic, genomic, and proteomic tools and resources available seems certain to ensure continuation of the current rapid rate of progress on these questions.

Acknowledgments. I apologize for the inability to cite many relevant original research articles due to space limitations. I thank the many colleagues who have contributed to the work from this laboratory cited here, and Ron Wells for manuscript preparation and editing. Research supported by NIH grant GM47475, DOE-BES grant DE-FG03-87ER13742, USDA CRIS 5335-21000-017-00D, and Torrey Mesa Research Institute, San Diego.

References

Alabadi D, Oyama T, Yanovsky MJ, Harmon FG, Mas P, Kay SA (2001) Reciprocal regulation between *TOC1* and *LHY/CCA1* within the Arabidopsis circadian clock. Science 293: 880–883

Bauer D, Viczian A, Kircher S, Nobis T, Nitschke R, Kunkel T, Panigrahi KCS, Adam E, Fejes E, Schäfer E, Nagy F (2004) Constitutive Photomorphogenesis 1 and multiple photoreceptors control degradation of Phytochrome Interacting Factor 3, a transcription factor required for light signaling in Arabidopsis. Plant Cell 16: 1433–1445

Bhoo SH, Davis SJ, Walker J, Karniol B, Vierstra RD (2001) Bacteriophytochromes are photochromic histidine kinases using a biliverdin chromophore. Nature 414: 776–779

Choi G, Yi H, Lee J, Kwon YK, Soh MS, Shin B, Luka Z, Hahn TR, Song PS (1999) Phytochrome signalling is mediated through nucleoside diphosphate kinase 2. Nature 401: 610–613

Clack T, Mathews S, Sharrock RA (1994) The phytochrome apoprotein family in *Arabidopsis* is encoded by five genes: the sequences and expression of *PHYD* and *PHYE*. Plant Mol Biol 25: 413–427

Fairchild CD, Schumaker MA, Quail PH (2000) *HFR1* encodes an atypical bHLH protein that acts in phytochrome A signal transduction. Genes Dev 14: 2377–2391

Fankhauser C (2000) Phytochromes as light-modulated protein kinases. Semin Cell Dev Biol 11: 467–473

Fankhauser C (2001) The phytochromes, a family of red/far-red absorbing photoreceptors. J Biol Chem 276: 11453–11456

Fankhauser C, Yeh KC, Lagarias JC, Zhang H, Elich TD, Chory J (1999) PKS1, a substrate phosphorylated by phytochrome that modulates light signaling in Arabidopsis. Science 284: 1539–1541

Franklin KA, Davis SJ, Stoddart WM, Vierstra RD, Whitelam GC (2003) Mutant analyses define multiple roles for phytochrome C in Arabidopsis photomorphogenesis. Plant Cell 15: 1981–1989

Furuya M (ed) (1987) Phytochrome and photoregulation in plants. Proceedings of the Yamada Conference XVI. Academic, New York

Goff SA, Ricke D, Lan TH, Presting G, Wang RL, Dunn M, Glazebrook J, Sessions A, Oeller P, Varma H, Hadley D, Hutchinson D, Martin C, Katagiri F, Lange BM,

Moughamer T, Xia Y, Budworth P, Zhong JP, Miguel T, Paszkowski U, Zhang SP, Colbert M, Sun WL, Chen LL, Cooper B, Park S, Wood TC, Mao L, Quail P, Wing R, Dean R, Yu YS, Zharkikh A, Shen R, Sahasrabudhe S, Thomas A, Cannings R, Gutin A, Pruss D, Reid J, Tavtigian S, Mitchell J, Eldredge G, Scholl T, Miller RM, Bhatnagar S, Adey N, Rubano T, Tusneem N, Robinson R, Feldhaus J, Macalma T, Oliphant A, Briggs S (2002) A draft sequence of the rice genome (*Oryza sativa* L. ssp. *japonica*). Science 296: 92–100

Gyula P, Schäfer E, Nagy F (2003) Light perception and signalling in higher plants. Curr Opin Plant Biol 6: 446–452

Henikoff S, Comai L (2003) Single-nucleotide mutations for plant functional genomics. Annu Rev Plant Biol 54: 375–401

Hisada A, Hanzawa H, Weller JL, Nagatani A, Reid JB, Furuya M (2000) Light-induced nuclear translocation of endogenous pea phytochrome A visualized by immunocytochemical procedures. Plant Cell 12: 1063–1078

Hudson ME, Quail PH (2003) Identification of promoter motifs involved in the network of phytochrome A-regulated gene expression by combined analysis of genomic sequence and microarray data. Plant Physiol 133: 1605–1616

Huq E, Quail PH (2002) PIF4, a phytochrome-interacting bHLH factor, functions as a negative regulator of phytochrome B signaling in *Arabidopsis*. EMBO J 21: 2441–2450

Huq E, Al-Sady B, Quail PH (2003) Nuclear translocation of the photoreceptor phytochrome B is necessary for its biological function in seedling photomorphogenesis. Plant J 35: 660–664

Huq E, Al-Sady B, Hudson M, Kim C, Apel K, Quail PH (2004) PHYTOCHROME-INTERACTING FACTOR 1 is a critical regulator of chlorophyll biosynthesis. Science 305: 1937–1941

Jones AM, Ecker JR, Chen JG (2003) A reevaluation of the role of the heterotrimeric G protein in coupling light responses in Arabidopsis. Plant Physiol 131: 1623–1627

Kendrick RE, Kronenberg GHM (eds) (1994) Photomorphogenesis in plants, 2nd edn. Kluwer, Dordrecht

Kim J, Yi H, Choi G, Shin B, Song PS, Choi G (2003) Functional characterization of phytochrome interacting factor 3 in phytochrome-mediated light signal transduction. Plant Cell 15: 2399–2407

Lariguet P, Boccalandro HE, Alonso JM, Ecker JR, Chory J, Casal JJ, Fankhauser C (2003) A growth regulatory loop that provides homeostasis to phytochrome A signaling. Plant Cell 15: 2966–2978

Ma L, Li J, Qu L, Hager J, Chen Z, Zhao H, Deng XW (2001) Light control of arabidopsis development entails coordinated regulation of genome expression and cellular pathways. Plant Cell 13: 2589–2607

Ma L, Gao Y, Qu L, Chen Z, Li J, Zhao H, Deng XW (2002) Genomic evidence for COP1 as a repressor of light-regulated gene expression and development in Arabidopsis? Plant Cell 14: 2383–2398

Martínez-García JF, Huq E, Quail PH (2000) Direct targeting of light signals to a promoter element-bound transcription factor. Science 288: 859–863

Mathews S, Sharrock RA (1997) Phytochrome gene diversity. Plant Cell Environ 20: 666–671

Matsushita T, Mochizuki N, Nagatani A (2003) Dimers of the N-terminal domain of phytochrome B are functional in the nucleus. Nature 424: 571–574

Millar AJ, McGrath RB, Chua NH (1994) Phytochrome phototransduction pathways. Annu Rev Genet 28: 325–349

Moller SG, Ingles PJ, Whitelam GC (2002) The cell biology of phytochrome signalling. New Phytol 154: 553–590

Monte E, Alonso JM, Ecker JR, Zhang Y, Li X, Young J, Austin-Phillips S, Quail PH (2003) Isolation and characterization of phyC mutants in Arabidopsis reveals complex crosstalk between phytochrome signaling pathways. Plant Cell 15: 1962–1980

Montgomery BL, Lagarias JC (2002) Phytochrome ancestry: sensors of bilins and light. Trends Plant Sci 7: 357–366

Nagy F, Schäfer E (2000a) Nuclear and cytosolic events of light-induced, phytochrome-regulated signaling in higher plants. EMBO J 19: 157–163

Nagy F, Schäfer E (2000b) Control of nuclear import and phytochromes. Curr Opin Plant Biol 3: 450–454

Nagy F, Schäfer E (2002) Phytochromes control photomorphogenesis by differentially regulated, interacting signaling pathways in higher plants. Annu Rev Plant Biol 53: 329–355

Nemhauser JL, Chory J (2004) BRing it on: new insights into the mechanism of brassinosteroid action. J Exp Bot 55: 265–270

Ni M, Tepperman JM, Quail PH (1998) PIF3, a phytochrome-interacting factor necessary for normal photoinduced signal transduction, is a novel basic helix-loop-helix protein. Cell 95: 657–667

Ni M, Tepperman JM, Quail PH (1999) Binding of phytochrome B to its nuclear signaling partner PIF3 is reversibly induced by light. Nature 400: 781–784

Quail PH (2000) Phytochrome interacting factors. Semin Cell Dev Biol 11: 457–466

Quail PH (2002a) Phytochrome photosensory signalling networks. Nat Rev Mol Cell Biol 3: 85–93

Quail PH (2002b) Photosensory perception and signalling in plant cells: new paradigms? Curr Opin Cell Biol 14: 180–188

Quail PH, Boylan MT, Parks BM, Short TW, Xu Y, Wagner D (1995) Phytochromes: photosensory perception and signal transduction. Science 268: 675–680

Sage LC (1992) Pigment of the imagination, a history of phytochrome research. Academic, San Diego

Saijo Y, Sullivan JA, Wang HY, Yang JP, Shen YP, Rubio V, Ma LG, Hoecker U, Deng XW (2003) The COP1–SPA1 interaction defines a critical step in phytochrome A-mediated regulation of HY5 activity. Genes Dev 17: 2642–2647

Sakamoto K, Nagatani A (1996) Nuclear localization activity of phytochrome B. Plant J 10: 859–868

Salter MG, Franklin KA, Whitelam GC (2003) Gating of the rapid shade-avoidance response by the circadian clock in plants. Nature 426: 680–683

Schmidt M, Schneider-Poetsch HAW (2002) The evolution of gymnosperms redrawn by phytochrome genes: the Gnetatae appear at the base of the gymnosperms. J Mol Evol 54: 715–724

Seo HS, Yang JY, Ishikawa M, Bolle C, Ballesteros ML, Chua NH (2003) LAF1 ubiquitination by COP1 controls photomorphogenesis and is stimulated by SPA1. Nature 423: 995–999

Serino G, Deng XW (2003) The COP9 signalosome: Regulating plant development through the control of proteolysis. Annu Rev Plant Biol 54: 165–182

Smith H (2000) Phytochromes and light signal perception by plants—and emerging synthesis. Nature 407: 585–591

Soh MS, Kim YM, Han SJ, Song PS (2000) REP1, a basic helix-loop-helix protein, is required for a branch pathway of phytochrome a signaling in Arabidopsis. Plant Cell 12: 2061–2073

Somers DE, Devlin PF, Kay SA (1998) Phytochromes and cryptochromes in the entrainment of the Arabidopsis circadian clock. Science 282: 1488–1490

Spiegelman JI, Mindrinos MN, Fankhauser C, Richards D, Lutes J, Chory J, Oefner PJ (2000) Cloning of the Arabidopsis *RSF1* gene by using a mapping strategy based on high-density DNA arrays and denaturing high-performance liquid chromatography. Plant Cell 12: 2485–2498

Sullivan JA, Shirasu K, Deng XW (2003) The diverse roles of ubiquitin and the 26S proteasome in the life of plants. Nature Rev Genet 4: 948–958

Tepperman JM, Zhu T, Chang HS, Wang X, Quail PH (2001) Multiple transcription-factor genes are early targets of phytochrome A signaling. Proc Natl Acad Sci USA 98: 9437–9442

Tepperman JM, Hudson ME, Khanna R, Zhu T, Chang H-S, Wang X, Quail PH (2004) Expression profiling of *phyB* mutant demonstrates substantial contribution of other phytochromes to red-light-regulated gene expression during seedling deetiolation. Plant J 38: 725–739

Toledo-Ortiz G, Huq E, Quail PH (2003) The Arabidopsis basic/helix-loop-helix transcription factor family. Plant Cell 15: 1749–1770

Wada M, Kanegae T, Nozue K, Fukuda S (1997) Cryptogram phytochromes. Plant Cell Environ 20: 685–690

Wada M, Kagawa T, Sato Y (2003) Chloroplast movement. Annu Rev Plant Biol 54: 455–468

Wang ZY, Tobin EM (1998) Constitutive expression of the *CIRCADIAN CLOCK ASSOCIATED 1* (*CCA1*) gene disrupts circadian rhythms and suppresses its own expression. Cell 93: 1207–1217

Wang H, Ma L, Habashi J, Li J, Zhao H, Deng XW (2002) Analysis of far-red light-regulated genome expression profiles of phytochrome A pathway mutants in Arabidopsis. Plant J 32: 723–733

Whitelam GC, Patel S, Devlin PF (1998) Phytochromes and photomorphogenesis in Arabidopsis. Phil Trans R Soc Lond B Biol Sci 353: 1445–1453

Yamaguchi R, Nakamura M, Mochizuki N, Kay SA, Nagatani A (1999) Light-dependent translocation of a phytochrome B-GFP fusion protein to the nucleus in transgenic Arabidopsis. J Cell Biol 145: 437–445

Yeh KC, Wu SH, Murphy JT, Lagarias JC (1997) A cyanobacterial phytochrome two-component light sensory system. Science 277: 1505–1508

Chapter 3

Evolutionary Selection of Phytochrome Chromophores

Takayuki Kohchi

Introduction

Phytochromes in plants are dimeric proteins that have a single linear tetrapyrrole molecule, phytochromobilin (PΦB), as chromophore in each protein molecule. Chromophore attachment to photoreceptor is essential for light perception. Although different types of phytochromes are encoded by a small multigene family in plants, the structure of chromophore molecules is common for all phytochromes. In prokaryotes, phytochrome-like proteins named bacteriophytochromes (Bphs), which have a different prosthetic group, biliverdin (BV) or phycocyanobilin (PCB), were discovered in the 1990s (Hughes and Lamparter 1999, Montgomery and Lagarias 2002). The genes and proteins for tetrapyrrole metabolism have been identified in plants by a molecular genetic approach with *Arabidopsis* photomorphogenic mutants (Muramoto et al 1999, Kohchi et al 2001) and in algae by a comparative genomics (Frankenberg et al 2001). Using mutants and biosynthetic genes of phytochrome chromophores as tools, a genetic system was developed allowing a structure–function assay of phytochrome chromophores in photochromic and physiological responses in plants (Kami et al 2004). Here we present a mini-review of phytochrome chromophore biosynthesis and structure from an evolutionary point of view.

Structure and Evolution of Phytochrome Chromophores

When the absorption spectrum of the red-light-absorbing form of phytochrome, Pr, was measured in the 1950s, its similarity to that of phycocyanin was noticed (Rüdiger and Thümmler 1992). Due to the abundance of phycocyanin as phycobiliprotein antennae, the structure of the phycocyanin chromophore was firstly identified as phycocyanobilin (PCB) composed of four tetrapyrrole rings, A, B,

Graduate School of Biostudies, Kyoto University, Kyoto 606-8502, Japan

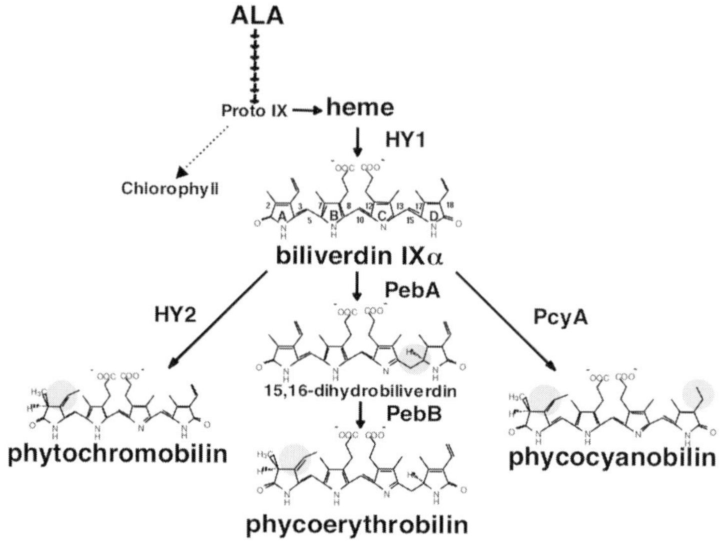

FIG. 1. Phytobilin biosynthesis pathway in photosynthetic organisms. Biliverdin (*BV*) is synthesized from 5-aminolevurinate (*ALA*) via heme. *HY1*, heme oxygenase; *HY2*, phytochromobilin synthase (phytochromobilin:ferredoxin oxidoreductase); *PcyA*, phycocyanobilin:ferredoxin oxidoreductase; *PebA*, 15,16-dihidrobiliverdin:ferredoxin oxidoreductase; *PebB*, phycoerythrobilin:ferredoxin oxidoreductase

C, and D (Figure 1). It took more than twenty years to determine the chromophore structure after the discovery of phytochromes (Rüdiger and Thümmler 1992). The chromophore of phytochrome is phytochromobilin (PΦB), which is closely related to PCB (Figure 1). The difference between PCB and PΦB is a substitution at ring D that consists of a vinyl group in PΦB and an ethyl group in PCB. The C3 ethylidene double bond of the chromophore is covalently linked to a Cys residue located in the N-terminal half of the protein.

In the mid 1990s, numerous phytochromes were discovered in algae and bacteria as well as in plants (Hughes and Lamparter 1999, Montgomery and Lagarias 2002). Cyanobacterial phytochrome 1 (Cph1) is an example of the bacteriophytochromes (Bphs) found by sequencing the *Synechocystis* PCC6803 genome (Kaneko et al 1996). The Cph1 protein expressed in *Escherichia coli* was a light-regulated histidine kinase that showed a typical red/far-red photoreversible spectral property in the presence of supplied PCB (Yeh et al 1997). Indeed, Cph1 proteins purified from cyanobacteria contained PCB as natural chromophore (Hübschmann et al 2001). In the cyanobacterium *Calothrix* PCC7601, there are two types of Bphs, CphA and CphB. CphA covalently binds PCB, whereas CphB that lacked the Cys residues required for covalent chromophore binding in plant phytochromes carries biliverdin (BV) as its chromophore (Quest and Gärtner 2004). Biliverdin is believed to be the chromophore for Bphs from nonphotosynthetic microbes (Bhoo et al 2001). In a broad sense, phytochrome families can be extended to a group of sensors with a bilin prosthetic group found in a wide range of organisms.

Biosynthetic Pathway of Phytochrome Chromophores

PΦB is synthesized from aminolevulinate through intermediates commonly observed during chlorophyll biosynthesis. Heme is synthesized by chelating iron into the closed tetrapyrrole ring of protoporphyrin IX, while chlorophyll is synthesized by chelating magnesium. Heme is not just a final product of tetrapyrrole metabolism, but is an intermediate for bilin biosynthesis in photosynthetic organisms. Although the PCB biosynthesis pathway in *Cyanidium caldarium* has become a model for bilin biosynthesis, the analytical methods used cannot easily be applied to study of PΦB biosynthesis owing to the low abundance of respective enzymes in plants. Enzyme purification and gene identification have been difficult even in studies of PCB biosynthesis.

Analysis of photomorphogenesis mutants proved powerful in research on phytochrome chromophore biosynthesis. Mutations of *HY1* and *HY2* genes in *Arabidopsis* cause deficiencies in light perception by phytochromes, resulting in mutants with photomorphogenetic phenotypes such as long hypocotyls, few and yellowish leaves, and early flowering due to a lack of photoactive phytochromes. Heme is first cleaved to BV by heme oxygenase, which is encoded by *HY1* in *Arabidopsis* (Figure 1). The *HY1* gene was first isolated by positional cloning (Davis et al 1999, Muramoto et al 1999) and shown to encode a functional heme oxygenase (Muramoto et al 1999, 2002). There are four heme oxygenase genes in the *Arabidopsis* genome, and the *HY1* gene that shows a clear photomorphogenic phenotype when mutated plays a major role in chromophore biosynthesis. But our understanding of the divergent roles of this gene family is far from complete (Terry et al 2002). In the next biosynthetic step, BV is reduced to PΦB by PΦB synthase encoded by *HY2* in *Arabidopsis* (Kohchi et al 2001). Both heme oxygenase and PΦB synthase in plants are ferredoxin-dependent and localized in plastids. PΦB biosynthesis occurs entirely in the plastids and the PΦB exported to the cytoplasm autocatalytically binds to phytochrome apoprotein.

Bilin Reductase Families in Photosynthetic Organisms

Identification of the *HY2* gene also shed light on phycobilin biosynthesis (Frankenberg et al 2001). In cyanobacteria and red algae, BV synthesized from heme is further reduced to PCB and phycoerythrobilin (PEB). A similarity search against the protein database identified the presence of HY2-related sequences in oxygenic photosynthetic organisms. Based on biochemical assays, the proteins that function during PCB and PEB biosynthesis were identified and named PcyA, PebA, and PebB (Figure 1). In cyanobacteria, PCB is the indispensable chromophore of phycobiliprotein and cyanobacterial phytochromes (Hughes and Lamparter 1999). The four-electron reduction of BV IXα at $2,3,3^1,3^2$-diene and C-$18^{1,2}$ to PCB is mediated by PcyA (Frankenberg et al 2001, 2003). The PCB product contains an ethyl group at C-18 where PΦB has a vinyl group. The pebA and pebB genes encode HY2-related bilin reductases involved in the two-step reduction of BV to PEB via 15,16-dihydrobiliverdin. These studies defined new bilin reductase families with different double bond specificities, and established links between phycobilin in microorganisms and PΦB biosynthesis in plants.

Structure–Function Assays of Phytochrome Chromophore

PΦB is believed to be the sole phytochrome chromophore in plants, although as described above, Bphs have BV and PCB chromophores. Therefore during evolution there could have been chromophores other than PΦB, which were potential phytochrome chromophores. PCB has been widely used in many experiments as a PΦB substitute since it is a ubiquitous and easily available photosynthetic pigment in algae. Recombinant oat apophytochrome from *E. coli* and yeast were shown to assemble efficiently with PCB as well as PΦB in vitro, and the holophytochrome obtained with PCB gave an absorption spectra blue-shifted by 10 to 15 nm as compared to the spectrum of the PΦB (Frankenberg et al 2001). Whether or not PΦB was selected as the phytochrome chromophore in plants as an adaptation to light environments on land is unclear. Hanzawa and colleagues (2001) established an approach for structurally analyzing structural relationship between chromophore and phytochrome. Chemically synthesized analogs of tetrapyrrole chromophores were reconstituted with PHYB, and used to evaluate ligation efficiency and spectral properties. The A-ring acts mainly as the anchor for protein attachment, while the B- and C-rings are crucial for the affinity of chromophore to protein and photoreversible spectra. Modification of the ring-D side chains resulted mainly in change in photoreversible spectra. Hanzawa et al (2002) also applied this system to feeding experiments in chromophore-deficient *Arabidopsis* mutants, *hy1* and *hy2*. Although R-HIR was efficiently rescued by feeding with PCB, far-red high-irradiance response (FR-HIR) sensitivity depended on the side chain structure of ring-D under certain light conditions.

To investigate the functional consequences of PΦB substitutions with PCB, Kami and colleagues (2004) developed a genetic chromophore modification system utilizing an *Arabidopsis* mutant and cyanobacterial bilin reductase genes. This genetic system has an advantage in evaluating the effect of chromophore substitutions on photobiological activities under different light conditions at any stages of development in planta. The chromophore substitution from PΦB to PCB was achieved by transgenic expression of a cyanobacterial pcyA gene for PCB:ferredoxin oxidoreductase in chromophore-deficient *hy2 Arabidopsis* mutants. In the *hy2* plants expressing *pcyA,* the phytochrome was present at wild type levels with similar photolabile properties, and showed a photoreversible but blue-shifted spectrum. This indicates that PCB produced by the pcyA transgene was efficiently incorporated into the holophytochrome in *Arabidopsis*. No obvious differences between the chromophore-modified plants and wild type were observed for the VLFR mediated by phyA, and for the LFR mediated by phyB. This indicated that the spectral changes that occurred by phytochrome modification with the PCB chromophore were relatively small for these responses. However, the phyA-mediated FR-HIR represented by hypocotyl elongation inhibition was impaired under far-red light conditions in the pcyA-expressing plants. Further monochromatic irradiations indicated that the response was deficient under far-red light of 730 nm but normal under that of 715 nm. These observations confirmed that phytochrome responses are mediated by different mode of action and that the selection of PΦB as a chromophore might have an ecological significance during physiological far-red responses in land plants.

Concluding Remarks

The recent discovery of functional cyanobacterial phytochromes has opened a new avenue of phytochrome research. With the numerous genetic resources and information in *Arabidopsis* and cyanobacteria, it is plausible to experimentally analyze the structural requirements of chromophores in plants. Since phyA with a PCB chromophore can sense shorter wavelengths (for example, 715 nm) during FR-HIRs, altered FR-HIR sensitivity was supposedly caused by the blue-shifted nature of the PCB chromophore. It should be noted, however, that a full FR-HIR was not achieved by the phytochrome with a PCB chromophore even if excess amounts of 730-nm light were given. Among the many phytochrome responses, the blue-shift spectra gave the most significant difference during FR-HIRs. It is hypothesized that the PΦB prosthetic group is required for plant-specific photobiological activities. In addition, it can be suggested that the co-evolution of phytochromes and chromophore biosynthesis occurred as an adaptation to the light environment.

References

Bhoo SH, Davis SJ, Walker J, Karniol B, Vierstra RD (2001) Bacteriophytochromes are photochromic histidine kinases using a biliverdin chromophore. Nature 414: 776–779

Davis SJ, Kurepa J, Vierstra RD (1999) The *Arabidopsis thaliana HY1* locus, required for phytochrome-chromophore biosynthesis, encodes a protein related to heme oxygenases. Proc Natl Acad Sci USA 96: 6541–6546

Frankenberg N, Mukougawa K, Kohchi T, Lagarias JC (2001) Functional genomic analysis of the HY2 family of ferredoxin-dependent bilin reductases from oxygenic photosynthetic organisms. Plant Cell 13: 965–978

Frankenberg N, Lagarias JC (2003) Phycocyanobilin:ferredoxin oxidoreductase of *Anabaena* sp. PCC 7120. Biochemical and spectroscopic characterization. J Biol Chem 278: 9219–9226

Hanzawa H, Inomata K, Kinoshita H, Kakiuchi T, Jayasundera KP, Sawamoto D, Ohta A, Uchida K, Wada K, Furuya M (2001) In vitro assembly of phytochrome B apoprotein with synthetic analogs of the phytochrome chromophore. Proc Natl Acad Sci USA 98: 3612–3617

Hanzawa H, Shinomura T, Inomata K, Kakiuchi T, Kinoshita H, Wada K, Furuya M (2002) Structural requirement of bilin chromophore for the photosensory specificity of phytochrome A and B. Proc Natl Acad Sci USA 99: 4725–4729

Hübschmann T, Börner T, Hartmann E, Lamparter T (2001) Characterization of the Cph1 holo-phytochrome from *Synechocystis* sp. PCC 6803. Eur J Biochem 268: 2055–2063

Hughes J, Lamparter T (1999) Prokaryotes and phytochromes. The connection to chromophore and signaling. Plant Physiol 121: 1059–1068

Kami C, Mukougawa K, Muramoto T, Yokota A, Shinomura T, Lagarias JC, Kohchi T (2004) Functional complementation of phytochrome chromophore-deficient *Arabidopsis* plants by expression of phycocyanobilin:Ferredoxin oxidoreductase. Proc Natl Acad Sci USA 101: 1099–1104

Kaneko T, Sato S, Kotani H, Tanaka A, Asamizu E, Nakamura Y, Miyajima N, Hirosawa M, Sugiura M, Sasamoto S, Kimura T, Hosouchi T, Matsuno A, Muraki A, Nakazaki N, Naruo K, Okumura S, Shimpo S, Takeuchi C, Wada T, Watanabe A, Yamada M, Yasuda M, Tabata S (1996) Sequence analysis of the genome of the unicellular cyanobacterium *Synechocystis* sp. strain PCC6803. II. Sequence determination of the entire genome and assignment of potential protein-coding regions. DNA Res 3: 109–136

Kohchi T, Mukougawa K, Masuda M, Yokota A, Frankenberg N, Lagarias JC (2001) The *Arabidopsis HY2* gene encodes phytochromobilin synthase, a ferredoxin-dependent biliverdin reductase. Plant Cell 13: 425–436

Montgomery BL, Lagarias JC (2002) Phytochrome ancestry: sensors of bilins and light. Trends Plant Sci 7: 357–66

Muramoto T, Kohchi T, Yokota A, Hwang I, Goodman HM (1999) The *Arabidopsis* photomorphogenic mutant *hy1* is deficient in phytochrome chromophore biosynthesis as a result of a mutation in a plastid heme oxygenase. Plant Cell 11: 335–348

Muramoto T, Tsurui N, Terry MJ, Yokota A, Kohchi T (2002) Expression and biochemical properties of a ferredoxin-dependent heme oxygenase required for phytochrome chromophore synthesis. Plant Physiol 130: 1958–1966

Quail PH (2002) Phytochrome photosensory signaling networks. Nat Rev Mol Cell Biol 3: 85–93

Quest B, Gärtner W (2004) Chromophore selectivity in bacterial phytochromes. Dissecting the process of chromophore attachment. Eur J Biochem 271: 1117–1126

Rüdiger W, Thümmler F (1992) The phytochrome chromophore. In: Kendrick RE, Kronenberg GHM (eds) Photomorphogenesis in plants. Kluwer Academic, Dordrecht

Terry MJ, Linley PJ, Kohchi T (2002) Making light of it: the role of plant haem oxygenase in phytochrome chromophore synthesis. Biochem Soc Trans 30: 604–609

Yeh KC, Wu SH, Murphy JT, Lagarias JC (1997) A cyanobacterial phytochrome two-component light sensory system. Science 277: 1505–1508

Chapter 4

Tertiary and Quaternary Structures of Phytochrome A

Masayoshi Nakasako[1,2] and Satoru Tokutomi[3]

Introduction

Phytochrome A (phyA) is a major member of phytochrome family and is photoconvertible between a red light (Pr) and a far-red light absorbing (Pfr) form. The Pfr form predominantly regulates downstream signaling cascades leading to a variety of photomorphogenic responses (Smith 2000, Fankhauser 2001, Quail 2002). In the last decade, molecular biological studies have identified functionally important residues and sequence motifs in phyA (Park et al 2000). However, the molecular structure of phyA and structural changes upon Pr-Pfr phototransformation are still unclear. Here, we describe our recent results on phyA structures with a brief review of structural studies on phytochromes.

Functional Domains and Motifs in the Primary Structure

PhyA comprises ca. 1100 amino acid residues and one chromophore, phytochromobiline, and folds into a photosensory region of the N-terminal half and a regulatory region of the C-terminal half (Figure 1). Molecular biological studies have revealed functionally important segments or motifs in each region (Figure 1).

The photosensory region is divided into a bilin lyase domain (BLD) (Wu and Lagarias 2000) and a phytochrome domain (PHY). BLD binds a phytochromobiline autocatalytically and shows reversible phototransformation, and PHY is requisite for the intactness of the absorption spectrum of Pfr. The primary structures of the BLD domains are well conserved among phyA–E, and the tertiary structure of its portion including the bilin-binding cysteine residue resembles those of GAF domains found in hundreds of signaling and sensory proteins

[1]Department of Physics, Faculty of Science and Technology, Keio University, 3-14-1 Hiyoshi, Kohoku-ku, Yokohama 223-8522, Japan
[2]RIKEN Harima Institute, Sayo, Hyogo 679-5148, Japan
[3]Research Institute for Advanced Science and Technology, Osaka Prefecture University, 1-2 Gakuen-cho, Sakai, Osaka 599-8570, Japan

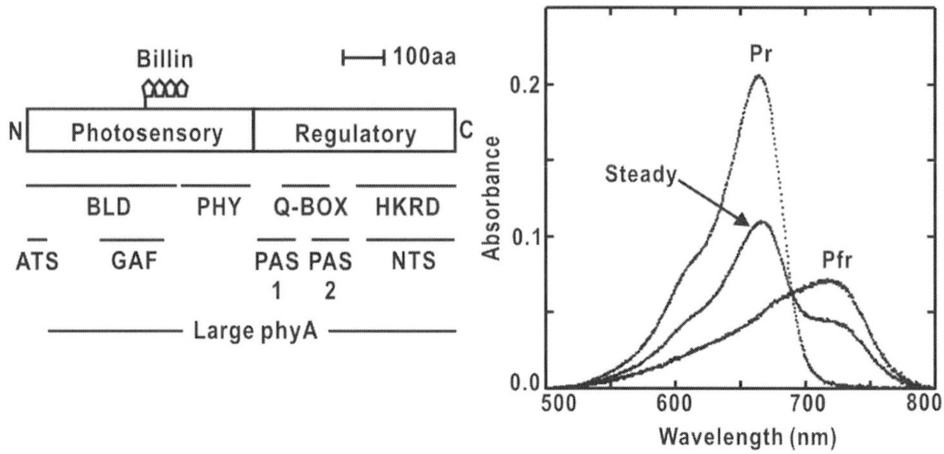

a

b

Fig. 1. **a** Locations of functional domains and segments in the primary structure of phyA. **b** Absorption spectra of pea large phyA in the dark (*Pr*) and under red light irradiation (*photostationary state*). The spectrum of Pfr is reconstructed by assuming the photostationary state to be composed of Pr (39%) and Pfr (61%)

(Montgomery and Lagarias 2002). The N-terminal segment of 6kDa (ATS) in BLD is shown to exhibit coil-to-helix transition upon Pr-to-Pfr transformation, and the conformational change is essential to stabilize Pfr (Deforce et al 1994). Serine residues located in ATS and near the C-terminal end of the photosensory region work as phosphorylation sites in down-regulation of phyA signaling (Park et al 2000).

The non-chromophoric region, named the regulatory region, is shown to be involved in signal transduction and dimerization. Two PAS (Per/Arnt/Sim)-related motifs are found in the N-terminal side of the regulatory region (Kay 1997). The motifs overlap with a cluster of residues named Quail-Box (QB) (Quail et al 1995), in which subtle conformational changes are observed upon Pr–Pfr phototransformation. The C-terminal segment of the regulatory region is called a histidine-kinase-related domain (HKRD), which is now considered to be a histidine kinase paralog with Ser/Thr kinase specificity (Yeh and Lagarias 1998). A nuclear localization signal (NLS) acting in the nuclear import of phyA (Kircher et al 1999) overlaps with HKRD.

Structural Studies on Phytochrome A

In contrast to the wealth of the information regarding the primary structure–function relationships of phytochromes, only limited and old information is available concerning their tertiary and quaternary structures. The molecular size and higher-order structures of phyA were investigated by using size exclusion chromatography (Lagarias and Mercurio 1985, Jones and Quail 1986, Tokutomi et al 1988), ultracentrifugation (Jones and Quail 1986), and quasi-elastic light scattering measurement (Sarker et al 1984). The estimated molecular mass of phyA in the range of 250–350 kDa suggested most likely dimeric association of phyA in solution. In addition, the estimated molecular mass of the chromophoric domain indicated its monomeric existence. Considering together the molecular mass and the estimated Stokes radius (ca. 56 Å), a molecular model composed of two subunits with an elongated tadpole-like shape was proposed for phyA: a head corresponding to the chromophoric region and a tail to the regulatory region.

Electron micrographs with rotary shadowing technique visualize the molecular shape of oat phyA (Jones and Erickson 1989) and pea large phyA, lacking ATS (Tokutomi et al 1989). Based on the images of oat phyA appearing as clusters of three domain-like structures, Jones and Erickson (1989) proposed the "Mickey Mouse" model to explain the structure–function relationships of phyA; the ears of "Mickey" sensing the environment correspond to the photosensory region, and the head communicating with others forms the regulatory region.

To visualize the molecular structure of phytochrome at an atomic resolution, a crystal structure analysis at a resolution better than 3.5 Å is required. Though phyA is the only isoform obtainable by purification from tissues on a large scale, it has been difficult to crystallize it due to its tendency to form amorphous aggregates, the heterogeneity that results from post-translational modifications, and so on. For studying the molecular structure of phyA, we applied the small-angle X-ray scattering (SAXS) technique. SAXS measurement provides molecular mass, radius of gyration, and maximum dimension of a protein, and enables estimation of low-resolution molecular structures of proteins in solution. About 10 years ago, we proposed a "Four Leaves" model based on SAXS and electron microscopy for Pr of pea large phyA (Tokutomi et al 1989, Nakasako et al 1990).

Molecular Structure of Phytochrome A in Pr

Since our previous studies, almost no progress has been made in structural studies on plant phytochromes. Recently, we have developed a new purification protocol for pea large phyA by modifying the previous one. The new protocol yields large phyA solution with little aggregation. Furthermore, the SAXS technique has been remarkably improved by utilizing highly brilliant synchrotron X-rays, detectors of high-sensitivity and ab initio molecular structure determination software. Now the software contributes significantly to provide unambiguous low-resolution molecular models of proteins from only their scattering profiles (Svergun 2001). Thus, we reinvestigated the molecular structures of Pr and Pfr by applying the modern SAXS techniques to pea large phyA prepared by the new protocol (Nakasako et al 2004).

In contrast to the previous study, the SAXS profile of Pr, recorded up to a resolution of ca. 8 Å, displays a good monodispersive property (Figure 2a). The

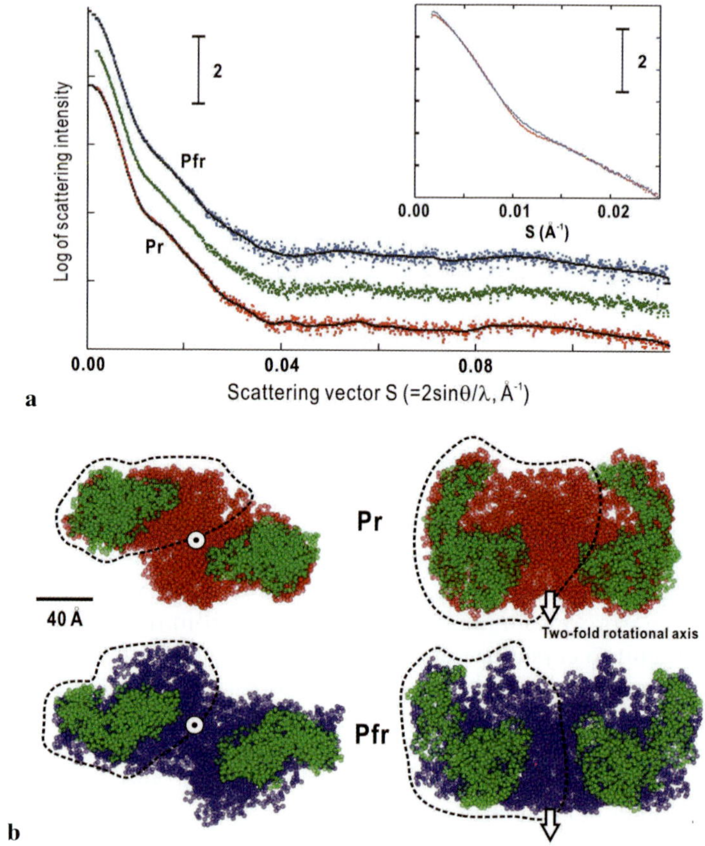

FIG. 2. **a** SAXS profiles of pea large phyA in Pr (*red dots*), under red light irradiation (*green dots*) and in Pfr (*blue dots*). The Pfr profile is calculated from the other two profiles by using the population ratio from Figure 1b. The profiles shown by *black lines* are those of the restored structural models in **b**. The profiles of Pr and Pfr in the *inset* are magnified in the region S < 0.025 Å^{-1}, where S is the length of scattering vector. **b** Molecular structures of Pr (*red-colored models in the upper panel*) and Pfr (*blue-colored models in the lower panel*) restored by the ab initio simulation for SAXS. Two views of molecular models are displayed as assemblies of dummy scatters used in the simulation. The *dashed lines* indicate plausible borders of two subunits. The *small circles in the left panel* and *arrows in the right* indicate the twofold rotational symmetry axes. The *green-colored model* of pea small phyA restored from the SAXS is superimposed on the models manually so as to maximize the overlapped area

characteristic enhancements appearing at around S = 0.015 Å^{-1}, 0.055 Å^{-1}, and 0.090 Å^{-1} indicate the presence of distinct structural domains and/or segments. The SAXS data confirm the dimeric association of Pr by determining the molecular weight and re-evaluate the radius of gyration (57.1 ± 0.1 Å) and the maximum dimension (183 ± 5 Å). The low-resolution molecular model predicted by the GASBOR program (Svergun 2001) is composed of two identical flat lobes stacked in an off-centered manner (Figure 2b, upper) and still has characteristics of the former "Four Leaves" model. One lobe plausibly assignable to one subunit has dimensions of 50 × 120 × 120 Å. A lobe is further divided into two portions: one facing to the twofold symmetry axis and the other located apart by ca. 60 Å from the axis. Taking the following two facts into account, the outer and the central lobes are possibly assignable to the photosensory and the regulatory regions, respectively. Firstly, the dimerization sites that are present in the central portion of the model exist only in the regulatory region (Edgerton and Jones 1993). Secondly, our previous study demonstrated that pea small phyA corresponding to the sensory region existed as a monomeric form in solution (Nakasako et al 1990). The structural model of small phyA predicted by the ab initio simulation (green-colored models in Figure 2b) is superimposable on the outer portion of a subunit lobe in the Pr model.

Red Light-Induced Structural Changes

It has been hypothesized that the light-induced conformational changes in BLD activate the regulatory domain through interdomain cross-talk, probably including the rearrangements and/or conformational changes of the functional segments. Light-induced changes in the physicochemical properties of phyA have been studied to understand the molecular mechanism initiating the phytochrome signaling cascades (Park et al 2000). Some biochemical assays revealed the changes in the surface properties upon Pr–Pfr conversion, and a few physicochemical experiments suggested possible global conformational changes by estimating molecular sizes of Pr and Pfr.

In the previous study, SAXS analyses on Pfr of pea large phyA could not be subjected to SAXS analysis owing to its severe aggregations under red light irradiation. In the present study, we have cleared the problem of the aggregation, and have detected successfully small but definite red light-induced changes of SAXS (Figure 2). Under red light, pea large phyA is in a photostationary state between Pr and Pfr, and the absorption spectrum is well simulated by the mixed spectra of 39% Pr and 61% Pfr (Figure 1). The populations of Pr and Pfr are the characteristic of pea large phyA. Thus, the scattering profile of 100% Pfr is reconstructed using the ratio. The reconstructed profile suggests strongly that the tertiary- and/or quaternary-structural changes occur predominantly in scales larger than 50 Å (Figure 2, inset). SAXS of Pfr are analyzed by the conventional methods applied to monodispersive solutions. The molecular weight of Pfr is the same as that of Pr, indicating no light-dependent dissociation into subunits or further association of dimers. The Rg and the maximum molecular dimension are determined to be 62.6 ± 0.4 Å and 211 ± 5 Å, respectively. These values suggest a significant expansion of the molecular size in the phototransformation to Pfr. In addition, more detailed analysis for SAXS shows that intermolecular interactions between large phyA molecules are significant only in Pfr. This may be advantageous for phyA in interacting with signal-mediating proteins under red light irradiation.

The predicted Pfr model has dimensions of 190 × 130 × 100 Å (Figure 2b) and is also approximated by a stack of two lobes. However, the lobes in the Pfr model display differences from those of Pr regarding their shapes and mutual orientations. The central lobe changes its shape and the outer lobe seems to rotate relative to the central lobe. This may be consistent with the exposure of a phosphorylation site in the "hinge" region between the sensory and the regulatory regions in Pfr of oat phyA. The structural changes may be correlated with the exposure of the NLS or the Quail-Box. Recently, rescue experiments of an *Arabidopsis* phyB mutant revealed that a combination of the photosensory domain with an artificial dimerization site and an NLS in its C-terminus is sufficient for phyB functions (Matsushita et al 2003). Because the arrangements of the functioning domains and segments in the primary sequence are similar among phyA, B and the artificial phyB, the artificial phyB may show similar conformational changes to large phyA.

Reversibility of the Red Light-Induced Structural Changes

Pfr is convertible to Pr by far-red light and is also shown to revert to Pr through a time-dependent thermal process termed dark reversion. The red light-induced SAXS changes are only partially reversed by a far-red light irradiation with a sufficient fluence and period to convert Pfr to Pr as judged from their absorption spectra. Thus, the SAXS results indicate that the absorption spectra do not reflect directly the overall structure of pea large phyA in the photoconversion by far-red light. When a red light-irradiated large phyA sample is kept in the dark, SAXS gradually goes back to that measured before light irradiation. In the dark reversion process, the SAXS changes couples with the fast component in the reversion of the absorption only within a time-scale of several minutes. This result indicates again that the absorption spectra of large phyA do not necessarily reflect the overall structure.

Concluding Remarks and Future Prospects

The modern SAXS measurements and analysis have provided reliable molecular models of pea large phyA and have revealed global conformational changes upon Pr–Pfr phototransformation. It should, however, be kept in mind that the present model for Pr is not a unique solution as would be obtained through X-ray crystallography, and this is clearly the next step.

References

Deforce L, Tokutomi S, Song PS (1994) Phototransformation of pea phytochrome A induces an increase in α-helical folding of the apoprotein: comparison with a monocot phytochrome A and CD analysis by different methods. Biochemistry 33: 4918–4922

Edgerton MD, Jones AM (1993) Subunit interactions in the carboxyl-terminal domain of phytochrome. Biochemistry 32: 8239–8245

Fankhauser C (2001) The phytochromes, a family of red/far-red absorbing photoreceptors. J Biol Chem 276: 11453–11456

Jones AM, Erickson HP (1989) Domain structure of 124-kilodulton phytochrome from *Avena sativa* visualized by electron micrograph. Photochem Photobiol 49: 479–483

Jones AM, Quail PH (1986) Quaternary structure of 124-kilodalton phytochrome from *Avena sativa* L. Biochemistry 25: 2987–2995

Kay SA (1997) PAS, present, and future: clues to the origins of circadian clocks. Science 276: 753–754

Kircher S, Kozma-Bognar L, Kim L, Adam E, Harter K, Schäfer E, Nagy F (1999) Light quality-dependent nuclear import of the plant photoreceptors phytochrome A and B. Plant Cell 11: 1445–1456

Lagarias JC, Mercurio FM (1985) Structure function studies on phytochrome. J Biol Chem 240: 2415–2423

Matsushita T, Mochizuki N, Nagatani A (2003) Dimers of the N-terminal domain of phytochrome B are functional in the nucleus. Nature 424: 571–574

Montgomery BL, Lagarias JC (2002) Phytochrome ancestry: sensors of bilins and light. Trends Plant Sci 7: 357–366

Nakasako M, Wada M, Tokutomi S, Yamamoto KT, Sakai J, Kataoka M, Tokunaga F, Furuya M (1990) Quaternary structure of pea phytochrome I dimmer studied with small-angle X-ray scattering and rotary-shadowing electron microscopy. Photochem Photobiol 52: 3–12

Nakasako M, Iwata T, Inoue K, Tokutomi S (2005) Light-induced conformational changes in phytochrome A regulating photomorphogenesis in plants. Eur J Biochem, in press.

Park CM, Bhoo SH, Song PS (2000) Inter-domain crosstalk in the phytochrome molecules. Cell Dev Biol 11: 449–456

Quail PH (2002) Phytochrome photosensory signaling networks. Nat Rev Mol Cell Biol 3: 85–93

Quail PH, Boylan MT, Parks BM, Short TW, Xu Y, Wagner D (1995) Phytochromes: photosensory perception and signal transduction. Science 268: 675–680

Sarker HK, Moon DK, Song PS, Chang T, Yu H (1984) Tertiary structure of phytochrome probed by quasi-elastic light scattering and rotational relaxation time measurements. Biochemistry 23: 1882–1888

Smith H (2000) Phytochromes and light signal perception by plants an emerging synthesis. Nature 407: 585–591

Svergun DI (2001) Determination of domain structure of proteins from X-ray solution scattering. Biophys J 80: 2946–2953

Tokutomi S, Kataoka M, Sakai J, Nakasako M, Tokunaga F, Tasumi M, Furuya M (1988) Small-angle X-ray scattering studies on the macromolecular structure of the red light-absorbing form of 121 kDa pea phytochrome and its 114 kDa chromopeptide. Biochim Biophys Acta 953: 297–305

Tokutomi S, Nakasako M, Sakai J, Kataoka M, Yamamoto KT, Wada M, Tokunaga F, Furuya M (1989) A model for the dimeric molecular structure of phytochrome based on small-angle X-ray scattering. FEBS Lett 247: 139–142

Wu SH, Lagarias JC (2000) Defining the bilin lyase domain: Lessons from the extended phytochrome superfamily. Biochemistry 39: 13487–13495

Yeh KC, Lagarias JC (1998) Eukaryotic phytochromes: light-regulated serine/threonine protein kinases with histidine kinase ancestry. Proc Natl Acad Sci USA 95: 13976–13981

Chapter 5

Interactions of the Arabidopsis Type II Phytochromes

Robert Sharrock

Introduction

The need to distinguish between two different "pools" or "types" of phytochrome in higher plants dates to the early 1960s, when efforts to correlate plant physiological light responses with the spectroscopic properties of phytochrome were sometimes confounding and inconsistent with the activity of a single molecular species of the photoreceptor (Pratt 1995). Moreover, it progressively became clear that phytochrome regulates responses to remarkably different parameters of the light environment (LFRs, VLFRs, HIRs and responses to ratios of R : FR) and the likelihood that all of these responses were mediated by one pigment seemed remote. In 1985, the three groups led by Furuya, Pratt, and Quail independently described forms of phytochrome that were present at low levels in extracts of both light-grown and dark-grown plants. These appeared to be distinct from the form of phytochrome that was abundant in etiolated plant tissues but rapidly degraded upon exposure to light (Furuya 1993). The distinction made by all three of these groups between a pool of phytochrome called "etiolated-tissue," "light-labile," or "type I" phytochrome and a pool called "green-tissue," "light-stable," or "type II" phytochrome has largely been supported by the identification and analysis of *PHY* gene families and the phytochrome apoproteins encoded by those genes, and by characterization of *phy* mutants that illuminate the individual activities of those genes. In the plants that have been examined, light-labile type I phytochrome appears to be a single molecular species, or a very highly related set of molecules, called phyA. Mutational loss of phyA results in disruption of a plant's ability to sense high fluences of continuous FR light and very low fluences of a broad spectrum of light wavelengths including R, FR, and B. In contrast, the pool of light-stable type II phytochrome in most plants appears to be heterogeneous, consisting of multiple divergent forms called phyB, phyC, and so on. Mutational loss of individual type II receptors results in partial reduc-

Department of Plant Sciences and Plant Pathology, Montana State University, Bozeman, MT 59717, USA

tion in a plant's ability to sense continuous R, with phyB forms having the strongest effects, and mutants lacking combinations of the type II forms are severely deficient in R sensing. Hence, the type I/type II distinction appears to meaningfully describe a fundamental difference in phytochrome structure and function.

Functions and Genetic Interactions of the Arabidopsis Phytochromes

Arabidopsis contains a single type I-encoding *PHYA* gene and four type II-encoding genes, *PHYB–PHYE*. The expression patterns and light-stabilities of these five proteins are consistent with these designations (Sharrock and Clack 2002) and the phenotypes of *phyA-phyE* null mutants demonstrate that they have distinct and differential activities (Whitelam and Devlin 1997). The phyA protein is 10–60-fold more abundant than any of the other forms in the dark and falls to a level 10-fold lower than the others in continuous white light. The *phyA* mutant shows a lack of response to continuous FR but very little effect on its morphology under white light conditions. The phyB–phyE type II phytochromes are present at low levels in the dark and fall in abundance only a few fold in continuous white or red light. The *phyB–phyE* mutants individually show varying defects in detection of continuous R, resulting in light-grown morphologies that mimic aspects of growth under a low R:FR ratio, the shade-avoidance responses. This is particularly evident in the *phyB* mutant, which shows a striking constitutive shade-avoidance phenotype (Whitelam and Devlin 1997). PhyB and phyD are the most closely related in sequence among the Arabidopsis phytochromes but, although the *phyD* mutant is defective in sensing the R:FR ratio, it shows a much weaker phenotype than the *phyB* mutant (Aukerman et al 1997). PhyE is somewhat more related in its sequence to phyB and phyD than to phyC or phyA and it also functions as an R sensor, prominently in controlling vegetative internode elongation in response to a low R:FR ratio (Devlin et al 1998) and in seed germination (Hennig et al 2002). PhyC is about equivalently related in sequence to phyA and phyB/D/E, and operates as a weak R sensor for seedling de-etiolation responses, rosette development, and flowering time (Franklin et al 2003a, Monte et al 2003). With respect to their photochemical properties, spectral analysis of plant-expressed phyA and phyB, and of phyC and phyE expressed in yeast and assembled with chromophore, indicate that these four Arabidopsis phytochromes undergo similar R/FR-reversible conformational changes but with significant differences in their absorption peak positions and dark reversion kinetics (Eichenberg et al 2000).

With the growing understanding of the individual functions of the phytochromes has come a realization that the activities of these receptors overlap and interact with each other. It is often seen that a given light-regulated plant trait, such as hypocotyl elongation, flowering time, and other characteristics, is altered in more than one of the single-gene *phy* mutants, indicating that multiple phytochromes co-regulate the response. In these cases, the activity of one R/FR photoreceptor or its signaling pathway may be affected by the activity of other photoreceptors. Construction of Arabidopsis lines with multiple *phy* mutations, that are deficient for various combinations of phytochromes, has allowed investigation of these regulatory interactions. The picture that has emerged is one of frequent and complex interactions not only among the different phytochromes but also between phytochromes and the blue light-sensing cryptochromes (Casal 2000). It has been observed that two or more phytochromes can have additive, synergistic, or antagonistic effects on each other's activities depending upon the response and the light conditions being studied. One example of this is the redundancy of phyB and phyD, which is revealed in the observation that the effects of a *phyD* mutation on hypocotyl and petiole elongation, leaf development, and flowering time are seen much more prominently in a *phyB* mutant background than in the presence of phyB (Aukerman et al 1997). Such redundancy is relatively common among the phyB/D/E phytochromes (Franklin et al 2003b). Combination of a *phyC* mutation with *phyA*, *phyB*, and *phyD* mutations has uncovered a large number of interactions between phyC and other members of the phytochrome family (Monte et al 2003). Synergy or antagonism of function between type I phyA and individual type II forms has also been observed (Casal 2000). In no case has the molecular mechanism of one of these regulatory interactions been identified and it is often proposed that these interactions occur at the level of cross-talk of signal transduction pathways. However, recent evidence suggests that direct physical interactions of type II phytochromes are occurring in plant cells and it is possible that these interactions play a role in modulating and integrating their activities.

Evidence for the Formation of Type II Phytochrome Heterodimers

The quaternary structure of phytochrome has been examined previously only in cases where one type of phytochrome is highly expressed. Experiments utilizing size-exclusion chromatography, analytical ultracentrifugation, and electron microscopy clearly showed that phytochrome purified from dark-grown oat tissue, phyA, is dimeric (Lagarias and Mercurio 1985, Jones and Quail 1986, Jones and Erickson 1989). The over-expressed phyB in transgenic plants containing the phyB coding region driven by the 35S promoter was also shown to be dimeric (Wagner and Quail 1995). Hence, it has often been assumed that all phytochromes are homodimers and that the number of types of mature active phytochrome present in a plant directly correlates with the number of *PHY* genes in that plant's genome. Recent results from immunoprecipitation experiments in Arabidopsis suggest that this may not be the case and that the phytochrome array likely includes multiple heterodimers of the type II forms (Sharrock and Clack

2004). In these experiments, an epitope-tagged phyB apoprotein, consisting of full-length phyB with six myc epitopes fused to its N-terminus, was shown to be fully active in complementing a *phyB* null mutation. When this tagged myc_6-phyB was immunoprecipitated from seedling extracts, phyC, phyD, and phyE were co-precipitated. This occurred irrespective of whether the seedlings were grown in the light or in the dark. Moreover, in the precipitations of dark-grown extracts, phyA was not pulled down by myc_6-phyB even though light-labile phyA is present at very high levels under those conditions. These results suggest that phyB physically interacts with the other type II phytochromes but not type I phyA and that it does so whether the molecules are in the Pr or Pfr conformation. These interactions were confirmed by expressing an epitope-tagged myc_6-phyD apoprotein in transgenic plants and showing co-immunoprecipitation of phyB and phyE with myc_6-phyD.

The interaction of phyB with the other type II phytochromes could occur as dimer/dimer structures (such as B/B binding to D/D), which would indicate that these receptors may influence each other's activities as higher order receptor complexes. Alternatively, the interactions could represent the formation of phytochrome heterodimers (such as B/C or B/D). To distinguish these possibilities, dark-grown seedling extracts were fractionated by size-exclusion chromatography under non-denaturing conditions and the migrations of the five phytochromes were assayed on immunoblots of the column fractions. In these experiments, all five of the Arabidopsis phytochromes migrated at masses characteristic of dimers and no evidence of higher molecular weight complexes was seen. In addition, when fractions from the dimer peak were immunoprecipitated with anti-myc antibody, phyC, phyD and phyE were observed to co-precipitate with myc_6-phyB. These results demonstrate that all five Arabidopsis phytochromes are in fact dimeric, as has often been assumed, but that it is likely that this dimer population includes B/C, B/D, and B/E heterodimers.

Implications of Phytochrome Heterodimer Formation

Currently, the evidence for heterodimer formation in Arabidopsis suggests that type II phytochromes have evolved as a set of proteins with dimerization affinities for each other but that type I phyA is possibly obligatorily homodimeric. Until the full complement of possible subunit interactions is assayed, this is a tentative conclusion, but it would correlate with the unique activity of phyA as a type I sensor of very low fluence light and continuous FR versus the overlapping and highly related activities of the type II forms as R sensors. If, indeed, phyA is present in plant cells only as a homodimer, analysis of *phyA* null mutants is an effective way to determine the in vivo activities of this phytochrome. On the other hand, if, for example, the phyB holoprotein can contribute to the formation of four different phytochrome types in plant cells (B/B, B/C, B/D, and B/E), a *phyB* null mutation will cause loss of all four of these forms and it will not be clear which aspects of the *phyB* mutant phenotype correspond with which form. Therefore, the identification of heterodimeric type II phytochromes necessitates a reinterpretation of the single and multiple *phyB–phyE* mutant phenotypes in Arabidopsis and in other plants. In at least some cases, the apparently overlapping, synergistic, or redundant activities of the type II phytochromes may in fact represent the differential activities of heterodimeric forms of the photoreceptor rather than, or in addition to, cross-talk of their downstream signaling mechanisms. It will likely be challenging to investigate the activities of the individual heterodimeric phytochromes because this will require their selective removal or perhaps their selective over-expression in plants. Moreover, their activities may differ in relatively subtle ways. Nevertheless, if the phytochrome array is more complex and heterogeneous than previously thought, the combinatorial assembly of phytochrome heterodimers must be considered as a potential mechanism for integration of receptor function and generation of photosensory versatility in plant R/FR sensing.

It is not yet known what fraction of phytochrome is present as heterodimers versus homodimers. The amounts of phyC, phyD, and phyE that co-immunoprecipitate with myc-tagged phyB are not greatly different in comparing dark-grown and light-grown seedling extracts (Sharrock and Clack 2004), so this fraction appears not to be strongly light-regulated and it is likely that heterodimeric phytochromes are present both in the cytosol and in the nucleus. Moreover, as determined by promoter fusions to the GUS coding sequence, the *PHYB*, *PHYD*, and *PHYE* genes are expressed in markedly different tissue-specific patterns in Arabidopsis (Goosey et al 1997), making it likely that different populations of homodimers and heterodimers of these phytochromes are present in different cells types and organs. If dimeric combinations of type II phytochrome chromoprotein subunits result in generation of many more different forms of phytochrome than the number of *PHY* genes, with the attendant possibility that each of those forms has unique photochemical properties and/or capacities to interact with downstream signaling pathways, more of the complexity and versatility of higher plant R/FR photosensing may occur at the level of the photoreceptor itself than has previously been recognized.

References

Aukerman MJ, Hirschfeld M, Wester L, Weaver M, Clack T, Amasino RM, Sharrock RA (1997) A deletion in the *PHYD* gene of the Arabidopsis Wassilewskija ecotype defines a role for phytochrome D in red/far-red light sensing. Plant Cell 9: 1317–1326

Casal JJ (2000) Phytochromes, cryptochromes, phototropin: photoreceptor interactions in plants. Photochem Photobiol 71: 1–11

Devlin PF, Patel SR, Whitelam GC (1998) Phytochrome E influences internode elongation and flowering time in Arabidopsis. Plant Cell 10: 1479–1487

Eichenberg K, Baurle I, Paulo N, Sharrock RA, Rüdiger W, Schäfer E (2000) Arabidopsis phytochromes C and E have different spectral characteristics from those of phytochromes A and B. FEBS Lett 470: 107–112

Franklin KA, Davis SJ, Stoddart WM, Vierstra RD, Whitelam GC (2003a) Mutant analyses define multiple roles for phytochrome C in Arabidopsis photomorphogenesis. Plant Cell 15: 1981–1989

Franklin KA, Praekelt U, Stoddart WM, Billingham OE, Halliday KJ, Whitelam GC (2003b) Phytochromes B, D, and E act redundantly to control multiple physiological responses in Arabidopsis. Plant Physiol 131: 1340–1346

Furuya M (1993) Phytochromes: their molecular species, gene families, and functions. Annu Rev Plant Physiol 44: 617–645

Goosey L, Palecanda L, Sharrock RA (1997) Differential patterns of expression of the Arabidopsis *PHYB*, *PHYD*, and *PHYE* phytochrome genes. Plant Physiol 115: 959–969

Hennig L, Stoddart WM, Dieterle M, Whitelam GC, Schäfer E (2002) Phytochrome E controls light-induced germination of Arabidopsis. Plant Physiol 128: 194–200

Jones AM, Erickson HP (1989) Domain structure of phytochrome from *Avena sativa* visualized by electron microscopy. Photochem Photobiol 49: 479–483

Jones AM, Quail PH (1986) Quaternary structure of 124 kilodalton phytochrome from *Avena sativa*. Biochemistry 25: 2987–2995

Lagarias JC, Mercurio FM (1985) Structure function studies on phytochrome. Identification of light-induced conformational changes in 124-kDa *Avena* phytochrome in vitro. J Biol Chem 260: 2415–2423

Monte E, Alonso JM, Ecker JR, Zhang Y, Li X, Young J, Austin-Phillips S, Quail PH (2003) Isolation and characterization of *phyC* mutants in Arabidopsis reveals complex crosstalk between phytochrome signaling pathways. Plant Cell 15: 1962–1980

Pratt LH (1995) Phytochromes: differential properties, expression patterns, and molecular evolution. Photochem Photobiol 61: 10–21

Sharrock RA, Clack T (2002) Patterns of expression and normalized levels of the five Arabidopsis phytochromes. Plant Physiol 130: 442–456

Sharrock RA, Clack T (2004) Heterodimerization of type II phytochromes in Arabidopsis. Proc Natl Acad Sci USA 101: 11500–11505

Wagner D, Quail PH (1995) Mutational analysis of phytochrome B identifies a small COOH-terminal-domain region critical for regulatory activity. Proc Natl Acad Sci USA 92: 8596–8600

Whitelam GC, Devlin PF (1997) Roles of different phytochromes in *Arabidopsis* photomorphogenesis. Plant Cell Environ 20: 752–758

Chapter 6

A Structure–Function Model Based on Inter-Domain Crosstalks in Phytochromes

Jeong-Il Kim[1,2] and Pill-Soon Song[1,2,3]

Introduction

The molecular mechanism for the on/off switching of the phytochromes is driven by the photochromic phototransformation between the two spectrally distinct forms of the phytochromes, a red light (R, 660 nm) absorbing Pr and a far-red light (FR, 730 nm) absorbing Pfr forms. The photoactivated Pfr signals are transduced by interacting with a wide array of downstream signaling components and finally regulate genes involved in photomorphogenesis (Quail 2002, Wang and Deng 2003 for reviews). Our understanding as to how phytochromes perceive light signals and how the light signals are transmitted intramolecularly within the phytochrome molecules is only beginning. The photoisomerization of the chromophore modulates apoprotein:chromophore interactions, and subsequently triggers conformational changes throughout the whole phytochrome molecule via intramolecular inter-domain crosstalks (Park et al 2000, Kim et al 2002b), like the well-characterized rhodopsin visual receptor in animals, as discussed in the next sections.

Structural Motifs and Domains in Phytochromes

The phytochrome molecule consists of two structural domains, the globular N-terminal chromophore-binding domain (~65 kDa) and the conformationally open or extended C-terminal domain (~55 kDa). The two domains are connected via a flexible hinge region. The N-terminal domain is necessary and sufficient for photoperception and possesses the bilin lyase domain (BLD) which allows the attachment of the chromophore to apo-phytochrome, while the C-terminal region is responsible for the transduction of the light signal (Fankhauser 2001)

[1] Kumho Life and Environmental Science Laboratory, Gwangju 500-712, South Korea
[2] Environmental Biotechnology National Core Research Center, Gyeongsang National University, Jinju 660-701, South Korea
[3] Department of Chemistry, University of Nebraska, Lincoln, NE 68588-0304, USA

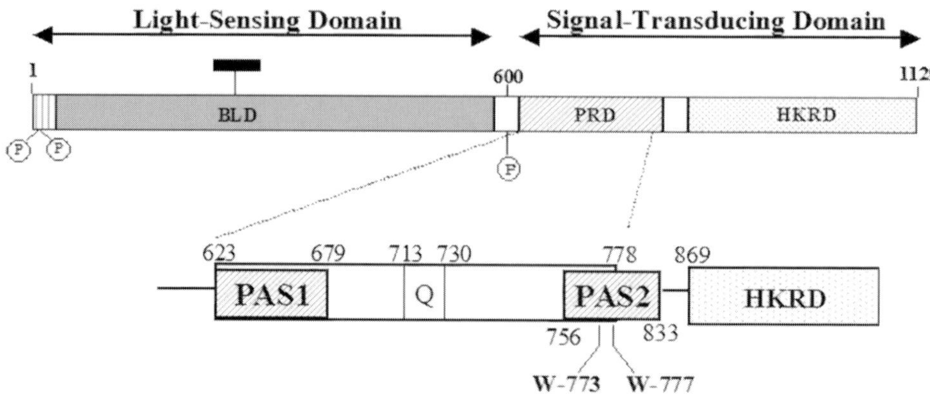

FIG. 1. Block diagram of a phytochrome monomer domain structure with conserved subdomains/motifs in the C-terminal domain. Diagram shows the bilin lyase domain (*BLD*) for chromophore attachment, the Per-Arnt-Sim (PAS)-related domain (*PRD*), and histidine kinase-related domain (*HKRD*). Three possible phosphorylation sites are also indicated. The PRD consists of two PAS repeats, PAS1 and PAS2, and overlaps with the regulatory core motif (Quail box, *Q*). The Trp-773 and Trp-777 residues reside near the PAS2 motif, and this peptide region undergoes detectable conformational changes in the phototransformation. Numbers are amino acid positions in the oat phyA and indicate the N- and C-terminal residues of each structural domain or motif

as shown schematically in Figure 1. Yeast two-hybrid screenings with a C-terminal segment as bait have revealed that the C-terminal domain interacts with the downstream phytochrome-interacting proteins (PIPs), such as phytochrome-interacting factor-3 (PIF3), nucleoside diphosphate kinase-2 (NDPK2), phytochrome kinase substrate-1 (PKS1), *EARLY FLOWERING 3* (ELF3), *ZEITLUPE* (ZTL), *ADAGIO1* (Ado1), and so on (Quail 2002, Matsushita et al 2003).

Several conserved subdomains/motifs have been identified in the phytochrome molecules (Figure 1). In the N-terminal light-sensing domain, the first 65 amino acids of the phyA protein are dispensable for chromophore binding, but they constitute the N-terminal extension (NTE) region that is necessary for a biological activity (Casal et al 2002). The serine-rich NTE of oat phyA helps regulate light responses and subnuclear localization of the photoreceptor, proposing that the NTE of phyA is involved in channeling downstream signaling in different cellular contexts. The N-terminal α-helix forming motif is likely to play a critical role here (Park et al 2000). The different NTE between phyA and phyB might explain the different photo-sensing specificities, and the longer NTE of phyB might induce a phyB-specific inter-domain crosstalk with the C-terminal domain. The importance of the C-terminal half of plant phytochromes is highlighted by numerous missense mutations affecting this part of the protein (Quail et al 1995, Krall and Reed 2000). In the signal-transducing C-terminal domain, there are Per-Arnt-Sim (PAS)-related domain (PRD) consisting of a pair of PAS repeats

(PAS1 and PAS2) and histidine kinase-related domain (HKRD) (see Figure 1). The PAS domain has been implicated in protein-protein interactions and inter-domain communications in some sensory proteins (Yeh and Lagarias 1998). The phytochrome PRD houses the regulatory core region ("Quail box") and is implicated as the functional domain for phytochrome dimerization and protein-protein interaction between phytochromes and PIPs. The HKRD is implicated as a protein kinase domain, but may not be a function kinase domain because the key conserved residues within the histidine kinase domain (HKD) are absent in the phytochrome HKRD (Quail 1997) and this domain is necessary but dispensable for phyB signaling (Krall and Reed 2000). Rather, it is more reasonable that the HKRD domain might have a regulatory role in the phytochrome signaling, because this domain interacts with PKS1 and the interaction negatively regulates phyB signaling (Fankhauser et al 1999).

Preparation for Crosstalks: Conformational Changes

What is the nature of the phytochrome-mediated signal transductions? The first step of phytochrome signaling is likely to be a conformational change upon absorbing light. Spectral and biochemical evidence suggest that the chromophore topography and the secondary and tertiary structures of the phytochromes are significantly changed through apoprotein:chromophore and inter-domain interactions (Park et al 2000 and references therein).

With native phyA, the Pfr-chromophore (chromophores in the Pfr forms) is more exposed than the Pr-chromophore. This preferential exposure of the Pfr-chromophore is modulated by the α-helix forming 6 kDa peptide in the N-terminal extension. On Pr→Pfr phototransformation, the N-terminal extension undergoes a conformational change from random coil to amphiphilic α-helix which interacts with the chromophore in the Pfr form (Park et al 2000). The 6 kDa peptide seems to directly interact with the chromophore and possibly with other structural motifs, causing a series of conformational changes. The N-terminal domain is more accessible in the Pr form than in the Pfr form (Lapko et al 1998). Also, the hinge region is preferentially exposed in the Pfr form. These data indicate that specific conformational changes occur in these regions during the phototransformation of phytochromes.

The surface topography of Pr and Pfr phytochromes includes differential exposure of tryptophan residues (Park et al 2000). Subtle conformational changes are detected in the region around Trp-569 and Trp-572. Ser-598 is located close to this regions (see below); this serine is preferentially phosphorylated in the Pfr form in vivo (Figure 3; Lapko et al 1999). In addition, Trp-773 and Trp-777 are preferentially exposed in the Pfr form, indicating that the peptide region containing these two Trp residues also undergoes significant changes in surface topography during the phototransformation. The two Trp residues are located within the PAS2 motif (Figure 1). These residues could be directly involved in the inter-domain interactions and/or in the protein-protein interactions. The preferential exposure of the Trp residues may arise from unmasking of the PAS domains in the Pfr form. The surface topography of phyA can also be monitored by probing the accessibility of Cysteine (Cys) residues with iodoacetamide (Lapko et al 1998). There are 22 Cys residues in oat phyA. Of the 11 Cys residues in the N-terminal domain, only one reactive Cys-311 near the chromophore shows a significant dependence of its surface exposure on the Pr→Pfr phototransformation, suggesting conformational changes near the chromophore binding pocket upon light absorption. Virtually all 11 Cys residues in the C-terminal domain are accessible to the –SH reagent in both Pr and Pfr forms of phyA. Thus, there is no distinct Pr- or Pfr-preferential accessibility of Cys residues in and around the Quail box, probably because the reagent is small enough to access the Quail box even in the Pr form. Cys-715 that is conserved among all members of the phytochrome family is fully exposed, suggesting that the "activation" of the Quail box may involve the Cys residue directly in inter-molecular crosstalks with signal transducers.

Taken together, conformational changes and differential surface topography of phytochromes are necessary for inter-domain crosstalks in the phytochrome photoactivation. The amphiphilic N-terminal α-helix appears to play a critical role in the N- to C-terminal/Q-box inter-domain crosstalk. The different size and amphiphilicity of the N-terminal α-helical chains may be recognized as phyA- and phyB-specific signals through the Quail box (Park et al 2000). The surface topography near to the PRD is also changed upon phototransformation, showing the presence of intramolecular inter-domain crosstalks. To prepare a phytochrome molecule for its signal transduction through a PIP, specific intermolecular inter-domain interactions are prerequisite for protein-protein interactions.

Inter-Domain Crosstalks

The "different" light signals perceived by phyA and phyB are transmitted from the N-terminal domain to the C-terminal regulatory domain through intramolecular inter-domain crosstalks in the Pfr form (Figure 2), where downstream signaling events begin. Since the N-terminal domains carry the determinants for the phytochrome individuality and the C-terminal domains are functionally interchangeable, it is suggested that phytochromes share common molecular mechanisms for the interaction with downstream signaling components (Wagner et al 1996). It is thus intriguing how C-terminal domain-interacting PIPs show the dependence of phytochrome species. Each phytochrome-interacting protein (PIP) is likely to interact with different structural motif or domain in the C-terminal domain. NDPK2 binds to the region of the Quail box preferentially in the Pfr-phytochrome (Choi et al 1999). This is consistent with the Pfr-dependent exposure of the PRD containing the Trp-773 and Trp-777. PIF3 also binds to phytochromes in a Pfr-preferential manner (Ni et al 1998). Both N-terminal and C-terminal domains are required for a full binding activity of Pfr-phytochromes to PIF3 (Zhu et al 2000). However, PIF3 also seems to bind to the Quail box

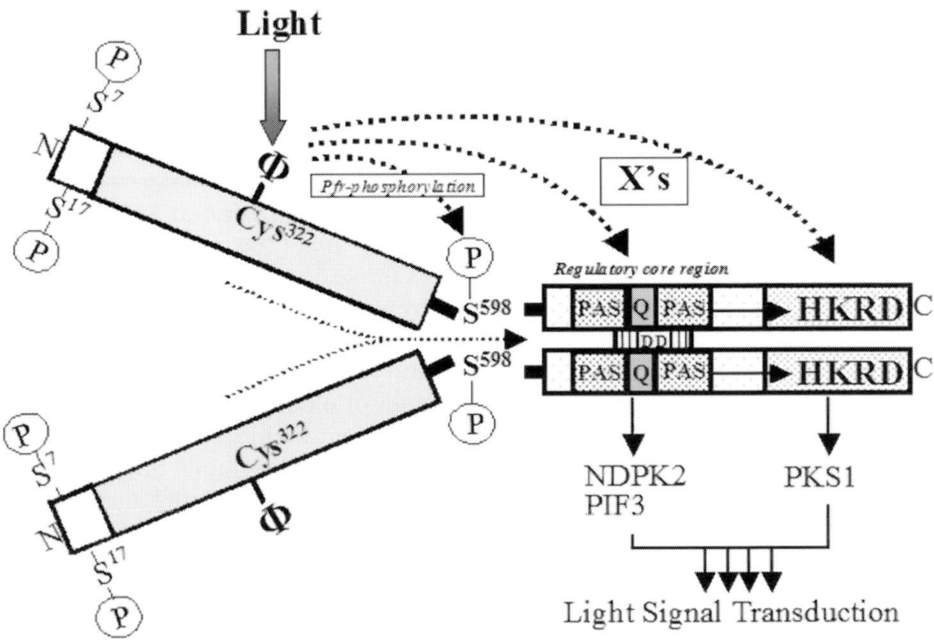

FIG. 2. Inter-domain signal transmission in a dimeric oat phyA. The Ser-598 is located in the hinge region and phosphorylated preferentially in the Pfr form. The Pfr-dependent conformational signals are generated through the chromophore photoisomerization by absorbing red light (*R*) in the N-terminal domain and trigger subtle conformational changes through the whole phytochrome molecule. The conformational signals are subsequently transmitted to the regulatory(Q/PAS)/dimerization domain (*DD*) either intramolecularly or intermolecularly (*bold dot arrows*) via the hypothetical signal transducers (*X*s). The photoactivated or uncovered Quail box (see Figure 3) interacts directly with the signal transducers such as NDPK2 and the PIF3, while the PKS1 associates with the HKRD. *N* and *C*, N- and C-terminal ends of the oat phyA; *Φ*, phytochromobilin; Cys^{322}, chromophore-binding residue. Numbers are the positions of amino acid residues in the oat phyA

specifically in the Pfr-phytochromes, because it was originally isolated through yeast two-hybrid screens using the C-terminal half as bait and some missense phytochrome mutants in the Quail box fail to bind with PIF3 (Ni et al 1998). On the other hand, PKS1 binds to the HKRD of phytochromes equally well in both Pr and Pfr forms, indicating that this motif is exposed in both spectral forms (Fankhauser et al 1999). Other PIPs such as ELF3 and ZTL/Ado1 also requires the C-terminal domain of phyB (Matsushita et al 2003). The only exception is ARR4 whose interaction is specifically accomplished with the extreme amino-terminus of phyB (Sweere et al 2001). These observations imply that differential inter-domain interactions activate a specific motif in the C-terminal domain and maybe in the NTE region to be recognized by different PIPs. It is thus apparent that the C-terminal domain takes a different conformation and/or surface topography in concert with the N-terminal domain through different inter-domain crosstalks, depending on light signals perceived by the phytochromes.

The photo-induced conformational signals could be further differentiated by inter-domain interactions through an intramolecular pathway (Figure 2). This pathway might be modulated by phytochrome phosphorylation/dephosphorylation at Ser-598 in the hinge region, since the phosphorylation at Ser-598 is Pfr-specific. An intermolecular pathway via putative signal transmitter proteins (e.g., Xs in Figure 2) could also be involved. For example, the SPA1 protein may serve this role as it is activated specifically by phyA (Hoecker et al 1999). Both the intermolecular and intramolecular pathways could be integrated to generate differential Pfr or conformational signals and to modulate the inter-domain crosstalks. Interestingly, the N-terminal domain of phyB by itself is shown to be functional in vivo when it could exist as dimers and be localized in the nucleus (Matsushita et al 2003). These results suggest that the C-terminal domain is necessary only for dimerization and nuclear localization. While the fact that N-terminal domain is apparently necessary and sufficient for phytochrome activity, it is puzzling if we consider that many of the missense mutants of phytochromes are due to mutations in the C-terminal regulatory domain (Quail et al 1995). Furthermore, many downstream phytochrome-interacting proteins physically interact with phytochromes through the C-terminal domain. In the case of phyA, 35 amino acid residues of the carboxy terminus are critical for the biological activity in plants. Although the N-terminal domain of phyB is functional under a specific condition (Matsushita et al 2003), its interaction with C-terminal domain may still be necessary for the completion of phytochrome signaling events such as signal attenuation and regulatory modulation.

Phosphorylation as a Switch for the Inter-Domain Crosstalks

Post-translational modification is important for the modulation of many signal transductions, e.g., rhodopsin is desensitized by phosphorylation by rhodopsin kinase (Sokal et al 2002). The three sites of phytochrome phosphorylation in vivo and in vitro have been identified with oat phyA (Lapko et al 1999); Serine-7 (Ser-7), Serine-17 (Ser-17) and Serine-598 (Ser-598). Circular dichroism analysis and proteolysis indicate that the phosphorylation of phytochromes by protein kinase A (PKA) induces subtle conformational changes near the hinge region containing Ser598. The PKA-catalyzed phosphorylation of oat phyA inhibited protease accessibility at the Lys536-Asn537 bond. These findings suggest that the phytochrome phosphorylation, especially at Ser-598 in the hinge region, could be a molecular mediator in the inter-domain interaction between the N- and C-terminal domains.

What is the in vivo function of this phosphorylation? Recently, we found phytochrome phosphorylation at Ser-598 in the hinge region is an inhibitory mechanism of phytochrome signaling (Kim et al 2004). The transgenic plants of mutant phyA with a Ser-598 to Ala substitution in *Arabidopsis* phyA null background exhibits hypersensitivity to far-red light and the phosphorylation at Ser598 prevents its interaction with putative signal transducers, NDPK2 and PIF3, suggesting that the Ser-598 phosphorylation has a negative regulatory role in photomorphogenesis. We also reported that an *Arabidopsis* serine/threonine specific protein phosphatase 2A (FyPP) interacts and dephosphorylates phyA (Kim et al 2002a). The FyPP-overexpressing transgenic plants stimulated phytochrome activity in flowering and hypocotyl shortening, whereas the anti-sense repression of FyPP transgenic plants displayed reduced phytochrome activity. These results are consistent with the negative regulation of phytochrome signaling in plants through protein phosphorylation. Since Ser-598 residue is phosphorylated only in the Pfr form that is considered the active form of phytochrome, the phosphorylation and dephosphorylation of Ser-598 may serve as a switch in phytochrome signaling (Kim et al 2002b). It is reminiscent of rhodopsin signaling, also modulated by phosphorylation and dephosphorylation. On stimulation, rhodopsin is phosphorylated at several sites on its C terminus as the first step in deactivation (Sokal et al 2002).

Another complex feature of phytochrome signaling is the autophosphorylating kinase activity of phytochromes (Yeh and Lagarias 1998). Phytochromes can be autophosphorylated and also phosphorylate several substrates such as cryptochromes, PKS1, and Aux/IAA proteins (Fankhauser et al 2001). What could be the function of this kinase activity of phytochromes in plants? The site for phytochrome autophosphorylation and kinase activity is suggested to be located in the N-terminal extension, either Ser7 or Ser17 (Kim et al 2002b, unpublished). The mutation of N-terminal serines to alanines including Ser-7 and Ser-17 results in an increased biological activity of phyA, suggesting the existence of desensitization regulation (Casal et al 2002). This desensitization might be due to the lack of phytochrome autophosphorylation or the kinase activity. Thus, phytochrome kinase activity might play a negative role in the signaling. Since phytochrome kinase activity is stimulated in the presence of histone H1 in a Pr-specific manner (Yeh and Lagarias 1998), it is suggested that phytochrome kinase activity is activated in the nucleus that contains cationinc molecules such as histones (Kim et al 2002b). Thus, phytochrome kinase activity may play a signal-attenuating role in the nucleus by phosphorylating PIPs. These indicate that conformational signals and accompanying inter-domain crosstalks are further diversified by protein phosphorylation and amplified through complex interactions with downstream signaling components.

A Schematic Model for the Inter-Domain Crosstalks

The conformational changes and inter-domain interactions involved in the phytochrome phototransformation are schematically summarized in Figure 3. Upon light absorption, the N-terminal extension undergoes a significant conformational changes from random coil to α-helix and the phytochrome molecule exhibits a

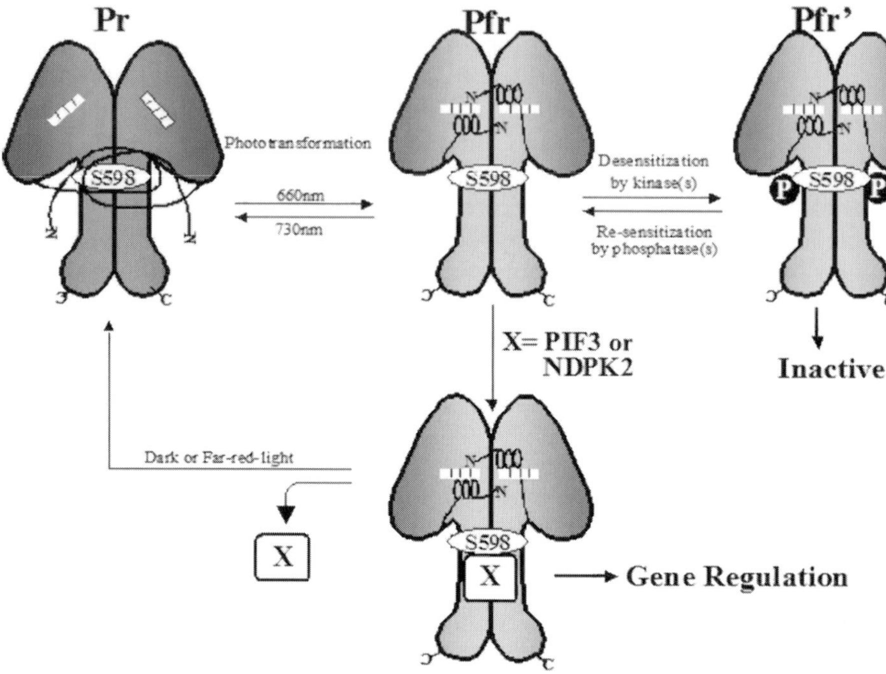

FIG. 3. A schematic model for the conformational changes and the inter-domain interactions in the phytochrome photoactivation. The photoactivation of the phytochrome includes changes in the domain conformation and in the surface topography triggered by apoprotein:chromophore interactions. The N-terminal 6 kDa peptide region forms an α-helical conformation in the Pfr but a random coil conformation in the Pr. The proximity between the N- and the C-termini in the Pr form is to depict the covered Quail box and shields the hinge region. Upon photoactivation, the hinge region is exposed, and the Ser-598 may be phosphorylated. The Pfr-phytochrome may phosphorylate the PKS1. The phospho-PKS1 is subsequently released from the photoactivated phytochrome, which can associate with the PIF3 or NDPK2. The Ser-598 phosphorylation is proposed as a mechanism to desensitize the Pfr activity. The desensitized Pfr (Pfr′) does not associate with PIPs in this working model. Alternatively, it may have a nongenomic regulatory function in the cytoplasm. A serine/threonine protein phosphatase such as FyPP is postulated in the resensitization step. *Open boxes*, chromophores; *X*, phytochrome interacting proteins (*PIPs*), such as PIF3 or NDPK2. This diagram is modified from Park et al (2000)

more exposed conformation in the C-terminal domain of the Pfr form than that of the Pr form, which prepares for the next inter-domain crosstalks. The photoactivated Pfr-phytochromes can then interact with their downstream signal transducers such as NDPK2 and PIF3. A key consequence of the crosstalk between the N- and C-terminal domains is closing (in Pr) and opening (in Pfr) of the Quail box and the hinge region. In the Pfr form, the two regions are exposed and the Ser-598 is phosphorylated. The C-terminal peptide region, without the N-terminus, is like a "photoactivated" Pfr form, which might explain why PIPs are positively screened with C-terminal domain of phytochromes by the yeast two hybrids methods. The phosphorylation of Ser-598 by protein kinase(s) desensitizes the Pfr-phytochrome, and the desensitized Pfr does not associate with PIPs. Phytochrome dephosphorylation by protein phosphatase(s) can re-sensitize phytochromes. The phytochrome signal transduction is thus modulated by protein phosphorylation and dephosphorylation; the phosphorylation blocks the interaction with its signal transducers while the dephosphorylation recovers the interaction. Phytochrome crosstalks with PIPs then regulate various light-regulated gene expression for plant's growth and development.

Concluding Remarks

Our current model qualitatively depicts the early event of phytochrome signaling. There are also other important mechanistic events to understand phytochrome signaling, including nuclear localization of phytochromes (Nagy and Schäfer 2000), regulated proteolysis (Wang and Deng 2003), and cytoplasmic events such as changes in ionic conductance across the plasma membrane.

Matsushita et al (2003) presented a model that the C-terminal domain is dispensable for phytochrome function except for dimerization and nuclear localization. Based on their model, the regulatory core region (Quail box) may not be crucial for some phytochrome functions, contradicting the Quail box missense mutation studies. It is difficult for this model to explain why most PIPs are obtained by yeast two-hybrid screens with the C-terminal domain as bait. C-Terminal domains are apparently necessary for the interaction with PIPs such as NDPK2, PKS1, ELF3 and Ado1/ZTL, even though N-terminal domain is sufficient for the interaction with ARR4 and PIF3 (Zhu et al 2000, Sweere et al 2001). The construct in their report contains GUS to force dimerization. Such a dimer may topographically simulate phytochrome dimers. The 3D-structures of all three *Arabidopsis* NDPKs share common electrostatic surface potentials in the upper side of hexamer where the enzymatic active site is located, while the distribution of electrostatic potentials at the lateral surface that houses the motif for phytochrome interaction is distinct for each type of NDPK isoforms (Im et al 2004). These findings suggest that electrostatic distribution of NDPK2 is responsible for the binding with phytochromes. Thus, if the GUS dimer mimics the surface charge distributions of phytochrome dimers, it can interact with PIPs such as NDPK2. More likely is that both N- and C-terminal domains possess dual signaling activities, with the former having a protein kinase function for a set of phytochrome-mediated signal transduction events. Phytochromes have pleiotropic effects on gene expression and development in plants. To control these multi-functional phytochromes, different modes of regulatory functions may originate from either or both N- and C-domains. The C-terminal domain may well be necessary for fine-tuning the regulatory mechanisms through inter-domain crosstalks.

Acknowledgments. The authors are supported by Kumho Petrochemical Co., Ltd. (KKPC, Publication No. 67), and in part by grants from National Research Laboratory/KISTEP, BioGreen 21 program of Rural Development Administration (RDA) (to P.-S.S.) and the KOSEF/MOST to the Environmental Biotechnology National Core Research Center (NCRC) (to J.-I.K. and P.-S.S; grant #R15-2003-002-01003-0).

References

Casal JJ, Davis SJ, Kirchenbauer D, Viczian A, Yanovsky MJ, Clough RC, Kircher S, Jordan-Beebe ET, Schäfer E, Nagy F, Vierstra RD (2002) The serine-rich N-terminal domain of oat phytochrome A helps regulate light responses and subnuclear localization of the photoreceptor. Plant Physiol 129: 1127–1137

Choi G, Yi H, Kwon YK, Soh MS, Shin B, Luka Z, Hahn TR, Song PS (1999) Phytochrome signaling is mediated through nucleoside diphosphate kinase 2. Nature 401: 610–613

Fankhauser C (2001) The phytochromes, a family of red/far-red absorbing photoreceptors. J Biol Chem 276: 11453–11456

Fankhauser C, Yeh KC, Lagarias JC, Zhang H, Elich TD, Chory J (1999) PKS1, a substrate phosphorylated by phytochrome that modulates light signaling in *Arabidopsis*. Science 284: 1539–1541

Hoecker U, Teppermann JM, Quail PH (1999) SPA1, a WD-repeat protein specific to phytochrome A signal transduction. Science 284: 496–499

Im YJ, Kim JI, Shen S, Na Y, Han YJ, Kim SH, Song PS, Eom SH (2004) Structural analysis of *Arabidopsis thaliana* nucleoside diphosphate kinase-2 for plant phytochrome signaling. J Mol Biol 343: 659–670

Kim DH, Kang JG, Yang SS, Chung KS, Song PS, Park CM (2002a) A phytochrome associated protein phosphatase 2A modulates light signals in flowering time control in *Arabidopsis*. Plant Cell 14: 3043–3056

Kim JI, Kozhukh GV, Song PS (2002b) Phytochrome-mediated signal transduction pathways in plants. Biochem Biophys Res Commun 298: 457–463

Kim JI, Shen Y, Han YJ, Kirchenbauer D, Park JE, Soh MS, Nagy F, Schäfer E, Song PS (2004) Phytochrome phosphorylation modulates light signaling by influencing the protein-protein interaction. Plant Cell 16: 2629–2640

Krall L, Reed JW (2000) The histidine kinase-related domain participates in phytochrome B function but is dispensable. Proc Natl Acad Sci USA 97: 8169–8174

Lapko VN, Jiang XY, Smith DL, Song PS (1998) Surface topography of phytochrome A deduced from specific chemical modification with iodoacetamide. Biochemistry 37: 12526–12535

Lapko VN, Jiang XY, Smith DL, Song PS (1999) Mass spectrometric characterization of oat phytochrome A: Isoforms and posttranslational modifications. Protein Sci 8: 1032–1044

Matsushita T, Mochizuki N, Nagatani A (2003) Dimers of the N-terminal domain of phytochrome B are functional in the nucleus. Nature 424: 571–574

Nagy F, Schäfer E (2000) Nuclear and cytosolic events of light-induced, phytochrome-regulated signaling in higher plants. EMBO J 19: 157–163

Ni M, Tepperman JM, Quail PH (1998) PIF3, a phytochrome-interacting factor necessary for normal photoinduced signal transduction, is a novel basic helix-loop-helix protein. Cell 95: 657–667

Park CM, Bhoo SH, Song PS (2000) Inter-domain crosstalk in the phytochrome molecules. Semin Cell Dev Biol 11: 449–456

Quail PH (1997) The phytochromes: A biochemical mechanism of signaling in sight? Bioassays 19: 571–579

Quail PH (2002) Phytochrome photosensory signaling networks. Mol Cell Biol 3: 85–93

Quail PH, Boylan MT, Parks BM, Short TW, Xu Y, Wagner D (1995) Phytochromes: Photosensory perception and signal transduction. Science 268: 675–680

Sokal I, Pulvermuller A, Buczylko J, Hofmann KP, Palczawski K (2002) Rhodopsin and its kinase. Methods Enzymol 343: 578–600

Sweere U, Eichenberg K, Lohrmann J, Mira-Rodado V, Baurle I, Kudla J, Nagy F, Schafer E, Harter K (2001) Interaction of the response regulator ARR4 with phytochrome B in modulating red light signaling. Science 294: 1108–1111

Wagner D, Fairchild CD, Kuhn RM, Quail PH (1996) Chromophore-bearing NH_2-terminal domains of phytochromes A and B determine their photosensory specificity and differential light lability. Proc Natl Acad Sci USA 93: 4011–4015

Wang H, Deng XW (2003) Dissecting the phytochrome A-dependent signaling network in higher plants. Trends Plant Sci 8: 172–178

Yeh KC, Lagarias JC (1998) Eukaryotic phytochromes: Light-regulated serine/threonine protein kinases with histidine kinase ancestry. Proc Natl Acad Sci USA 95: 13976–13981

Zhu Y, Tepperman JM, Fairchild CD, Quail PH (2000) Phytochrome B binds with greater apparent affinity than phytochrome A to the basic helix-loop-helix factor PIF3 in a reaction requiring the PAS domain of PIF3. Proc Natl Acad Sci USA 97: 13419–13424

Chapter 7

Functions of Different Domains of Phytochrome

Akira Nagatani

Introduction

Phytochrome is a photoreceptor regulating various aspects of plant photomorphogenesis. Phytochrome affects plant development by regulating gene expression. Recent analysis using the DNA microarray technique revealed many genes that are under the control of phytochrome (Chapter 2 by Quail). To regulate the gene expression, phytochrome translocates from the cytoplasm to the nucleus upon light-activation (Nagatani 2004). In the nucleus, phytochrome physically interacts with specific transcription factors such as PIF3 (Ni et al 1998). However, details of the signal transduction process that takes place in the nucleus remains obscure.

The phytochrome molecule consists of two major domains, N-terminal chromophoric and C-terminal dimerization domains (Figure 1). The N-terminal domain covalently binds the open tetrapyrrole chromophore, phytochromobilin (Chapter 3 by Kohchi) and exhibits reversible photoconversion between two spectrally distinct forms, Pr and Pfr. The C-terminal domain, through which phytochrome dimerizes, contains motifs such as Per-Arnt-Sim (PAS) domains and a histidine kinase-related domain but these are not required for the photoconversion.

Phytochrome is encoded by a small multigene family (Chapter 5 by Sharrock). Among them, phytochrome A (phyA) and phytochrome B (phyB) are the two major molecular species. In vivo, phyB exists at relatively low levels irrespective of light conditions. By contrast, phyA accumulates at a high level in darkness and is degraded rapidly upon the light activation. The loss of phyB causes long hypocotyl phenotype under continuous red light, whereas the *phyA* deficient mutant fails to respond to continuous far-red light. Further analysis of these mutants have established the view that phyB acts as a major photoreceptor for the red/far-red reversible low fluence responses, whereas phyA mediates atypical very low fluence and far-red high irradiance responses.

Department of Botany, Graduate School of Science, Kyoto University, Sakyo-ku, Kyoto 606-8502, Japan

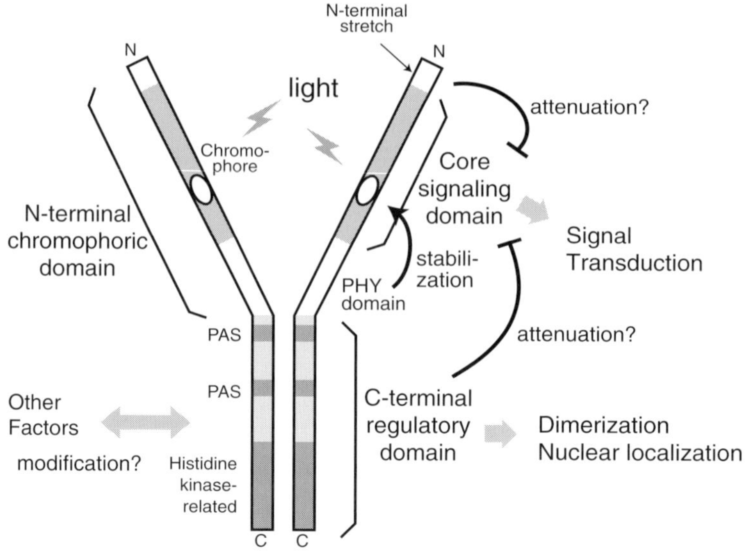

Fig. 1. A schematic diagram of the phyB domain structure and possible functions of each domain. PhyB monomer consists of the N-terminal chromophoric and C-terminal dimerization domains. For details, see text. *PAS*, Per-Arnt-Sim

In the present review, I summarize recent data on how the different domains of phytochrome contribute to the functions of the whole molecule.

Functions of the N-Terminal Domain

The N-terminal domain of phytochrome consists of 600–650 amino acids and binds one phytochromobilin molecule at a cysteine residue residing at the center of this domain. Proteolytic fragments as well as recombinant proteins corresponding to this domain exhibit almost normal Pr/Pfr photoconversion in vitro and in vivo. It is intriguing here that proteins that exhibit homology to this domain have been found in cyanobacteria and some other bacteria, which suggests that the origin of this domain can be traced back to an ancient photosynthetic organisms (Chapters 35, 36).

For many years, the C-terminal domain but not the N-terminal domain has been believed to be involved in the interaction of phytochrome with downstream signal transduction partners. Indeed, the N-terminal fragments of phytochrome expressed without the C-terminal domain do not exhibit the biological activity. However, more detailed analysis has revealed that the N-terminal domain alone can transduce the signal in response to light under certain conditions (see below).

Phytochrome translocates into the nucleus upon light activation. We examined whether the N-terminal domain by itself has nuclear localization activity (Matsushita et al 2003). Although the N-terminal fragment of phyB fused to GFP (N650G) localized both in the nucleus and cytoplasm, that fused to β-glucuronidase (GUS) and green fluorescent protein (GFP) (N650G-GUS) was not imported into the nucleus. The former observation can be explained by passive entry into the nucleus, which is often observed for relatively small proteins. In conclusion, the N-terminal domain is not actively transported into the nucleus.

During the above analysis, we noticed that the N-terminal fragment of phyB exhibited some biological activity. The seedlings of transgenic lines expressing N650G were shorter than the parental *phyB* mutant under continuous red light. By contrast, N650G-GUS, which is not localized in the nucleus, did not show the biological activity with respect to this phenotype. Hence, we examined whether N650G-GUS was active when it is targeted into the nucleus with the aid of the nuclear localization signal (NLS). To our surprise, the fusion protein with NLS (N650G-GUS-NLS) acted almost normally as a photoreceptor (Matsushita et al 2003). The *phyB* mutant expressing N650G-GUS-NLS responded to red light with a higher sensitivity to light. Hence, we concluded that the N-terminal domain alone can transduce the signal in the nucleus. This conclusion is consistent with a former report that the phyB without the kinase domain retains some biological activity (Krall and Reed 2000).

The Core Signal Transduction Domain of phyB

As discussed above, importance of the N-terminal domain of phytochrome for its signaling activity has been highlighted. The N-terminal domain includes the N-terminal stretch, a chromophore-bearing core region, and the GAF-related phytochrome (PHY) domain (Montgomery and Lagarias 2002). The N-terminal stretch, which is highly divergent compared with other domains, exhibits dramatic structural changes upon the photoconversion (Chapter 6 by Kim and Song). Phosphorylation of serine residues in this domain is suggested to result in the desensitization of phytochrome signal transduction activity (Stockhaus et al 1992, Jordan et al 1996). However, this domain is dispensable for the phyB signal transduction because phyB lacking the N-terminal stretch of 103 amino acid residues retains partial biological activity (Wagner et al 1996).

We examined the biological activity of the N-terminal domain without the PHY domain. The fragment corresponding to amino acid positions 1–451 of Arabidopsis phyB (N451) was fused to GFP, GUS, and NLS (N451G-GUS-NLS) and expressed in the *phyB* mutant of Arabidopsis (Oka et al 2004). The hypocotyl assay of the resulting plants indicated that N451G-GUS-NLS retains the activity to respond to continuous red light. Further analysis revealed that the Pfr form of N451 is less stable and reverts back more rapidly to the Pr form in darkness, which is consistent with the proposed role of the PHY domain to maintain the integrity of Pfr (Montgomery and Lagarias 2002). Together with the observation on phyB lacking the N-terminal stretch (see above), a core region for the phyB signal transduction activity was defined as amino acid positions 104–451. Since no clear motifs that are presumed to be involved in any type of signal transduction is found within this region, the mechanism by which this region transduces the signal to downstream components remains a mystery.

Factors Interacting with the N-Terminal Domain of Phytochrome

The above observation is quite surprising because only few proteins are known to interact with the N-terminal domain of phytochrome. One class of such proteins is the phytochrome-interacting-protein 3 (PIF3) and its relatives (PIL). PIF3 was originally identified as a protein recognizing the C-terminal domain of phyB (Ni et al 1998) but later it was shown to interact with the N-terminal domain as well (Ni et al 1999). The PIF3 is a basic-helix-loop-helix type transcription factor. It recognizes *cis*-elements found in light-responsive genes (Martinez-Garcia 2000). Hence, PIF3 is a good candidate of the downstream signal transduction targets for the N-terminal domain of phytochrome. However, recent analysis on the *pif3* loss-of-function mutants has suggested that the biological function of this protein is complicated and that this class of protein may not be the primary target of the phytochrome signal transduction (Kim et al 2003).

Arabidopsis response regulator 4 (ARR4) is another example of the proteins that interact with the N-terminal domain of phytochrome (Sweere et al 2001). Response regulators act as signal transduction components in the bacterial two-component signal transduction system. In plants, they are involved in the cytokinin signal transduction. Interestingly, ARR4 binds to the N-terminal domain of phyB and increase the stability of its Pfr form in vivo. Hence, it is speculated that ARR4 plays a role in the cross-talk between phytochrome and cytokinin. However, it remains unclear whether this factor is involved in the main signal transduction pathway of phytochrome.

Functions of the C-Terminal Domain

Phytochrome exists as a homodimer through the interaction between the C-terminal domains of each monomer (Chapter 4 by Nakasako and Tokutomi). By contrast, the N-terminal domain does not have this activity. Although the significance of dimerization with respect to the biological function has not been fully understood, dimerization may be required for efficient signal transduction. The N650 fragment of phyB, which exists as a monomer, exhibits weak but measurable signaling activity (Matsushita et al 2003). This observation indicates that dimerization is not essential for the phyB signaling. By contrast, N650 fused to

GUS exhibits much higher activity. GUS is known to multimerize a protein attached to it. Hence, the dimer form of N650 appears to be more active than the monomer. However, it remains unclear why the dimer is more active. Analysis of N650 fused to various dimerizing protein is awaited.

In addition to dimerization activity, the C-terminal domain appears to confer nuclear localization activity to phytochrome. The C-terminal domain fused to GFP localizes exclusively in the nucleus and form speckles regardless of the light conditions (Nagy et al 2000). This is in striking contrast to the N-terminal domain, which never enters the nucleus by itself even after the light activation. Hence, the N-terminal domain in Pr form appears to have an activity to retain phytochrome in the cytoplasm. The molecular mechanism of this effect should be elucidated in future studies.

It is intriguing here that N650G-GUS-NLS was two orders of magnitude more sensitive to light than the full-length phyB (Matsushita et al 2003). Hence, the C-terminal domain appears to attenuate the signaling activity of the N-terminal domain of phyB rather than enhance it in the nucleus. Possible mechanisms explaining this effect of the C-terminal domain are as follows: (1) The C-terminal domain may chemically modify the N-terminal domain to reduce its signaling activity. Phosphorylation would be a good candidate mechanism for this effect (Chapter 6 by Kim and Song). (2) The quaternary structure determined by the C-terminal domain may not be optimal for the signaling reaction compared with that determined by GUS. (3) The C-terminal domain may withdraw phytochrome from its reaction site by forming aggregates (or speckles) in the nucleus. Indeed, the C-terminal domain by itself forms speckles in the nucleus, which may be the site of sequestering.

In the C-terminal domain of phytochrome, a region homologous to bacterial histidine kinase resides at its C-terminal end. Biochemical analysis has suggested that this domain indeed has serine/threonine kinase activity (Yeh and Lagarias 1998). However, evidence indicates that this domain is dispensable for the signal transduction activity of phyB (see above). Nevertheless, it may have activity to phosphorylate the N-terminal domain of phytochrome to modify its signaling activity. Proteins that are reported to be phosphorylated by phytochrome include cry1 (Ahmad et al 1998), PKS1 (Fankhauser et al 1999) and IAA (Colon-Carmona et al 2000).

Amino acid substitutions that affect the signal transduction of phytochrome but not the spectral activity of phytochrome reside within a relatively small region, which is often referred to as Q-box, within the C-terminal domain. Overlapping with this domain, PAS domains are recognized. PAS domain is found in many signal sensor proteins in archaea, eubacteria, and eukarya. Some of these mutations are known to reduce the affinity of phytochrome to PIF3 (Ni et al 1998) and/or to reduce the speckle formation in the nucleus (Chen et al 2003). Hence, it is likely that this domain is involved in the regulation of phytochrome signal transduction. Nevertheless, it should be emphasized here that the PAS domains are not required for the signaling activity of phytochrome in the nucleus (see above).

Factors Interacting with the C-Terminal Domain of Phytochrome

Several proteins are known to physically interact with the C-terminal domain of phytochrome. PIF3 has been identified as a protein interacting with the C-terminal domain of phyB through the yeast two-hybrid screening of the Arabidopsis cDNA library (Ni et al 1998). As mentioned above, PIF3 recognizes both the C-terminal and N-terminal domains of phytochrome (Ni et al 1999). Since the binding is stronger with the full-length phyB compared with that with the N-terminal or C-terminal domain, the two binding domains appear to act synergistically in the full-length context. It is intriguing here that the C-terminal domain may attenuate the signaling activity of phytochrome in the nucleus (Matsushita et al 2003). The interaction between the C-terminal domain and PIF3 may be involved in this attenuation process.

NDPK2 (Choi et al 1999) and PKS1 (Fankhauser et al 1999) have been identified as factors that interact with the C-terminal domain of phytochrome through the yeast two-hybrid screening. The *NDPK2* gene encodes a nucleoside diphosphate kinase, which acts as a tumor suppressor in animal cells. The enzymatic activity of NDPK2 is regulated by phyA in vitro. The *ndpk2* mutant exhibits partial defects in the regulation of hypocotyl elongation and cotyledon opening in response to continuous far-red light. PKS1 is a cytoplasmic protein that is phosphorylated by phytochrome both in vitro and in vivo. Recent analysis using mutants suggests that PKS1 and PKS2, a closest homologue of PKS1, are involved in a growth regulatory loop that provides homeostasis to the phyA signaling (Lariguet et al 2003).

Besides the above proteins, a few factors are known to physically interact with the C-terminal domain of phytochrome. These include ADO1/ZTL/LKP1, ELF3, and COP1. ADO1/ZTL/LKP1 is a protein containing the PAS domain. The deficiency in this protein results in malfunctioning of the biological clock. The C-terminal domain of phyB interacts with ADO1/ZTL/LKP1 in the yeast-two hybrid system and in vitro (Jarillo et al 2001). Since phytochrome regulates and is regulated by the circadian clock, this interaction may define the link between these two systems. ELF3 is a nuclear protein whose abundance is regulated by the circadian clock. The C-terminal domain of phyB and ELF3 interact in the yeast two-hybrid system and in vitro (Liu et al 2001). The *elf3* mutants exhibit early flowering phenotype although this phenotype is independent of the presence of phyB. COP1, which has E3 ubiquitin ligase activity, is a key regulator of photomorphogenesis (Chapter 29 by Yanagawa et al). The C-terminal domain of phyB interacts with COP1 in the yeast two-hybrid system (Yang et al 2001). More recently, the phyA PAS domain has been shown to interact with the COP1 WD40 domain (Seo et al 2004).

In summary, diverse factors interacts with the C-terminal domain of phytochrome. It should be noted here that none of the mutations in these factors has ever been shown to completely mimic the phytochrome deficient mutants such as *hy1*, *hy2*, *phyA*, and *phyB*. This may be explained by the redundant functions of related genes. More likely, these factors may modify the functions of phytochrome rather than mediate the phytochrome signal directly. These two possibilities should be tested experimentally in the future study.

A Proposed Model of the Phytochrome Signal Transduction

Our current model on the functions of different domains of phyB is shown in Figure 1. As suggested by the analysis of N451G-GUS-NLS plants, the core domain for phyB signaling has been defined to be amino acid positions 104–451 (Oka et al 2004). The N-terminal stretch is required for the phyA (Jordan et al 1996) but not phyB (Wagner et al 1996) functions. This domain may be a target of phosphorylation and be involved in the signal attenuation (Chapter 6 by Kim and Song). The PHY domain is required to stabilize the Pfr form of phyB. The C-terminal domain appears to have multiple functions. It is required for the nuclear localization of phyB. It dimerizes phyB, which enhances the signaling activity of the N-terminal domain substantially. Nevertheless, the C-terminal domain appears to attenuate the signaling activity of phytochrome in the nucleus by an unknown mechanism. The N651G-GUS-NLS plant is much more sensitive than the plant expressing the full-length phyB fused to GFP. In addition, this domain interacts with various factors, which may lead to fine-tuning of the phyB function.

It has been under debate whether phytochrome transduces the signal only in the nucleus. Recent analysis using phyB derivatives attached to NLS, NES (Matsushita et al 2003) or a cytoplasmic retention protein (Huq et al 2003) supports that nuclear localization is essential for the function of phyB. Nevertheless, it is possible that phyA and other phytochromes may transduce the signal in the cytoplasm. For example, phytochrome mediates a rapid change in the cytoplasmic motility in response to light (Chapter 9 by Takagi). The phyA and phyB responses are different in many aspects regardless of the fact that these molecular species exhibit similar spectral characteristics in vitro. Only phyA mediates the very low fluence response and far-red high-irradiance response, both of which are triggered by very low levels of phyA Pfr. Hence it is possible that phyA utilizes a totally different mechanism to transduce the signal for these specific responses. It is also intriguing that most of the phytochrome-interacting factors reside in both the nucleus and the cytoplasm. PKS1 exists exclusively in the cytoplasm. Phytochrome may transduce the signal in the cytoplasm through the interaction with these factors.

References

Ahmad M, Jarillo JA, Smirnova O, Cashmore AR (1998) The CRY1 blue light photoreceptor of Arabidopsis interacts with phytochrome A in vitro. Mol Cell 1: 939–948

Chen M, Schwab R, Chory J (2003) Characterization of the requirements for localization of phytochrome B to nuclear bodies. Proc Natl Acad Sci USA 100: 14493–14498

Choi G, Yi H, Lee J, Kwon YK, Soh MS, Shin B, Luka Z, Hahn TR, Song PS (1999) Phytochrome signalling is mediated through nucleoside diphosphate kinase 2. Nature 401: 610–613

Colon-Carmona A, Chen DL, Yeh KC, Abel S (2000) Aux/IAA proteins are phosphorylated by phytochrome in vitro. Plant Physiol 124: 1728–1738

Fankhauser C, Yeh KC, Lagarias JC, Zhang H, Elich TD, Chory J (1999) PKS1, a substrate phosphorylated by phytochrome that modulates light signaling in Arabidopsis. Science 284: 1539–1541

Huq E, Al-Sady B, Quail PH (2003) Nuclear translocation of the photoreceptor phytochrome B is necessary for its biological function in seedling photomorphogenesis. Plant J 35: 660–664

Jarillo JA, Capel J, Tang RH, Yang HQ, Alonso JM, Ecker JR, Cashmore AR (2001) An Arabidopsis circadian clock component interacts with both CRY1 and phyB. Nature 410: 487–490

Jordan ET, Cherry JR, Walker JM, Vierstra RD (1996) The amino-terminus of phytochrome A contains two distinct functional domains. Plant J 9: 243–257

Kim J, Yi H, Choi G, Shin B, Song PS, Choi G (2003) Functional characterization of phytochrome interacting factor 3 in phytochrome-mediated light signal transduction. Plant Cell 15: 2399–2407

Krall L, Reed JW (2000) The histidine kinase-related domain participates in phytochrome B function but is dispensable. Proc Natl Acad Sci USA 97: 8169–8174

Lariguet P, Boccalandro HE, Alonso JM, Ecker JR, Chory J, Casal JJ, Fankhauser C (2003) A growth regulatory loop that provides homeostasis to phytochrome a signaling. Plant Cell 15: 2966–2978

Liu XL, Covington MF, Fankhauser C, Chory J, Wagner DR (2001) ELF3 encodes a circadian clock-regulated nuclear protein that functions in an Arabidopsis PHYB signal transduction pathway. Plant Cell 13: 1293–1304

Martinez-Garcia JF, Huq E, Quail PH (2000) Direct targeting of light signals to a promoter element-bound transcription factor. Science 288: 859–863

Matsushita T, Mochizuki N, Nagatani A (2003) Dimers of the N-terminal domain of phytochrome B are functional in the nucleus. Nature 424: 571–574

Montgomery BL, Lagarias JC (2002) Phytochrome ancestry: sensors of bilins and light. Trends Plant Sci 7: 357–366

Nagatani A (2004) Light-regulated nuclear localization of phytochromes. Curr Opin Plant Biol 7: 708–711

Nagy F, Kircher S, Schäfer E (2000) Nucleo-cytoplasmic partitioning of plant photoreceptors phytochromes. Semin Cell Dev Biol 11: 505–510

Ni M, Tepperman JM, Quail PH (1998) PIF3, a phytochrome-interacting factor necessary for normal photoinduced signal transduction, is a novel basic helix-loop-helix protein. Cell 95: 657–667

Ni M, Tepperman JM, Quail PH (1999) Binding of phytochrome B to its nuclear signalling partner PIF3 is reversibly induced by light. Nature 400: 781–784

Oka Y, Matsushita T, Mochizuki N, Suzuki T, Tokutomi S, Nagatani A (2004) Functional analysis of a 450-amino acid N-terminal fragment of phytochrome B in Arabidopsis. Plant Cell 16: 2104–2116

Seo HS, Watanabe E, Tokutomi S, Nagatani A, Chua NH (2004) Photoreceptor ubiquitination by COP1 E3 ligase desensitizes phytochrome A signaling. Genes Dev 18: 617–622

Stockhaus J, Nagatani A, Halfter U, Kay S, Furuya M, Chua NH (1992) Serine-to-alanine substitutions at the amino-terminal region of phytochrome A result in an increase in biological activity. Genes Dev 6: 2364–2372

Sweere U, Eichenberg K, Lohrmann J, Mira-Rodado V, Baurle I, Kudla J, Nagy F, Schäfer E, Harter K (2001) Interaction of the response regulator ARR4 with phytochrome B in modulating red light signaling. Science 294: 1108–1111

Wagner D, Koloszvari M, Quail PH (1996) Two small spatially distinct regions of phytochrome B are required for efficient signaling rates. Plant Cell 8: 859–871

Yang HQ, Tang RH, Cashmore AR (2001) The signaling mechanism of Arabidopsis CRY1 involves direct interaction with COP1. Plant Cell 13: 2573–2587

Yeh KC, Lagarias JC (1998) Eukaryotic phytochromes: light-regulated serine/threonine protein kinases with histidine kinase ancestry. Proc Natl Acad Sci USA 95: 13976–13981

Chapter 8

Light Regulation of Intracellular Localization of Phytochrome B

Eberhard Schäfer[1] and Ferenc Nagy[2]

Introduction

The nucleo/cytoplasmic distribution of members of the phytochrome photoreceptor family is regulated by light. Light quality- and quantity-dependent nuclear import of these photoreceptors plays a critical role in mediating light-induced signalling. In this chapter we discuss results obtained by analysing the kinetics of changes in the light-dependent intracellular localization of wild-type and mutant PHYB:YFP fusion proteins.

To monitor the red/far-red part of the solar spectrum, plants make use of the red/far-red photoreversible photoreceptors phytochromes. In *Arabidopsis thaliana* phytochromes are encoded by a small gene family consisting of five members, named *PHYA-D* (Clack et al 1994). Phytochromes are chromoproteins with a covalently linked open-chain tetrapyrrol as chromophore. The spectral properties of all phytochromes are very similar (Eichenberg et al 2000); based on their kinetic properties and physiological roles, however, they can be divided to two classes: light-labile or type I and light-stable or type II phytochromes (Furuya and Schäfer 1996). The light-labile PHYA controls VLFR (very low fluence responses) and far-red HIR (high irradiance responses). PHYA is rapidly degraded in its physiologically active Pfr (far-red absorbing) form. In contrast to PHYA, the light-stable phytochromes—of which PHYB is the best analysed—control the red/far-red photoreversible and continuous red light responses. The active Pfr form of PHYB is inactivated by fast dark reversion and the response regulator ARR4, interacting with the extreme N-terminal part of PHYB, modulates this process both in yeast cells and in planta (Sweere et al 2001).

[1]Institute of Botany, University of Freiburg, Schaenzlestrasse 1, D-79104 Freiburg, Germany
[2]Plant Biology Institute, Biological Research Center, Temesvari krt. 62, H-6726 Szeged, Hungary

Nucleo/Cytoplasmic Partitioning of Phytochromes

For many years it was generally assumed that phytochromes are localized in the cytosol. In their pioneering work Yamaguchi et al (1999) and Kircher et al (1999, 2002) showed that phytochromes A–D are transported into the nucleus in a light-dependent manner. With the first observations of light-dependent nuclear transport of phytochrome–GFP fusion proteins, it became possible to study how the differences in the signal transduction properties of the various phytochromes manifest themselves.

Kinetics of Nuclear Translocation of PHYA and PHYB

Kircher et al (1999) were the first to describe the striking difference in kinetics and wavelength dependence of the nuclear import of PHYA:GFP and PHYB:GFP. These authors showed that the import of PHYB:GFP was rather slow and could only be induced by either continuous red or multiple red light pulses. The effect of the red light pulses could be reverted by subsequent far-red light pulses. Thus the nuclear transport of PHYB:GFP showed the characteristics of a classical LFR. After being transported from the cytosol into the nucleus, PHYB:GFP forms characteristic complexes called speckles (Kircher et al 1999). In a subsequent paper Gil et al (2000) demonstrated that nuclear PHYB:GFP speckles disappear slowly after a light-to-dark transition. This reaction can be accelerated by a far-red light pulse applied before transferring the seedlings to darkness. More interestingly, the sensitivity of PHYB:GFP speckle formation to a subsequent light pulse was strongly enhanced after the first light-to-dark transition (i.e., light treatment is followed by a dark period, allowing the speckles to dissolve). This sensitivity amplification depends on the length of the dark interval and thus resembles the responsiveness amplification described previously for several photomorphogenetic responses (Drumm and Mohr 1978).

In contrast to PHYB:GFP, PHYA:GFP exhibits significant nuclear import even after a single light pulse (Kim et al 2000). In this case, a far-red light pulse is sufficient to induce nuclear import of the PHYA:GFP fusion protein, and continuous far-red light treatment evokes enhanced import of PHYA:GFP. More detailed analysis showed that this transport is both wavelength and fluence rate dependent (Kim 2002). It follows that light-dependent nuclear transport of PHYA:GFP shows the characteristics of both the VLFR and HIR and therefore it reflects the well-known properties of PHYA-mediated physiological responses. Similarly to PHYB:GFP, a red light pulse or continuous light treatment induces not only nuclear transport but also formation of nuclear PHYA:GFP speckles (Kircher et al 1999, Kim et al 2000). However, in sharp contrast to PHYB:GFP, illumination also leads to the formation of PHYA:GFP speckles in the cytosol. These cytosolic PHYA:GFP complexes are probably the equivalent of the previously described cytosolic SAPs (sequestered areas of phytochrome; Speth et al

1986). Those complexes have been believed to be the place of PHYA degradation in the cytosol, but the critical experimental tests of this hypothesis are still missing. In this context it is worth noting that in a very recent paper, Seo et al (2004) showed that COP1 (constitutive photomorphogenesis), an important negative regulator of photomorphogenesis, may play multiple roles in desensitizing PHYA signalling. These authors showed that COP1 acting as an E3 ubiquitin ligase ubiquinates multiple proteins in vitro. These proteins include not only those mediating PHYA signalling such as HY5 (Osterlund et al 2000) and LAF1 (Seo et al 2003), but also the PHYA photoreceptor itself. The light-regulated nucleo/cytoplasmic partitioning of COP1 and that of PHYA show opposite patterns and PHYA degradation is a quick process. To determine how and to what extent COP1 contributes to PHYA degradation in the nucleus and cytosol in planta will remain a challenging task.

Nucleo/Cytoplasmic Partitioning of Other Phytochromes

It has been established that PHYA:GFP and PHYB:GFP fusion proteins show a striking difference regarding the light dependence of their nuclear transport: as for PHYB:GFP, significant nuclear transport was only observed in continuous illumination or after several pulses of red light (Kircher et al 1999, Gil et al 2000), whereas nuclear transport of PHYA:GFP is detected during continuous far-red illumination or even after a single pulse of far-red light (Kim et al 2000). Other members of the phytochrome family show additional subtle differences regarding the kinetics of their nuclear translocation after light treatment (Nagy and Schäfer 2002). A comparative analysis of light-dependent intracellular partitioning of all five members of the phytochrome family showed that only PHYA:GFP was exclusively cytosolic in dark-grown tissues (Kircher et al 2002). This means that no PHYA:GFP signals could be detected in the nucleus in dark-grown seedlings. In contrast, for all the other PHY:GFP species a diffuse nuclear staining was already detectable in the dark. This can be caused by the fact that the expression of these transgenes was driven by the strong viral 35S promoter, thus over expression of the fusion protein may result in their nuclear translocation in the dark. This hypothesis assumes that the cytoplasmic localization of these photoreceptors in the dark is controlled by an active cytosolic retention mechanism, which can be saturated by the appearance of an excessive amount of substrate (PHY:GFP fusion proteins). The effectiveness of light-dependent nuclear transport of phytochromes supports the existence of such a saturable retention mechanism. Nevertheless, it is still not clear whether the weak nuclear staining observed in dark-grown transgenic *Arabidopsis* seedlings expressing PHYB-D:GFP is due to over-expression, or whether a limited nuclear transport of these PHY:GFP fusion proteins indeed takes place in the dark. Independently of the diffuse nuclear staining in the dark, all phytochromes show light-dependent nuclear import and formation of nuclear complexes. Clearly, both the wavelength and fluence rate dependences of these processes are quite different for the different phytochromes. PHYA is the only member, which shows a VLFR and a far-red HIR. All other phytochromes show red light dependence and PHYB is clearly the most light-sensitive species (Kircher et al 2002). Thus, light-dependent nuclear transport of PHYA and PHYB seems to reflect their previously reported, physiologically analysed spectral sensitivity. Our knowledge about the physiological function of PHYC-E is rather limited, yet the light-dependent nuclear transport of the different phytochromes reflects, at least qualitatively, the differences in light responsiveness described for the phytochromes and it can therefore been assumed that this process is an essential step for the determination of the specificity of the action of the different phytochromes. The verification of this statement clearly needs further study; however, the recently described fact that the overwhelming majority of the target genes of PHYA and PHYB are identical supports this view (Tepperman et al 2004).

Search for Mutants Displaying Aberrant Intracellular Distribution of PHYB:GFP

The primary function of phytochromes in plants is still not known. Bacterial phytochromes are light-dependent histidine kinases and function as input receptors of a two-component signalling system. Thus activation of bacterial phytochromes leads to a change in the phosphorylation state of their cognate response regulators. Although the phytochromes of higher plants contain a histidine kinase-like domain in the C-terminal part of the molecule, this domain was shown to be dispensable for their physiological function (Matsushita et al 2003). The same authors demonstrated that (i) the N-terminal part of the PHYB molecule is sufficient to induce light signalling when it is driven into the nucleus by the SV40 NLS, and (ii) the C-terminal domain probably plays a role in the regulation of the light-dependent nuclear import of the photoreceptor in vivo. It has recently become evident that phytochromes localized in the nucleus, including PHYA, PHYB and PHYB:GFP, specifically accumulate at sub-nuclear foci and form large nuclear complexes. The absence or perturbation of PHYA- or PHYB-containing complex formation led to impaired signalling in numerous cases (Nagy and Schäfer 2002). To characterize these sub-nuclear complexes at the molecular level, we purified phytochrome B (phyB) speckles from adult *Arabidopsis* plants and determined their protein composition by in-gel digestion and mass spectrometry. We also investigated the localization of some of their protein components by immunogold electron microscopy. We found that phyB speckles vary in size and are primarily localized in the interchromatin space of the nucleus. They are resistant to high concentrations of salts and non-ionic detergents, and to DNAseI treatment. They are enriched in Ser-Arg (SR) rich proteins and proteins involved in pre-mRNA processing. These features indicate that they are similar to **I**nterchromatin **G**ranule **C**lusters (IGCs) described in animal cells. The precise function of mammalian and plant IGCs is not yet clear, but these results indicate for the first time specific association of a photoreceptor with IGC-like structures (KSC Panigrahi et al unpublished).

To gain further insight into the mechanism regulating nucleo/cytoplasmic partitioning of PHYA and PHYB, we performed several screens to identify mutants displaying impaired light-dependent nuclear import and/or complex formation of PHYA:GFP and PHYB:GFP within the nucleus. In the past, most screens performed to isolate photomorphogenic mutants used either hypocotyl length or cotyledon phenotype as read-outs, which are obviously late phytochrome-mediated responses. Thus we expected that by employing the same strategy, only a low number of mutants affecting early response elements that play a role in controlling nuclear translocation of phytochromes can be isolated. Nevertheless, in the absence of a specific screen we decided to use light-dependent inhibition of hypocotyl growth as a read-out. We used low fluence rate to screen for hypersensitive lines and strong irradiation with continuous red light to screen for hyposensitive lines. Transgenic seeds expressing a single copy of 35S:PHYB:GFP or PHYA:GFP in a wild-type background were mutagenized by EMS.

The number of mutants obtained in the screens for both PHYA:GFP and PHYB:GFP lines are given in Table 1. In this report we will concentrate only on the mutants isolated in the PHYB:GFP screen.

Our data show that the number of mutants that exhibit aberrant growth phenotype and nuclear import and/or formation of PHYB:GFP speckles is surprisingly high. Characterization of the majority of isolated mutants is still in progress, but in some cases physiological and microscopical studies have indicated that the mutation may have occurred within the transgene. Indeed, members of this class of mutations were mapped mostly to the transgene locus. Sequencing of the PHYB:GFP transgene in these lines identified several mutations, which displayed hypo- or hypersensitive phenotype for hypocotyl growth and aberrant intracellular localization of PHYB:GFP.

The hyposensitive mutants characterized so far had mutations either in the C-terminal domain (Quail-box), or in the hinge region between the N-terminal chromophore-binding and C-terminal dimerization domains. Most of our work has been concentrated on the two hypersensitive alleles. One shows a G to S transition at aa565, the other a G to N transition at aa515.

Both mutants showed a phenotype similar to that of the previously described *phyB-401* mutant, which had a G to E transition at aa564 (Kretsch et al 2000).

Table 1. Number of mutants obtained in the screens for PHYA:GFP and PHYB:GFP lines

	PHY A	PHY B
No. of individuals screened	40000	150000
Isolated "mutants"	74	400
Growth and localization phenotype	21	18
Hyposensitive ones	8	9
Hypersensitive ones	9	2

The main characteristic of this mutant is that it has no far-red reversibility of hourly given R pulses. Biochemical analysis of the mutated form expressed in yeast cells showed that it had normal spectral properties but it lacked dark reversion of Pfr to Pr. The analysis of the dark reversion of the new phyB mutant allele is still in progress, but physiological data clearly indicate that dark reversion is lost also in this mutant. Analysis of light-dependent nuclear translocation indicates that for these three mutated versions a single light pulse is sufficient to induce strong nuclear translocation.

Furthermore, immunolocalization data indicate that in the *phyB-401* mutant the peripheral cytosolic localization is lost (T. Kunkel et al unpublished). Surprisingly, the phenotypes of mutations in the hinge region were not restricted to hypersensitivity: several mutations leading to hyposensitive phenotype were also found. In two cases point mutations in the hinge region affected amino acids separated only by a short distance, yet these mutations resulted in contrasting phenotypes. Work is in progress to test whether these hyposensitive mutations lead to the opposite kinetic and localization properties relative to the hypersensitive mutations, i.e., enhanced dark reversion and retardation in the periphery of the cytosol.

Acknowledgments. Work in Freiburg, Germany, was supported by the Wolfgang Paul Award to F.N. and SFB 592 and Fonds Chemical Industry grants to E.S. Work in Szeged, Hungary, was supported by OTKA Grant T032565 and HHMI Grant #55000325 to F.N.

References

Clack T, Matthews S, Sharrock RA (1994) The phytochrome apoprotein family in *Arabidopsis* is encoded by five genes: the sequence and expression of *PHYD* and *PHYE*. Plant Mol Biol 25: 413–417

Drumm H, Mohr H (1978) The mode of action in between blue (UV) light photoreceptors and phytochrome in anthocyanin formation of the *Sorghum* seedling. Photochem Photobiol 27: 241–248

Eichenberg K, Bäurle I, Paulo N, Sharrock RA, Rüdiger W, Schäfer E (2000) *Arabidopsis* phytochromes C and E have different spectral characteristics from those of phytochromes A and B. FEBS Lett 470: 107–112

Furuya M, Schäfer E (1996) Photoperception and signalling of induction reactions by different phytochromes. Trends Plant Sci 1: 301–307

Gil P, Kircher S, Adam E, Bury E, Kozma-Bognar L, Schäfer E, Nagy F (2000) Photocontrol of subcellular partitioning of phytochrome-B:GFP fusion protein in tobacco seedlings. Plant J 22: 135–145

Kim L (2002) Analysen zur intrazellulären Lokalisation von Phytochrom A. PhD Thesis, Universität Freiburg

Kim L, Kircher S, Toth R, Adam E, Schäfer E, Nagy F (2000) Light-induced nuclear import of phytochrome-A:GFP fusion proteins is differentially regulated in transgenic tobacco and *Arabidopsis*. Plant J 22: 125–134

Kircher S, Kozma-Bognar L, Kim L, Adam E, Harter K, Schäfer E, Nagy F (1999) Light quality-dependent nuclear import of the plant photoreceptors phytochrome A and B. Plant Cell 11: 1445–1456

Kircher S, Gil P, Kozma-Bognar L, Fejes E, Speth V, Husselstein-Muller T, Bauer D, Adam E, Schäfer E, Nagy F (2002) Nucleo-cytoplasmic partitioning of the plant photoreceptors phytochrome A, B, C, D and E is differentially regulated by light and exhibits a diurnal rhythm. Plant Cell 25: 1222–1232

Kretsch T, Poppe C, Schäfer E (2000) A new type of mutation in the plant photoreceptor phytochrome B causes loss of photoreversibility and an extremely enhanced light sensitivity. Plant J 22: 177–186

Matsushita T, Mochizuki N, Nagatani A (2003) Dimers of the N-terminal domain of phytochrome B are functional in the nucleus. Nature 424: 571–574

Nagy F, Schäfer E (2002) Phytochromes control photomorphogenesis by differentially regulated, interacting signalling pathways in higher plants. Annu Rev Plant Biol 53: 329–355

Osterlund MT, Hardtke CS, Wie N, Deng XW (2000) Targeted destabilization of HY5 during light-regulated development of *Arabidopsis*. Nature 405: 462–466

Seo HS, Yang JY, Ishikawa M, Bolle C, Ballasteros ML, Chua NH (2003) LAF1 ubiquination by COP1 controls photomorphogenesis and is stimulated by SPA1. Nature 424: 995–999

Seo HS, Watanabe E, Tokutomi S, Nagatani A, Chua NH (2004) Photoreceptor ubiquitination by COP1 E3 ligase desensitizes phytochrome A signaling. Genes Dev 18: 617–622

Speth V, Otto V, Schäfer E (1986) Intracellular localisation of phytochrome in oat coleoptiles by electron microscopy. Planta 168: 299–304

Sweere U, Eichenberg K, Lohrmann J, Mira-Rodado V, Bäurle I, Kudla J, Nagy F, Schäfer E, Harter K (2001) Interaction of the response regulator ARR4 with phytochrome B in modulating red light signalling. Science 942: 1108–1111

Tepperman JM, Hudson ME, Khanna R, Zhu T, Chang SH, Wang X, Quail PH (2004) Expression profiling of phyB mutant demonstrates substantial contribution of other phytochromes to red-light-regulated gene expression during seedling de-etiolation. Plant J 38: 725–739

Yamaguchi R, Nakamura M, Mochizuki N, Kay SA, Nagatani A (1999) Light-dependent translocation of a phytochrome B:GFP fusion protein to the nucleus in transgenic *Arabidopsis*. J Cell Biol 145: 437–445

Chapter 9

Photoregulation of Cytoplasmic Motility

Shingo Takagi

Introduction

Intracellular movement is ubiquitously observed in the plant kingdom and it plays a pivotal role in the regulation of a wide spectrum of activities in plant cells. Well-organized motile apparatuses, which are predominantly composed of actin filaments, microtubules, and their associated proteins, drive respective unique types of intracellular movement. In several types of plant cell, intracellular movements are known to be under the control of light. In many cases, a new pattern of movement of cell organelles and/or smaller cytoplasmic particles is induced in response to light stimuli.

For intracellular movement to be induced, a motive force must be generated to drive the movement of cytoplasm. In addition, the cytoplasmic matrix, in which the cell organelles and cytoplasmic particles are buried, must become mobile simultaneously with the generation of motive force. Consequently, high cytoplasmic motility is achieved through fulfilling at least two conditions; the generation of motive force and the appropriate mechanical properties of cytoplasmic matrix. The motive force for intracellular movement is generated through the interaction of ATP-hydrolyzing motor proteins, such as myosins (Shimmen et al 2000), kinesins and dyneins (Asada and Collings 1997), with actin filaments or microtubules, respectively. Although the mechanochemical processes associated with the motor proteins have been extensively investigated in biophysical and molecular biological approaches, we have only a small amount of knowledge on the nature of mechanical properties of cytoplasmic matrix. This review aims first to briefly summarize the dissection of light-dependent cytoplasmic motility in an aquatic angiosperm *Vallisneria gigantea*, and next to discuss possible modes of regulation of the cytoplasmic motility.

Department of Biology, Graduate School of Science, Osaka University, Toyonaka, Osaka 560-0043, Japan

Cytoplasmic Streaming in *Vallisneria* Mesophyll Cells

In mesophyll cells of *Vallisneria*, unidirectional streaming of the cytoplasm along the anticlinal walls is induced by continuous illumination, which has been known as "photodinesis." Several aspects of photodinesis have been extensively investigated; however, most of classic studies focused merely on the relationships between the nature of actinic illumination and the numbers of cells in which the cytoplasmic streaming was induced. Through precise observation of the initial process of induction of cytoplasmic streaming, we noticed that the cytoplasm never exhibits the typical unidirectional movement from the beginning. The cytoplasm of dark-adapted mesophyll cells, which includes the nucleus, cell organelles, and cytoplasmic particles, is apparently quiescent. The initial event observed in the illuminated cells is acceleration of movement of cytoplasmic particles in random directions. In time, the cytoplasmic particles make local streamlets, but their movement is still haphazard. After a few minutes of illumination, a continuous, smooth movement of the cytoplasm commences. Between the start of illumination and the commencement of unidirectional movement, the pattern of movement of cytoplasm dynamically changes from a motionless state, via an accelerated state in random directions, and then to a directed streaming. These changes are transient, and moreover, observed only in an early phase after the start of illumination. We can assume that light first increases the motility of cytoplasm, which may correspond to the acceleration of random movement. The more ordered pattern of movement seems to be established after a certain period of lag time.

We succeeded in separating the effects of light on the induction of cytoplasmic streaming into those of phytochrome and of photosynthesis. Photosynthesis accelerates the enzymatic activity of plasma membrane H^+-ATPase (Harada et al 2002b) to generate the H^+ motive force (Harada et al 2002a), which provides a driving force for Ca^{2+} efflux. Since Ca^{2+} is a general regulatory factor of intracellular movement, we have proposed that Ca^{2+} fluxes across the plasma membrane are involved in regulation of the cytoplasmic level of Ca^{2+}. These aspects were briefly summarized in another review (Takagi 1997). On the other hand, the effect of phytochrome in the regulation of cytoplasmic streaming was ambiguous. Possible phytochrome-dependent changes in the mechanical properties of cytoplasmic matrix were examined by centrifugation methods. Using a centrifuge microscope of stroboscopic type (Hiramoto and Kamitsubo 1995), we detected rapid, partially reversible effects of red and far-red light on the resistance of chloroplasts to centrifugal force. However, the resistance of chloroplasts to centrifugal force does not necessarily represent the mechanical properties of cytoplasmic matrix, because it has become evident that the intracellular positioning of chloroplasts is determined through complex interactions of numerous extracellular factors (Wada et al 2003), and that the actin cytoskeleton is involved in the intracellular anchoring of chloroplasts (Takagi 2003). In the cytoplasmic matrix of living cells, chloroplasts do not behave like falling balls of the sedimentation assay.

Cytoplasmic Motility in *Vallisneria* Epidermal Cells

The light-dependent regulation of cytoplasmic motility seems to occur rapidly and only transiently. Moreover, there was no sensible method to analyze a random movement of cytoplasmic particles in a quantitative manner. Consequently, the initial process of light-induced changes in the cytoplasmic motility has been unable to be dissected by conventional light microscopy. To address these problems, we applied a dynamic image processing technique (Mineyuki et al 1983) to a microscopic analysis of cytoplasmic motility. The cytoplasmic motility was quantified by monitoring changes in the brightness of individual pixels on digitized images sequentially captured under infrared light. The method enabled us to effectively separate a random movement of the cytoplasm from the directional movement. We selected epidermal cells of *Vallisneria* as the materials, which have flat outer periclinal walls, so that we could obtain much better optical images of the cytoplasm in the focal plane of a light microscope.

The results obtained by these procedures are summarized as follows (Takagi et al 2003). (1) The cytoplasmic motility is attenuated in darkness and accelerated upon illumination. (2) The acceleration and deceleration of cytoplasmic motility is reversibly regulated by low-fluence red and far-red light. (3) The acceleration of cytoplasmic motility does not occur uniformly but in a patchy manner over the cytoplasmic layer along the outer periclinal walls. (4) The acceleration of cytoplasmic motility is detectable only a few seconds after the start of pulse illumination with red light. (5) The acceleration of cytoplasmic motility is a subcellularly localized response. (6) The cytoplasmic motility is actin-dependent and can be accelerated by treatment with Ca^{2+}-chelating reagents. On the basis of these findings, we proposed a working hypothesis on the photoregulation of cytoplasmic motility in *Vallisneria* epidermal cells (Figure 1). Type II phytochrome (Furuya 1993) in the cytoplasm matrix functions as the photoreceptor. Phototransformation of the red light-absorbing form to the far-red light-absorbing form of the type II phytochrome produces a change in the cytoplasmic level of Ca^{2+}. This change provokes modulation in the actin cytoskeleton, such as an acceleration of the activity of motor protein and changes in the mechanical properties of cytoplasmic matrix. Transduction of the type II phytochrome-dependent signals onto the actin cytoskeleton is completed within a few seconds after the start of illumination only in the illuminated region. In the following sections, several conspicuous features of this response will be discussed, namely, the rapid and localized regulation and factors involved in the Ca^{2+} regulation.

Rapid, Localized Regulation of Cytoplasmic Motility

The acceleration of cytoplasmic motility was detectable as early as a few seconds after the start of illumination and thus it can be categorized as one of the most rapid responses under the control of type II phytochrome to date. This time

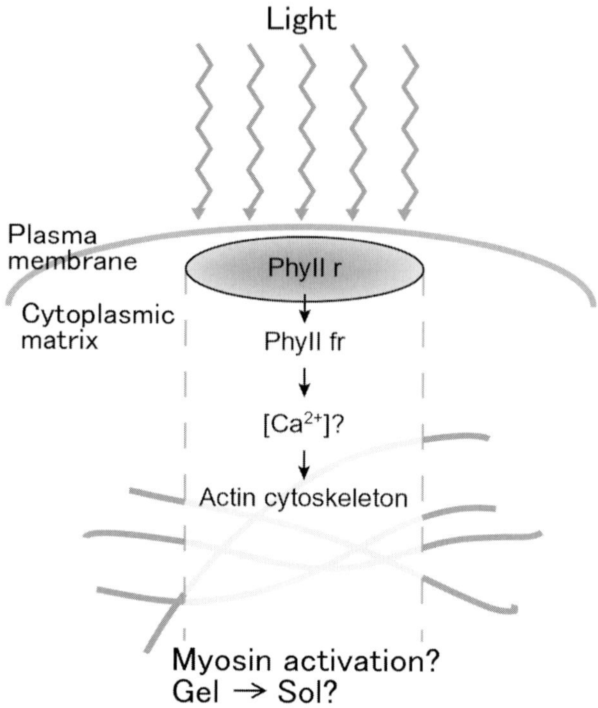

FIG. 1. A schematic model for rapid photoregulation of actin-dependent cytoplasmic motility in *Vallisneria* epidermal cells. Phototransformation of the red light-absorbing form of type II phytochrome (*PhyIIr*), which is located in the cytoplasm matrix, to the far-red light-absorbing form (*PhyIIfr*) transmits the signal onto the actin cytoskeleton via a pathway that might include a change in the cytoplasmic Ca^{2+} (*[Ca^{2+}]*). Modulation in the actin cytoskeleton, which is demonstrated by faded filaments, may include an acceleration of the myosin activity and/or a solation of the cytoplasmic matrix. All the processes are completed within a few seconds after the start of illumination, and furthermore, only in the light-exposed cytoplasmic region, which is indicated as a *column*

course may exclude a possible involvement of newly expressed gene products in the response. Consequently, type II phytochrome(s) in epidermal cells of *Vallisneria* may function, at least partly, in the cytoplasm. Although a possible cytoplasmic partner for phytochrome in this response is unknown, the cytoplasmic localization of phytochrome molecules was confirmed by immunocytochemistry (Takagi et al 2003). Light-dependent rapid responses in plant cells have been identified using electrophysiological procedures. In a couple of cases, a possible association of phytochrome molecules with the plasma membrane was proposed. Since the acceleration of cytoplasmic motility occurred only in the cytoplasmic region exposed to red light, some interaction of the phytochrome-dependent signals with the plasma membrane might be suggested.

In the case of whole illumination, the acceleration of cytoplasmic motility occurred not uniformly in the cytoplasm but in a patchy manner. On the other hand, microbeam illumination did not always induce the acceleration of cytoplasmic motility. When the cytoplasmic motility was accelerated by microbeam illumination, however, the response was limited exclusively in the illuminated cytoplasm. These results might indicate that there is an uneven distribution of the relevant components required for the response in the dark-adapted cytoplasm, and that phytochrome molecules and/or the other signal components become associated with the plasma membrane or the cortical cytoplasm only after the phototransformation of phytochrome. The molecular basis for the localized transmission of signals from the photoreceptors is an open question.

Ca^{2+} Regulation of Cytoplasmic Motility

Treatment of dark-adapted epidermal cells with the actin-depolymerizing reagents produced a substantial decline in the cytoplasmic motility. In those cells, light never accelerated the cytoplasmic motility. The effects of the actin-depolymerizing reagents were reversible. In contrast, microtubule-depolymerizing reagents neither affected the cytoplasmic motility in dark-adapted cells nor its acceleration by illumination. The most plausible downstream mechanism for the light-induced increase in cytoplasmic motility is the acceleration of motor protein activity and/or the reorganization of actin filaments. In dark-adapted epidermal and mesophyll cells, Ca^{2+}-chelating reagents, such as EGTA and BAPTA, can accelerate the cytoplasmic motility in darkness. The effects of exogenously applied EGTA on the mode of movement of chloroplasts were also reported in the fern *Adiantum capillus-veneris* (Kadota and Wada 1992). The chloroplasts exhibited jerky movement, even after the completion of photorelocation movement. These results may indicate that the cytoplasmic motility is under the control of Ca^{2+}, and that lower concentrations of Ca^{2+} are favorable for the motility.

We have already succeeded in biochemically isolating the Ca^{2+}-sensitive motor activity from *Vallisneria* leaves that can interact with actin filaments in an ATP-dependent manner in vitro (Takagi 1997). The activity was accelerated in the presence of Ca^{2+} lower than 1 μM, whereas it was suppressed at higher concentrations of Ca^{2+}. Consequently, one of the candidates that provide Ca^{2+} sensitivity for cytoplasmic motility is the motor protein myosin. On the other hand, if this activity is solely responsible for the light-dependent acceleration of cytoplasmic motility, the level of Ca^{2+} in the dark-adapted cytoplasm should be higher than 1 μM, which is not likely. Other possible candidates responsible for the Ca^{2+}-sensitive cytoplasmic motility are gelation and solation of the cytoplasmic matrix, which have been known to participate in the amoeboid movement. These changes in the mechanical properties of cytoplasmic matrix are attributable, at least partly, to reorganization of actin filaments. There have been several reports that demonstrate that actin-binding proteins function in plant cells in response to a variety of signals, regulating dynamic reorganization of actin cytoskeleton (Staiger 2000). Recently, plant villin (Yokota et al 2000) and gelsolin (Huang et al 2004) were identified in pollen tubes of lily and field poppy, respectively. The former exhibits the bundling activity and the latter exhibits the severing activity of actin filaments, both in Ca^{2+}-sensitive manners. Such Ca^{2+}-sensitive actin-binding proteins might play essential roles in regulation of the mechanical properties of cytoplasmic matrix.

Actin-Independent Regulation of Cytoplasmic Motility

Although the mechanical properties of cytoplasmic matrix are generally supposed to depend on the mode of assembly of actin filaments, a possible involvement of actin-independent factors is also reported. Using optical tweezers to trap and manipulate statoliths in living rhizoid cells of *Chara vulgaris*, Leitz et al (1995) measured the resistance of cytoplasm to displacement of statoliths. The force required to displace the statoliths in acropetal or basipetal directions was larger than that in lateral directions. In the presence of the actin-depolymerizing reagent, the force required to displace the statoliths in all directions became the same level, which was obtained in lateral directions in the absence of the actin-depolymerizing reagent. These results suggest that there are actin-dependent and -independent factors that can affect the mechanical properties of cytoplasmic matrix. In preliminary experiments, the resistance of cytoplasm to centrifugal force was markedly reduced by treatment with EGTA of mesophyll cells of *Vallisneria* pretreated with the actin-depolymerizing reagent (Takagi unpublished). This also raised a possibility that the mechanical properties of cytoplasmic matrix can be changed in an actin-independent, though in a Ca^{2+}-sensitive manner.

Physiological Significance of Cytoplasmic Motility

In epidermal cells of *Vallisneria,* a light-induced rapid increase in the cytoplasmic motility seems to be indispensable for an efficient accumulation of chloroplasts along the outer periclinal wall, a process accomplished through a dynamic rearrangement of actin filaments (Dong et al 1998). Kagawa and Wada (2000) found that photorelocation of chloroplasts in *Arabidopsis thaliana* was efficiently induced by blue light only under background illumination with red light. By image analysis, Kagawa and Wada (2000) revealed that red light enhances the cytoplasmic motility. Together with the results obtained in the mesophyll cells of *Vallisneria,* it seems likely that in early processes of the induction of intracellular movement, high cytoplasmic motility is a prerequisite for the subsequent establishment of more ordered patterns of movement brought about by cytoskeletal reorganization.

On the other hand, the cytoplasmic streaming in several kinds of plant cell, which include mesophyll cells of *Vallisneria* (Hayashi and Takagi 2003), is transiently inhibited upon mechanical stimuli such as touch and extracellular osmotic changes. In many cases, a transient cessation of cytoplasmic streaming is accompanied by an increase in the cytoplasmic level of Ca^{2+}. Since mechanical stimuli are known to induce a variety of downstream responses, for example, turgor movement, thigmotropism, and chloroplast relocation (Sato et al 2003), not only the cytoplasmic level of Ca^{2+} but also the level of cytoplasmic motility is postulated to transmit signals mediating between the mechanoperception and the downstream responses. Furthermore, a possible involvement of activities of certain kinds of kinases and phosphatases in regulation of the tension and organization of plant actin cytoskeleton has been suggested by optical displacement assay (Grabski et al 1998). All together, we presume that modulation in the cytoplasmic motility imposed by diverse external stimuli may function as one of the integral components constituting the signal transduction pathways.

Concluding Remarks

Light-regulated cytoplasmic motility seems to play crucial roles in early steps in the transduction pathway of light signals. We can infer that phytochrome functions in the cytoplasmic matrix as the photoreceptor that mediates the response. The cytoplasmic partner for phytochrome molecules and the regulatory mechanism of actin cytoskeleton should be urgently clarified. Other open questions are which kind of light dependent developmental processes are downstream of the cytoplasmic motility and what kind of regulation operates in those processes.

Acknowledgments. I express my sincere thanks to Dr. Masaki Furuya for his hearty and patient encouragement throughout the study of *Vallisneria*. Thanks are also due to all colleagues and friends involved in the study.

References

Asada T, Collings D (1997) Molecular motors in higher plants. Trends Plant Sci 2: 29–37

Dong XJ, Nagai R, Takagi S (1998) Microfilaments anchor chloroplasts along the outer periclinal wall in *Vallisneria* epidermal cells through cooperation of Pfr and photosynthesis. Plant Cell Physiol 39: 1299–1306

Furuya M (1993) Phytochromes: their molecular species, gene families, and functions. Annu Rev Plant Physiol Plant Mol Biol 44: 617–645

Grabski S, Arnoys E, Busch B, Schindler M (1998) Regulation of actin tension in plant cells by kinases and phosphatases. Plant Physiol 116: 279–290

Harada A, Okazaki Y, Takagi S (2002a) Photosynthetic control of the plasma membrane H^+-ATPase in *Vallisneria* leaves. I. Regulation of activity during light-induced membrane hyperpolarization. Planta 214: 863–869

Harada A, Fukuhara T, Takagi S (2002b) Photosynthetic control of the plasma membrane H^+-ATPase in *Vallisneria* leaves. II. Presence of putative isogenes and a protein equipped with a C-terminal autoinhibitory domain. Planta 214: 870–876

Hayashi T, Takagi S (2003) Ca^{2+}-dependent cessation of cytoplasmic streaming induced by hypertonic treatment in *Vallisneria* mesophyll cells: possible role of cell wall-plasma membrane adhesion. Plant Cell Physiol 44: 1027–1036

Hiramoto Y, Kamitsubo E (1995) Centrifuge microscope as a tool in the study of cell motility. Int Rev Cytol 157: 99–128

Huang S, Blanchoin L, Chaudhry F, Franklin-Tong VE, Staiger CJ (2004) A gelsolin-like protein from *Papaver rhoeas* pollen (PrABP80) stimulates calcium-regulated severing and depolymerization of actin filaments. J Biol Chem 279: 23364–23375

Kadota A, Wada M (1992) Photoorientation of chloroplasts in protonemal cells of the fern *Adiantum* as analyzed by use of a video-tracking system. Bot Mag Tokyo 105: 265–279

Kagawa T, Wada M (2000) Blue light-induced chloroplast relocation in *Arabidopsis thaliana* as analyzed by microbeam irradiation. Plant Cell Physiol 41: 84–93

Leitz G, Schnepf E, Greulich KO (1995) Micormanipulation of statoliths in gravity-sensing *Chara* rhizoids by optical tweezers. Planta 197: 278–288

Mineyuki Y, Yamada M, Takagi M, Wada M, Furuya M (1983) A digital image processing technique for the analysis of particle movements: its application to organelle movements during mitosis in *Adiantum* protonemata. Plant Cell Physiol 24: 225–234

Sato Y, Kadota A, Wada M (2003) Chloroplast movement: Dissection of events downstream of photo- and mechano-perception. J Plant Res 116: 1–5

Shimmen T, Ridge RW, Lambiris I, Plazinski J, Yokota E, Williamson RE (2000) Plant myosins. Protoplasma 214: 1–10

Staiger CJ (2000) Signaling to the actin cytoskeleton in plants. Annu Rev Plant Physiol Plant Mol Biol 51:257–288

Takagi S (1997) Photoregulation of cytoplasmic streaming: cell biological dissection of signal transduction pathway. J Plant Res 110: 299–303

Takagi S (2003) Actin-based photo-orientation movement of chloroplasts in plant cells. J Exp Biol 206: 1963–1969

Takagi S, Kong SG, Mineyuki Y, Furuya M (2003) Regulation of actin-dependent cytoplasmic motility by type II phytochrome occurs within seconds in *Vallisneria* epidermal cells. Plant Cell 15: 331–345

Wada M, Kagawa T, Sato Y (2003) Chloroplast movement. Annu Rev Plant Biol 54: 455–468

Yokota E, Muto S, Shimmen T (2000) Ca^{2+}-calmodulin suppresses the F-actin-binding activity of a 135-kDa actin-bundling protein isolated from lily pollen tubes. Plant Physiol 123: 645–654

CHAPTER 10

Polymorphism of Phytochrome A and Its Functional Implications

VITALY A. SINESHCHEKOV

Introduction

One of the most important achievements in phytochrome (phy) research is the discovery of a number of its isoforms, with phyA and phyB as major ones (Smith 1997). They perform complementary and contrasting photophysiological functions. These functions are distinguished by fluence and spectral light requirements and attributed to different modes of action—very low fluence responses (VLFR) and high irradiance responses (HIR) mediated by phyA under far-red light (FR) and the "classical" red light (R)-induced/FR-reversible low fluence responses (LFR) mediated by phyB. The phytochrome system has, however, further levels of complexity (Sineshchekov 1995, 1998, 1999, 2004). There exist two distinct isoforms of phyA (second level) and different conformers within one molecular species (third level). In this review, we concentrate on the two native populations of phyA which may account, at least partially, for the complexity of its action.

Two Phenomenological Phy Pools

Heterogeneity of phytochrome in the cell was directly shown by the detection of (1) its light-labile and light-stable pools, (2) immunologically distinguishable phys, and (3) different phy genes and their products (Furuya 1993). In our group, different phy species in vivo were found based on their distinct spectral and photochemical properties. Fluorescence of phy in the cell was detected and a highly sensitive method of its in situ fluorescence assay was developed (Figure 1) (Sineshchekov 1995). Using this approach, phy in the cell was characterized by a number of key parameters (Figure 1a): (1) phy content ($[P_{tot}]$); (2) position of its fluorescence emission and excitation (absorption) spectra (λ_{max}); (3) extent of Pr phototransformation into lumi-R at low temperature (γ_1) and into Pfr at ambient temperature (γ_2); and (4) activation barrier (E_a) of the Pr $\rightarrow$ lumi-R photoreaction.

Biology Faculty, Chair of Physicochemical Biology, M.V. Lomonosov Moscow State University, Moscow 119899, Russia

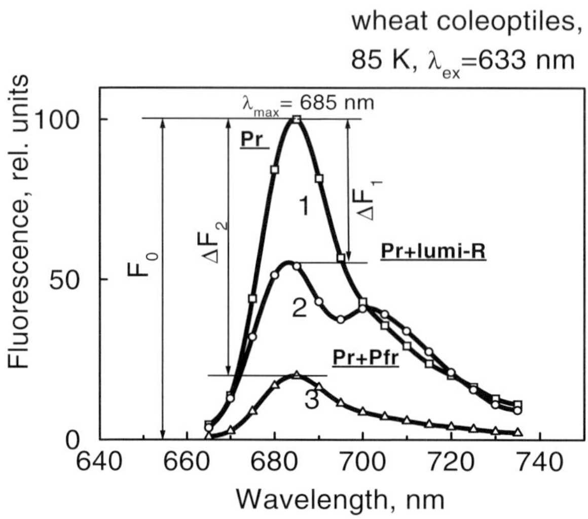

a

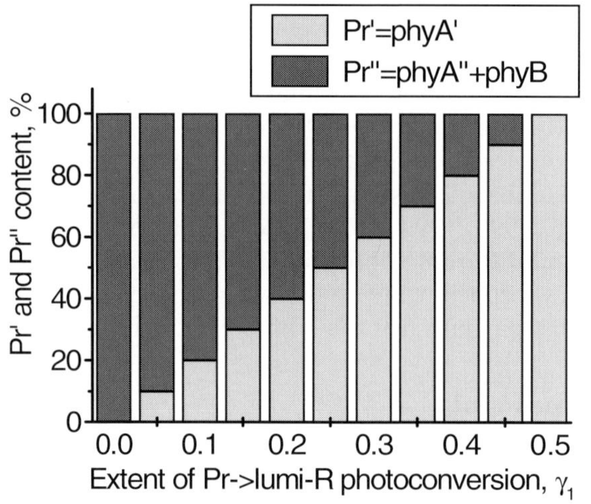

b

FIG. 1. In planta fluorescence assay of phytochrome. **a** Low-temperature emission spectra of phytochrome in etiolated tissue (*1*) and after R-induced Pr conversion into lumi-R (at 85 K) (*2*) and into Pfr (at 273 K) (*3*). The spectra provide information on λ_{max}, $[P_{tot}] \approx F_0$, and extent of the Pr photoconversion into lumi-R ($\gamma_1 = \Delta F_1/F_0$) and into Pfr ($\gamma_2 = \Delta F_2/F_0$). **b** Evaluation of the proportion of Pr′ (phyA′) and Pr″ (phyA″ + phyB) in plant tissues from experimental γ_1 values (see text)

Investigations of phy in different tissues and organs of a number of monocots and dicots have shown that all the above parameters varied: the most pronounced were changes in γ_1 and E_a, and there was a correlation between them and $[P_{tot}]$ (Sineshchekov 1998, 1999). These variations were interpreted as a manifestation of the existence of two distinct phy species: Pr′—longer-wavelength (λ_{max} = 685–687/671–673 nm) with high photochemical activity at low temperature ($\gamma_1' = 0.5$) and relatively low E_a and Pr″—shorter-wavelength (λ_{max} = 680–682/666–668 nm), inactive at low temperature ($\gamma_1'' = 0$) and with high E_a. At ambient temperature, the two species did not differ much in their photochemical activity ($0.7 \leq \gamma_2 \leq 0.85$). Based on γ_1 of Pr′ and Pr″, experimental γ_1 values and $[P_{tot}]$, their proportion and content in the cell were evaluated (Figure 1b) and

Pr′ was shown to be the major species in growing etiolated tissues, variable and light-labile, whereas Pr″ was the minor, more evenly distributed species in tissues, conserved and relatively light-stable.

Two Distinct PhyA Molecular Species

To attribute the two phenomenological Pr species to the different phys we turned to phy mutants and transgenic plants (Sineshchekov 1995, Sineshchekov et al 1996, 1997, 1999a, b, 2001b, c). In the double *phyAphyB* mutants of *Arabidopsis* and pea, almost no phy fluorescence was observed suggesting that the emission in the respective wild types belonged to phyA and phyB. In the single *phyA* mutants of *Arabidopsis* and pea, the detected fluorescence species was entirely of the Pr″ type and phyB was therefore attributed to this pool. In favor of this was also the fact that rice and authentic phyB overexpressed in *Arabidopsis* and *Arabidopsis* phyB overexpressed in transgenic potato belonged to Pr″. Besides, phyB, obtained by a light-induced depletion of phyA in transgenic potato overexpressing phyB, also possessed the properties of Pr″. In the *phyB* mutants of cucumber, *Arabidopsis* and pea, however, both Pr′ and Pr″ were detected and their content was almost identical to that in the wild type (WT). Moreover, oat phyA overexpressed in *Arabidopsis*, authentic phyA in transgenic potato and oat phyA in transgenic wheat were also found to comprise both Pr′ and Pr″. Thus, it was proved that phyA in the cell is represented by two distinct phenomenological types, phyA′ belonging to Pr′ and phyA″, which together with phyB was attributed to Pr″. Interestingly, cryptogam (fern phy1 and moss CP2) and cyanobacterial (Cph1) phys also revealed photochemical properties of the Pr″ type (see for review Sineshchekov 2004).

The contrasting activity of Pr′ (phyA′) and Pr″ (phyA″ + phyB) at low temperature were interpreted in terms of an energy level scheme whose major distinctive feature is an activation barrier (E_a) in the Pr excited state for the Pr → lumi-R photoreaction (Sineshchekov 1995). This barrier is relatively low for Pr′ (hundreds $\mathrm{J\,mol^{-1}} \leq E_a \leq 3\text{–}4\,\mathrm{kJ\,mol^{-1}}$) and it can be overcome at low temperature. For Pr″, it is higher by more than a factor of 10, what slows down the photoreaction and results in a shift of the Pr ↔ lumi-R photoequilibrium towards Pr. At ambient temperature, this barrier is easily overcome in both Pr species with the formation of a "hot" intermediate (prelumi-R?) and the yield of the initial photoreaction is determined by the partitioning coefficients of the conversion of the latter to lumi-R (complete photoreaction) or back to Pr (incomplete photoreaction).

The Nature of the Two PhyA Pools

The photochemical differences between phyA′ and phyA″ are obviously connected with their structural distinctions. Our data indicate that they are due to post-translational modification of the pigment, possibly, phosphorylation of the molecule at the 10kDa N-terminal segment. This is pointed at by a number of experimental facts. Firstly, gene-engineered phyA (rice and *Arabidopsis*) in transgenic yeast assembled with the chromophore (phytochromobilin or phycocyanobilin) in vivo was entirely represented by a species close to but not identical with phyA″ (Sineshchekov et al 2001a). This suggests that the minor phyA″ is the product of the same gene as the major phyA′ and that plant-specific modification is needed to form phyA′ and phyA″. Secondly, the equilibrium between phyA′ and phyA″ is connected with phosphorylation/dephosphorylation of the pigment. Dephosphorylation of oat phyA in vitro shifted the phyA′/phyA″ ratio more than twofold towards phyA″ (Sineshchekov and Lamparter unpublished). On the contrary, inhibition of phosphatases with an unspecific inhibitor phenylmethylsulfonyl fluoride, PMSF, drastically increased the content of phyA′ at the expense of phyA″ in maize roots (Koppel and Sineshchekov 2003).

The sites of the pigment molecule implicated in the phyA′ and phyA″ formation are at the N-terminus within the 6–69 amino acid stretch. This is evidenced by the facts that (i) Δ7–69 truncated oat phyA in tobacco was of the phyA″ type (Sineshchekov et al 1999a) and (ii) Δ6–12 oat phyA expressed in the *phyA* mutant of *Arabidopsis* belonged on the contrary to phyA′ (Sineshchekov et al unpublished). The 13–69 and 6–12 stretches are thus critical for the formation of phyA′ and phyA″, respectively. Taking into consideration the data that phyA is phosphorylated at serine 7 and serine 17 (Lapko et al 1999), we hypothesize that these serines may be critical for the formation of its isoform(s). This is currently verified in our work with the use of point-mutated phyA.

It is of interest to note that the phytochrome kinase substrates, PKS1 and PKS2, which interact with phyA in the cytosol (see Lariguet et al 2003 and the literature cited therein), may have a relation to the formation of phyA′ because in the double *pks1pks2* mutant we observed a considerable redistribution of the pigment towards phyA″ (Sineshchekov and Fankhauser 2004). In the single *pks1* or *pks2* mutants no such redistribution was found, suggesting that PKS1 and PKS2 can be redundant in this respect. Also, the interaction of phyA with cry1, whose activity is enhanced by phy induced phosphorylation (Ahmad et al 1998), may affect the phyA state in the cell. In *HY4* mutant lacking cry1, we observed a shift of the phyA′/phyA″ equilibrium towards phyA′ (Koppel and Sineshchekov unpublished). All this implicates phosphorylation of phyA as a likely source of structural distinctions between phyA′ and phyA″.

Nuclear-Cytoplasmic Partitioning of PhyA′ and PhyA″

Recent findings suggest that the nuclear-cytoplasmic partitioning of phyA and phyB is a prerequisite of their functions (see Casal et al 2002 and the literature cited therein). With the use of chimeric phyA:GFP it was shown that after FR illumination part of the phyA pool translocates from cytoplasm to the nucleus forming there speckles of different types. In the context of our research, it was of interest to find out if the two phyA isoforms take place in this crucial process.

For this, we turned to transgenic plants expressing chimeric phyA:GFP. Rice phyA:GFP overexpressed in tobacco and authentic phyA:GFP overexpressed in *Arabidopsis* comprised both phyA′ and phyA″, suggesting that either of them is a potential participant in this process (Sudnitsin et al 2003). More specifically, experiments with full-length (FL) and Δ6–12 truncated oat phyA:GFP expressed in phyA-deficient *Arabidopsis* revealed that FL phyA:GFP contained both phyA species in comparable amounts, characteristic of WT *Arabidopsis*, whereas Δ6–12 phyA:GFP was represented only by phyA′ (Sineshchekov et al unpublished). Proceeding from this and from the fact that nuclei with many tiny spots were observed in the case of Δ6–12 phyA:GFP (Casal et al 2002), we can conclude that phyA′ participates in the nuclear translocation with this type of speckle formation. In the case of FL phyA:GFP consisting of phyA′ and phyA″, both nuclei with many tiny spots and nuclei with few small spots were found (Casal et al 2002). Thus, the latter type may be associated with phyA″. This assumption needs, however, direct experimental verification with transgenic plants expressing mutant phyA:GFP comprising only phyA″.

Functions of PhyA′ and PhyA″

We tried to follow possible functional implications of the detected phyA heterogeneity by investigating transgenic plants with modified phyA′ and phyA″ content and altered phenotypes. Experiments with potato over- and underexpressing authentic phyA (*BIN PS* and *AP* lines, respectively) have shown that [phyA′] in these species varied by ≈50-fold whereas alterations of [phyA″] were an order of magnitude lower (Sineshchekov et al 1996). This suggests that the exaggerated and suppressed de-etiolated phenotypes of *BIN PS* and *AP* plants under FR (HIR) are likely to be due to the changes in the phyA′ content although the input of phyA″ in these effects cannot be ruled out. More decisive were results on transgenic wheat overexpressing oat phyA (Sineshchekov et al 2001b). This wheat acquired HIR (inhibition of growth, leaf unrolling, anthocyanin accumulation), which are not characteristic of WT. These responses were attributed to phyA′ because the transgenic wheat overexpressed primarily this pool and retained this excessive phyA′ content under FR.

The role of phyA and its isoforms was also verified in experiments on the FR-induced phyA-mediated regulation of the active protochlorophyllide, Pchlide^{655}. We have found up-regulation of Pchlide^{655} in tobacco and pea in contrast to the well known its down-regulation in *Arabidopsis*, tomato and barley (see Sineshchekov et al 2004a and the literature cited therein). A reversion of the sign of the effect was also observed in jasmonate-free rice *hebiba*—from negative in WT to positive in the mutant (Sineshchekov et al 2004b). Moreover, in WT rice, the sign depended on the lighting conditions—it was positive under pulsed FR (VLFR) and negative under constant FR (HIR). Experiments on transgenic tobacco imply that phyA′ is the likely participant in these FR effects. Under constant FR, a considerable induction of Pchlide^{655} was observed in wild-type tobacco and its transgenic lines overexpressing FL oat phyA, which comprises both phyA species, and in Δ7–69 phyA consisting of only phyA″. However, the effect was much lower in the phyA″ overexpressor suggesting that phyA″ is inactive under FR and inhibits the action of phyA′ (Sineshchekov et al 2004a).

Participation of phyA′ in VLFR is pointed at by the fact that Δ6–12 oat phyA, which according to our data is represented entirely by the phyA′ pool (see above), mediates VLFR (growth effects) when overexpressed in *Arabidopsis* (Casal et al 2002). Thus, in agreement with Casal et al (2002) and Yanovsky et al (2002), one molecular phyA species could be responsible for the two modes of phyA action (HIR and VLFR) and this species is likely to be phyA′. PhyA″, on the other hand, may contribute to the R-induced responses (Sineshchekov 2004). In the case of cotyledon unfolding, R had more profound effects on the phyA″ tobacco overexpressor (NA) than on WT whereas FR was completely inactive in this effect in NA in contrast to FL and WT. Interestingly, the effect of phyA″ overexpression was also found in the dark: the level of the active $Pchlide^{655}$ was much higher in the NA tobacco line overexpressing phyA″ than in WT (Sineshchekov et al 2004a).

Light Regulation of PhyA′ and PhyA″

The specificity of the phyA′ and phyA″ action is likely to be connected with their turnover under different light conditions. We have observed two modes of the light-induced phyA decline—with and without a phyA′ → phyA″ shift in their equilibrium. These effects depend on the plant species and character of illumination. The former dominates under R whereas the latter, in FR-grown seedlings. The first mode was absent or suppressed in transgenic wheat overexpressing oat phyA (Sineshchekov et al 2001b), in *phyA-3D* and *phyA-3DphyB* pea with point-mutated phyA (A194V) (Sineshchekov and Weller 2004) and *hebiba* mutant of rice (Sineshchekov et al 2004b). These plant species were characterized by higher light stability and activity of phyA in comparison with respective wild types. These data allowed a conclusion that the first component of the phyA decline (with the phyA′ → phyA″ shift) is due to the phyA destruction primarily in the more light-labile phyA′ form whereas the second (without the shift), to the negative autoregulation of the phyA biosynthesis. Regulation of the phyA′/phyA″ balance under certain light conditions may thus be part of the mechanism of the fine-tuning of the phyA functioning.

Conclusions

PhyA is represented in the cell by two isoforms, phyA′ and phyA″, whose content depends on plant species and tissues and physiological conditions. They are the products of one and the same gene and differ by post-translational modification, possibly, phosphorylation, at the N-terminus. Both of them are likely to translocate to the nucleus upon illumination forming there speckles of different types. The light-labile phyA′ is responsible for de-etiolation under FR (HIR and VLFR) whereas the relatively light-stable phyA″ is active under R and could be functional together with phyB throughout the whole plant life cycle. PhyA″ suppresses the action of phyA′ and the regulation of the ratio between phyA′ and phyA″ could be important for the phyA functioning. The nature and mode of action of phyA′ and phyA″ is not, however, fully understood and investigations of transgenic plants with point-mutated phyAs comprising only one of the two phyA species seem to be promising in this respect.

Acknowledgments. This work was supported by The Russian Foundation for Fundamental Investigations, grant no. 02-04-49516 to V.S.

References

Ahmad M, Jarillo JA, Smirnova O, Cashmore AR (1998) The CRY1 blue light photoreceptor of *Arabidopsis* interacts with phytochrome A in vitro. Mol Cell 1: 939–948

Casal JJ (2000) Phytochromes, cryptochromes, phototropin: Photoreceptor interactions in plants. Photochem Photobiol 71: 1–11

Casal JJ, Davis SJ, Kirchenbauer D, Viczian A, Yanovsky MJ, Clough RC, Kircher S, Jordan-Beebe ET, Schäfer E, Nagy F, Vierstra RD (2002) The serine-rich N-terminal domain of oat phytochrome A helps regulate light responses and subnuclear localization of the photoreceptor. Plant Physiol 129: 1127–1137

Furuya M (1993) Phytochromes: their molecular species, gene families, and functions. Annu Rev Plant Physiol Plant Mol Biol 44: 617–645

Koppel L, Sineshchekov V (2003) Effect of a non-specific phosphatase inhibitor, PMSF, on the two phyA populations in roots of maize. Plant Photobiology Meeting, Marburg, Abstract P10

Lapko VN, Jiang XY, Smith DL, Song PS (1999) Mass spectrometric characterization of oat phytochrome A: isoforms and posttranslational modifications. Protein Sci 8: 1032–1044

Lariguet P, Boccalandro HE, Alonso JM, Ecker JR, Chory J, Casal JJ, Fankhauser C (2003) A growth regulatory loop that provides homeostasis to phytochrome A signaling. Plant Cell 12: 2966–2978

Sineshchekov VA (1995) Photobiophysics and photobiochemistry of the heterogeneous phytochrome system. Biochim Biophys Acta 1228: 125–164

Sineshchekov VA (1998) The system of phytochromes: photobiophysics and photobiochemistry in vivo. Membr Cell Biol 12: 691–720

Sineshchekov VA (1999) Phytochromes: molecular structure, photoreceptor process and physiological function. In: Singhal GS, Renger G, Sopory SK, Irrgang K-D, Govindjee (eds) Concepts in photobiology: photosynthesis and photomorphogenesis. Kluwer, Boston/Narosa, Delhi, pp 755–795

Sineshchekov VA (2004) Phytochrome A: functional diversity and polymorphism. In: Braslavsky S (ed) Biological photosensors, special issue. Photochem Photobiol Sci 3: 596–607

Sineshchekov V, Fankhauser C (2004) PKS1 and PKS2 affect phyA state in etiolated *Arabidopsis* seedlings. In: Braslavsky S (ed) Biological photosensors, special issue. Photochem Photobiol Sci 3: 608–611

Sineshchekov VA, Weller JL (2004) Two modes of the light-induced phytochrome A decline—with and without changes in the proportion of its isoforms (phyA′ and phyA″): evidence from fluorescence investigations of mutant *phyA-3D* pea. J Photochem Photobiol B Biol 75: 127–135

Sineshchekov VA, Heyer AG, Gatz C (1996) Phytochrome states in transgenic potato plants with altered phyA levels. J Photochem Photobiol B Biol 34:137–142

Sineshchekov VA, Ogorodnikova OB, Devlin PF, Whitelam GC (1997) Fluorescence spectroscopy and photochemistry of phytochromes A and B in wild-type, mutant and transgenic strains of *Arabidopsis thaliana*. J Photochem Photobiol B Biol 42: 133–142

Sineshchekov VA, Clough RC, Jordan-Beebe ET, Vierstra RD (1999a) Fluorescence analysis of oat phyA deletion mutants expressed in tobacco suggests that the N-terminal domain determines the photochemical and spectroscopic distinctions between phyA′ and phyA″. Photochem Photobiol 69: 728–732

Sineshchekov VA, Ogorodnikova OB, Weller JL (1999b) Fluorescence and photochemical properties of phytochromes A and B in etiolated pea seedlings. J Photochem Photobiol B Biol 49: 204–211

Sineshchekov V, Hennig L, Lamparter T, Hughes J, Gärtner W, Schäfer E (2001a) Recombinant phytochrome A in yeast differs by its spectroscopic and photochemical properties from the major phyA′ and is close to the minor phyA″: evidence for post-translational modification of the pigment in plants. Photochem Photobiol 73: 692–696

Sineshchekov V, Koppel L, Shlumukov L, Barro F, Barcelo P, Lazzeri P, Smith H (2001b) Fluorescence and photochemical properties of phytochromes in wild-type wheat and a transgenic line over-expressing an oat phytochrome A (*PHYA*) gene: functional implications. Plant Cell Environ 24: 1289–1297

Sineshchekov V, Ogorodnikova O, Thiele A, Gatz C (2001c) Fluorescence and photochemical characterization of phytochromes A and B in transgenic potato expressing *Arabidopsis* phytochrome B. J Photochem Photobiol B Biol 59: 139–146

Sineshchekov V, Belyaeva O, Sudnitsin A (2004a) Phytochrome A positively regulates biosynthesis of the active protochlorophyllide in dicots under far-red light. J Photochem Photobiol B Biol 74: 47–54

Sineshchekov VA, Loskovich AV, Riemann M, Nick P (2004b) The jasmonate-free rice mutant *hebiba* is affected in the response of phyA′/phyA″ pools and protochlorophyllide biosynthesis to far-red light. Photochem Photobiol Sci 3: 1058–1062

Smith H (ed) (1997) Photomorphogenesis. Special issue. Plant Cell Environ 20: 657–844

Sudnitsin A, Adam E, Nagy F, Schäfer E, Sineshchekov V (2003) phyA-GFP is spectroscopically and photochemically close to phyA and comprises both phyA′ and phyA″. Plant Photobiology Meeting, Marburg, Abstract P9

Yanovsky MJ, Luppi JP, Kirchbauer D, Ogorodnikova OB, Sineshchekov VA, Adam E, Kircher S, Staneloni RJ, Schäfer E, Nagy F, Casal JJ (2002) Missense mutation in the PAS2 domain of phytochrome A impairs subnuclear localization and a subset of responses. Plant Cell 14: 1591–1603

Chapter 11

Phytochromes and Functions: Studies Using Gene Targeting in *Physcomitrella*

Jon Hughes[1], Gerhard Brücker[2], Alexander Repp[2], Mathias Zeidler[1], and Franz Mittmann[1]

Introduction

The lowly moss *Physcomitrella patens* is an excellent model organism for modern molecular physiology. The protonemal filaments are haploid, displaying the phenotype of a genetic lesion immediately after mutagenesis (selfing is possible but unnecessary). Filament cells are amenable to a wide range of cell biological methods including microinjection and, uniquely amongst plants, gene targeting via homologous recombination. Thus a *Physcomitrella* gene can be cloned, disrupted in situ and the mutant filament—together with its phenotype—regenerated within a couple of weeks. Protonemata are especially interesting photobiologically as they use phytochrome to steer their direction of growth in relation to light (phototropism). As we shall see, according to the fashionable view of phytochrome molecular action, this is simply not possible. We hope to resolve this self-contradictory situation using the power of the *Physcomitrella* system.

In this chapter we first describe the specific paradoxes which arise from potential explanations of the data, then summarise what is known about the moss phytochrome system, particularly recent work in which the phenotypes of targeted phytochrome knockouts have been studied.

Three Paradoxes

For many years a battle raged over whether phytochrome works via gene expression *or* by some effect on membranes or in the cytoplasm. It was quite clear all along that both ideas were correct—but arguing was more fun than research.

[1] Plant Physiology, Justus Liebig University, Zeughaus, Senckenbergstrasse 3, D-35390 Giessen, Germany
[2] Plant Physiology, Free University, Königin-Luise-Strasse 12–16, D-14195 Berlin, Germany

With the rise of molecular genetics, the fascinating mechanism by which phytochrome controls gene expression is being revealed. Very little effort has been devoted to elucidating the role of phytochrome in the cytoplasm, it being tacitly assumed that transcriptional control in the nucleus somehow organises the whole thing. This is impossible as three significant paradoxes arise.

Paradox 1

Although the phototropic response in *Physcomitrella* is actually more complex (see below), at appropriate fluence rates the tip cells of dark-adapted caulonemal filaments rebuild their actin cytoskeleton (Meske and Hartmann 1995, Meske et al 1996) under the influence of light to shift their growth direction towards that of the light beam. In *Ceratodon* the most effective spectral region is red light (R) around 660 nm, and the response is negated by far-red (FR) around 730 nm (Hartmann et al 1983). This behaviour implies the involvement of the R/FR photochromic photoreceptor, phytochrome. Although the optics of the response are little known, microbeam irradiations show that growth reorientation is positively directed by a higher fluence rate at the side of the cell closest to the light source (rather than a growth away from an area of the cell which might be more strongly illuminated due to refractive effects). So, Pfr is formed maximally in this region of the cell and somehow reorganises the actin microfilaments to promote tip growth in this direction.

- In all phytochromes known, Pfr in vivo is quite stable ($\tau > 10$ min), so in monochromatic light the level of Pfr rapidly becomes uniform throughout the cell. The directional vector of the incoming light can, at best, be transmitted transiently via local differences in [Pfr].
- Perhaps moss phytochromes have a special, short-lived signalling state (Px* or something equally hypothetical) connected to its own signal transduction system.

Paradox 2

Remarkably, protonemal growth direction can also be steered by the polarisation of R, as indeed is chloroplast rotation in the algae *Mesotaenium* and *Mougeotia*. In the Jaffe/Etzhold/Haupt model these responses derive from phytochrome molecules fixed anisotropically to a structure near the plasmamembrane with their dipole-moments in the Pr form parallel to the cell surface. The dipolemoment of Pfr formed following photon absorbance is rotated by 90°, perpendicular to the plasmamembrane (see Figure 1). This allows Pfr to accumulate specifically at the site of filament tip growth (Haupt and Häder 1994).

- Biochemical and cell biological studies of phytochrome in vivo show that Pr is cytosolic. Furthermore, current notions of phytochrome action see newly formed Pfr entering the nucleus according to an unknown mechanism, there perhaps acting as a kinase to activate transcription factors such as PIF3. The Jaffe/Etzhold/Haupt model cannot be reconciled with this system as the response looses all directional information provided by the stimulus.

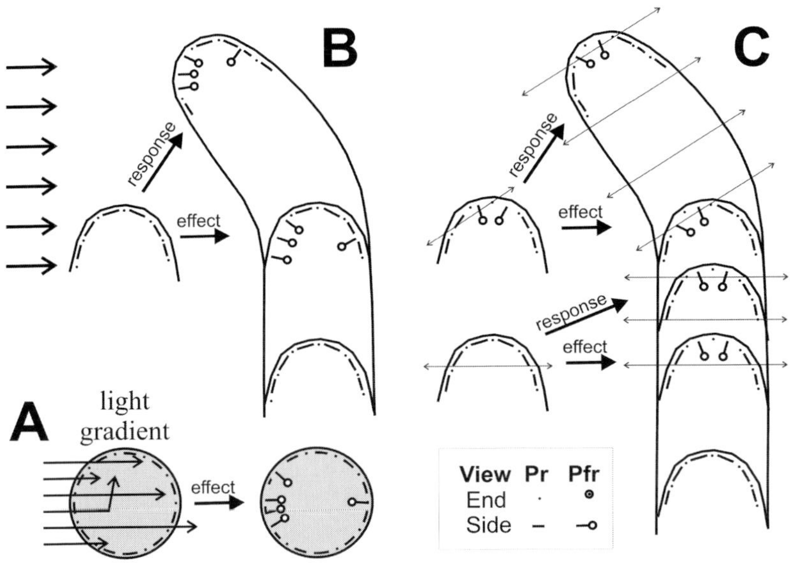

FIG. 1. The Jaffe/Etzhold/Haupt model of phytochrome-mediated phototropism (**A**, **B**) and polarotropism (**C**).
Hypothetical protonemal tip cell is shown in transverse (**A**) and longitudinal (**B**, **C**) section with phytochrome molecules attached near the plasma membrane. The microfilaments are reorganised to the site of Pfr formation and direct cell-wall outgrowth.
→; unidirectional red light, ↔; electrical vector of polarised red light.

- Perhaps a (small) proportion of Pr molecules associate with a factor attached to the plasmalemma (at least in lower plants). This additional primary mode of phytochrome action would be coupled to an appropriate transduction system. We note that even in the case of higher-plant phyA, less than half the phytochrome relocates to the nucleus.

Paradox 3

Lest it be forgotten, a number of phytochrome effects in both higher and lower plants occur within minutes or even seconds (Quail 1983). R/FR reversible changes in the surface potential of oat coleoptiles or barley and mung-bean root caps arise within a minute (the Tanada effect). Phytochrome-controlled potassium redistribution during leaf movement in *Samanea* begins within a few minutes as does the first phase of moss photo- and polarotropism. Moreover, the local [Ca^{2+}] changes which might initiate to these effects occur within seconds of irradiation. Cytoplasmic streaming in *Valissneria* responds to R within 2.5 s (Takagi et al 2004). In the case of phytochrome acting in the nucleus as a regulator of transcription, Pfr must first be transported to and then accumulate in the nucleus before anything happens. PhyB requires hours for this, phyA only perhaps 5 min—but that is still a long way from the *physiological* response. Numerous phytochrome-regulated responses in both higher and lower plants are much too fast to be explained by transcriptional regulation. Phytochrome must posses a much faster signal transduction system within the cytoplasm—but what form does this take?

Summa summarum, there exists at least one hitherto unknown mode of phytochrome signalling. This is fast, associated with a phytochrome-binding protein fixed isotropically at the periphery of the cytoplasm and able to transmit a vectorial light stimulus to a vectorial physiological response.

We are using *Physcomitrella* to elucidate this signalling system, while fully expecting the results to be relevant to the situation in higher plants: the fact that directional, cell-autonomous phytochrome responses in higher plants are not known does not mean that plant phytochromes only regulate transcription. The known phytochrome-interacting proteins PKS1 and NPDK2 as well as Ca^{2+} and cGMP (if not G-proteins, see Jones et al 2003) all play still unknown roles in light signalling, and are at least partially cytoplasmic. Similarly, at least 50% of the phytochrome molecules in the cell are always in the cytoplasm—irrespective of light treatments. Clearly, some phytochrome molecules enter the nucleus to regulate genes, but what does Cinderella do during the party?

Phytochromes in Mosses

The first biochemical evidence that lower plants contained canonical phytochromes came in the form of Western data with monoclonal antibodies which bound epitopes representing the conserved phytochrome subdomains LIPPIFGADE #848–#858[1] and KYIECLLS #934–#941 in a wide range of lower and higher plants. The Schneider–Poetsch laboratory cloned canonical phytochrome cDNAs from the clubmoss *Selaginella* and the moss *Physcomitrella*, *PP1*[2] (Hanelt et al 1992, Kolukisaoglu et al 1993), both showing subdomains similar to those recognised by the two monoclonals. Subsequently we cloned two canonical phytochrome genes from *Ceratodon*, *CP2* and *CP3*, and four from *Physcomitrella*, *PP1–4* (Lamparter et al 1995, Hughes et al 1996, Mittmann et al 2004)—see Figure 2. Earlier, a divergent phytochrome-like gene, *CP1*, had been cloned from *Ceratodon* (Thümmler et al 1992). The C-terminal half of the hypothetical peptide showed no homology to other phytochromes but rather to tyrosine kinases, hitherto unknown in plants. *CP1* remains an enigma: as it is conserved in *Ceratodon* lines from five continents (Mittmann and Hughes unpublished), it is unlikely to be a classical pseudogene. On the other hand, it seems to be unique

---

[1]The #-numbering system used refers to the alignment www.uni-giessen.de/~gf1251/Phytochrome/align2x.htm.

[2]The abbreviated gene names used here are, for example, CP2=Cerpu;PHY0;2 and PP2=Phypa;PHY0;2.

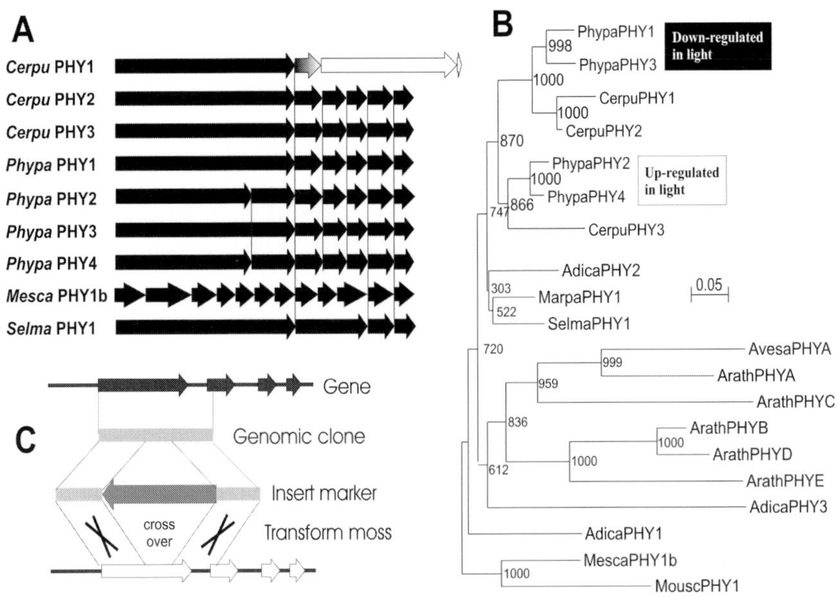

FIG. 2. The phytochrome family. *Adiantum capillus-veneris* PHY1-3, *Arabidopsis thaliana* PHYA-E, *Avena sativa* PHYA, *Ceratodon purpureus* PHY1-3 (CP1-3), *Marchantia paleacea* PHY1, *Mesotaenium caldorum* PHY1b, *Mougeotia scalaris* PHY1, *Physcomitrella patens* PHY1-4 (PP1-4), *Selaginella martensi* PHY1. (**A**) Structure of lower plant phytochrome gene homologs. (**B**) Phylogenetic analysis of the N-terminal 600 amino acids of higher and lower plant phytochromes (Phylip neighborhood joining method with Kimura correction. For alignment see www.uni-giessen.de/~gf1251/Phytochrome/align2x.htm; in this analysis gaps were ignored). The numbers near the nodes are bootstrap values (identities per 1000). (**C**) Phytochrome gene targeting via homologous recombination

to *Ceratodon* and there is no evidence that it encodes a photoreceptor. Neither mosses nor higher plants possess homologues of the phytochrome–phototropin chimera "superchrome" discovered in the fern *Adiantum* (Nozue et al 1998).

The *Physcomitrella PHY* genes now to hand have allowed us to begin a penetrating study of their biology especially in relation to directional responses. We expect these four genes and cDNAs as well as the two from *Ceratodon* represent the entire phytochrome family in each species (ignoring *CP1*). All currently known moss ESTs as well as Southern data can be correlated with these six genes. The conclusion is also consistent with the notion that the *Physcomitrella* genome has been duplicated at some point. The *PHY* gene structures and phylogeny are illustrated in Figure 2.

In higher plants *PHYA* is strongly down-regulated by light while the other four family members are transcribed more-or-less constitutively, PHYB predominating. Protein levels reflect the transcriptional differences. In *Ceratodon* total phytochrome levels increase when filaments are transferred from light to darkness

(Lamparter et al 1995). In *Physcomitrella* PP1 and PP3 transcripts are down-regulated whereas PP2 and PP4 are up-regulated in light (Mittmann et al 2004), correlating with their phylogeny and intron structure as well as implying different functions in photoperception.

The phylogenetic tree in Figure 2b implies that moss phytochromes might have unique functions distinct from those of any higher plant phytochrome. One probe for such differences is photochemistry. We showed that difference spectra of crude total phytochrome preparations from dark-grown *Ceratodon* (Lamparter et al 1995) and *Physcomitrella* (Mittmann et al 2004) resembled those of phytochrome from etiolated seedlings of higher plants (i.e. the native phyA–PΦB adduct). We also overproduced CP2 apoprotein in yeast and autoassembled the PΦB-adduct (Zeidler et al 1998). This too was spectrally similar to phyA–PΦB. Moreover, the Pr→Pfr photolysis intermediates resembled those known for plant phytochromes. Thus mosses seem not only to use the same PΦB chromophore as higher plants but also, despite the peptide sequence divergence, to provide it with a similar molecular environment. Thus, photochemically at least, moss phytochromes are functionally similar to those of higher plants.

Phytochrome Functions in *Physcomitrella*: Wild-Type and Targeted Knockout Mutants

Mutational studies in *Physcomitrella* are not only attractive because of its haploid genetics but because its high rate of homologous recombination allows chromosomal genes to be knocked out accurately (Schaefer and Zryd 1997). As we cloned each of the *Physcomitrella* phytochrome genes we set about targeting it, genetically characterising the resulting knockout lines and, finally, studying the phenotype relative to that of the wild type (Mittmann et al 2004).

Our studies are most advanced in relation to phototropism in dark-grown caulonemal filament tips cells. Unlike *Ceratodon*, *Physcomitrella* caulonemata show a complex fluence rate/response behaviour. Below 5 nmol $m^{-2} s^{-1}$ no response is seen, whereas between 15 and 150 nmol $m^{-2} s^{-1}$ most filaments grow away from the light source (negatively phototropically). At around 500 nmol $m^{-2} s^{-1}$ individual tip-cells display a clear tendency to grow *either* away from *or* towards the light, whereas at 1.5 µmol $m^{-2} s^{-1}$ almost all are strongly positively phototropic. Above 5 µmol $m^{-2} s^{-1}$, however, the filaments bend away from the light (Mittmann et al 2004).

Does each of these responses derive from a separate phytochrome or from a different signalling mode of one phytochrome (or both)? To answer this we repeated the analysis for each knockout (Mittmann et al 2004). Both *pp1* and *pp2* show reduced sensitivity to R around 1 µmol $m^{-2} s^{-1}$ but otherwise behave similarly to the wild type. *pp3*, on the other hand, shows an attenuated avoidance response at high irradiances. The "opposite" phenotype is shown by *pp4*: here the positive phototropism shown by the wild type is lost completely, the filaments seeming blind to R in that fluence rate range, while the avoidance response at higher light levels is retained.

Other defects are evident in *pp4*. It is polarotropically insensitive too, fitting in nicely with the Jaffe/Etzold/Haupt model. Chloroplast movement is also impaired. Kadota showed that polarised R presented with the E-vector parallel or perpendicular to the filament access leads to chloroplast accumulation along the side- and crosswalls, respectively (Kadota et al 2000), whereas we see this in the wild type and in *pp1*, *pp2*, and *pp3*, we have never observed an equivalent relocation in any *pp4* cell.

Summary and Conclusions

Studies in lower plants including mosses indicate that phytochrome possesses a rapidly acting cytoplasmic signalling system able to transmit directional information. A phytochrome-binding molecule in or anchored near the plasmalemma is probably involved, and a local interaction with actin filaments is likely at least in some cases, but apart from these hypotheses, the molecular nature of the system is unknown. A related although possibly non-identical cytoplasmic pathway is likely to exist in higher plants too. The system probably operates in parallel with transcriptional regulation via Pfr imported into the nucleus, although which phytochromes are attached to which transduction pathway remains to be discovered. The four *PHY* genes in *Physcomitrella* form a clade distinct from that of higher plant phytochromes. Their transcript levels are differentially up- and down-regulated by light. Gene-specific targeted knockouts show specific lesions in the phototropic behaviour of caulonemal tip cells. In particular, *pp4* shows no positive phototropism in R. These lines also show defective polarotropism and chloroplast relocation in polarised red light. It would thus seem that the different phytochromes have different signalling properties and perhaps distinct cytoplasmic signalling routes. Given the differential light-dependent transcript levels, the predominant photoreceptors are likely to be different in light- (chloronemal) and dark- (caulonemal) adapted filaments. We plan to use appropriate tools to detect potential interacting partners of each phytochrome in the cytoplasm, then to target the cognate genes to identify their physiological roles.

References

Hanelt S, Braun B, Marx S, Schneider-Poetsch HA (1992) Phytochrome evolution: a phylogenetic tree with the first complete sequence of phytochrome from a cryptogamic plant (*Selaginella martensii* Spring). Photochem Photobiol 56: 751–758

Hartmann E, Klingenberg B, Bauer L (1983) Phytochrome-mediated phototropism in protonemata of *Ceratodon purpureus*. Photochem Photobiol 38: 599–603

Haupt W, Häder DP (1994) Photomovement. In: Kendrick RE, Kronenberg GHM (eds) Photomorphogenesis in plants. Kluwer, Dordrecht, pp 707–732

Hughes J, Lamparter T, Mittmann F (1996) Cerpu;PHY0;2, a "normal" phytochrome in *Ceratodon* (accession no. U56698) (PGR 96–067). Plant Physiol 112: 446

Jones AM, Ecker JR, Chen JG (2003) A reevaluation of the role of the heterotrimeric G protein in coupling light responses in Arabidopsis. Plant Physiol 131: 1623–1627

Kadota A, Sato Y, Wada M (2000) Intracellular chloroplast photorelocation in the moss *Physcomitrella patens* is mediated by phytochrome as well as by a blue-light receptor. Planta 210: 932–937

Kolukisaoglu HÜ, Braun B, Martin WF, Schneider-Poetsch HA (1993) Mosses do express conventional, distantly B-type-related phytochromes. Phytochrome of *Physcomitrella patens* (Hedw.). FEBS Lett 334: 95–100

Lamparter T, Podlowski S, Mittmann F, Hartmann E, Schneider-Poetsch HA, Hughes J (1995) Phytochrome from protonemal tissue of the moss *Ceratodon purpureus*. J Plant Physiol 147: 426–434

Meske V, Hartmann E (1995) Reorganisation of microfilaments in protonemal tip cells of the moss *Ceratodon purpureus* during the phototropic response. Protoplasma 188: 59–69

Meske V, Rupert V, Hartmann E (1996) Structural basis for the red light induced repolarisation of tip growth in caulonemal cells of *Ceratodon purpureus*. Protoplasma 192: 189–198

Mittmann F, Brücker G, Zeidler M, Repp A, Abts T, Hartmann E, Hughes J (2004) Targeted knockout in *Physcomitrella* reveals direct actions of phytochrome in the cytoplasm. Proc Natl Acad Sci USA 101: 13939–13944

Nozue K, Kanegae T, Imaizumi T, Fukuda S, Okamoto H, Yeh KC, Lagarias JC, Wada M (1998) A phytochrome from the fern *Adiantum* with features of the putative photoreceptor NPH1. Proc Natl Acad Sci USA 95: 15826–15830

Quail PH (1983) Rapid action of phytochrome in photomorphogenesis. Encycl Plant Physiol 16A: 178–212

Schaefer DG, Zrÿd JP (1997) Efficient gene targeting in the moss *Physcomitrella patens*. Plant J 11: 1195–1206

Takagi S, Kong SG, Mineyuki Y, Furuya M (2004) Regulation of actin-dependent cytoplasmic motility by type II phytochrome occurs within seconds in *Valissneria gigantea* epidermal cells. Plant Cell 15: 331–345

Thümmler F, Dufner M, Kreisl P, Dittrich P (1992) Molecular cloning of a novel phytochrome gene of the moss *Ceratodon purpureus* which encodes a putative light-regulated protein kinase. Plant Mol Biol 20: 1003–1017

Zeidler M, Lamparter T, Hughes J, Hartmann E, Remberg A, Braslavsky S, Schaffner K, Gärtner W (1998) Recombinant phytochrome of the moss *Ceratodon purpureus*: heterologous expression and kinetic analysis of Pr–>Pfr conversion. Photochem Photobiol 68: 857–863

Chapter 12

Distinct Functions of Phytochromes on the Photomorphogenesis in Rice

Makoto Takano[1], Xianzhi Xie[1], Noritoshi Inagaki[1], and Tomoko Shinomura[2]

Introduction

Phytochromes are the most extensively characterized photoreceptors in plants; however, their diverse functions in the regulation of plant development have been characterized mainly in dicots. Very little information in this regard is available in monocots principally due to the unavailability of phytochrome mutants.

Rice is an excellent model for monocot plant systems to explore complex genetic and physiological phenomena because a wealth of information is available on its genomic structure and functionality (Shimamoto 1995). The task of designing and executing complex genetic studies in rice has been further facilitated by the availability of discrete research materials and tools, such as whole genome sequence, the high-density genetic linkage map and thousands of expressed sequence tag (EST) clones made available by the Rice Genome Research Program (Yamamoto and Sasaki 1997). Among them, the most useful tool is a gene knockout system available in rice for the reverse genetics where a retrotransposon named *Tos17* makes it possible to isolate mutants for genes of interest (Hirochika et al 1996, Hirochika 1997, 1999). In this system, tentatively activated *Tos17* can be used to easily generate a large number of *Tos17*-tagged mutant lines, and the mutants of a specific gene can be identified from the large mutant population by polymerase chain reaction (PCR)-based screening.

The generation and characterization of phytochrome mutants of rice can greatly enhance the existing knowledge of phytochrome function in plants, especially in terms of differentiating the phytochrome functionality between monocots and dicots. Phytochromes in higher plants are encoded by a small gene family. While Arabidopsis has five members in phytochrome family (*PHYA* to *PHYE*; Sharrock and Quail 1989, Clack et al 1994), rice has only three members, namely *PHYA*, *PHYB*, and *PHYC* (Kay et al 1989, Dehesh et al 1991, Tahir et

[1] Department of Plant Physiology, National Institute of Agrobiological Sciences, Tsukuba, Ibaraki 305-8602, Japan
[2] Hitachi Central Research Laboratory, Hatoyama, Saitama 350-0395, Japan

al 1998, Basu et al 2000). Therefore, rice has a distinct advantage for studying the functions of individual phytochromes by the isolation and analysis of the mutants.

Isolation of Phytochrome Mutants from Rice

We screened a large population of *Tos17* insertional mutants of rice to isolate rice *phyA* mutants. DNAs isolated from the mutant plants were organized as super pools and correspondingly named as "Mutant panel" (Hirochika 1997, 1999). The Mutant panels were efficiently screened by PCR using *PHY*- and *Tos17*-specific primers. After an extensive screening of Mutant panels (more than 50,000 mutant lines) we isolated more than 10 different alleles for the *phyA* mutation (Figure 1) and only one mutant line for phyC, but we could not isolate any *phyB* mutant. Since we could not get any mutant lines for phyB from the Mutant panels, we screened the M_3 generation of rice population mutagenized by γ-ray irradiation for isolating *phyB* mutants phenotypically. Using this forward genetics screening, we isolated four alleles of *phyB* mutants including one line with Nipponbare background and three lines with Norin-8 background.

Characterization of Rice Phytochrome Mutants

All of the rice phytochrome genes are located on chromosome 3 where *PHYB* is on the short arm, and *PHYA* and *PHYC* are present on the long arm in close proximity (10 cM). To confirm that the isolated lines are mutants, we raised antibodies against PHYB and PHYC proteins. For PHYA, we used a monoclonal antibody against rye PHYA, donated by Hitachi's group. The results of Western blotting indicate that each antibody is specific to the individual phytochromes

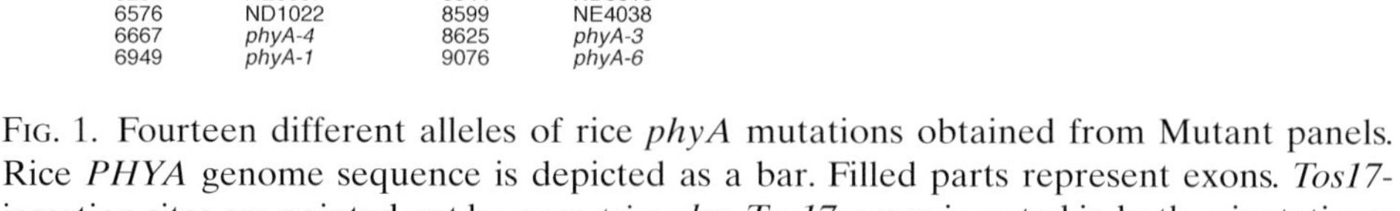

Insertion site	line name	Insertion site	line name
5718	NE1031	7748	NF6831
6092	*phyA-5*	8125	NE1038
6223	NG3530	8166	*phyA-2*
6234	NE0007	8544	ND5013
6576	ND1022	8599	NE4038
6667	*phyA-4*	8625	*phyA-3*
6949	*phyA-1*	9076	*phyA-6*

FIG. 1. Fourteen different alleles of rice *phyA* mutations obtained from Mutant panels. Rice *PHYA* genome sequence is depicted as a bar. Filled parts represent exons. *Tos17*-insertion sites are pointed out by *open triangles*. *Tos17*s were inserted in both orientations; inserted *Tos17*s with forward orientation are *inverted triangles* above the bar and those with reverse orientation are *upright triangles* below the bar. Translation start and stop sites are indicated by an *arrow* and *cross*, respectively. Nucleotide numbers are based on the rice phy18 gene sequence (accession no. X14172)

and that the isolated mutant lines are null mutants. One thing to be noted here is that the content of PHYC protein is greatly reduced in the *phyB* mutant, a phenomenon that has also been observed in Arabidopsis.

Physiological analyses of the *phyA* mutants at the young seedling stage revealed that phyA is responsible for the inhibition of coleoptile elongation, a phenomenon which is induced by irradiation with a pulse of very low-fluence red light (VLF-R) or far-red light (FR). Moreover, phyA is also involved in the inhibition of mesocotyl elongation and the induction of the gravitropic response in crown roots under continuous FR. These observations support the idea that rice phyA mediates light signals via VLFR mode as well as via HIR mode as also observed in Arabidopsis (Furuya and Schäfer 1997). When grown under natural light conditions, *phyA* mutants develop normally and display a vegetative phenotype and flowering time that are indistinguishable from the wild type. Therefore, the defect of phyA in rice appears to cause phenotypical changes that are mainly restricted to the de-etiolation process.

Phenotypic characterization of seedlings of rice *phyB* mutants showed that phyB is responsible for the response to R in rice as well. However, it was also revealed that phyB is not a sole photoreceptor for R, because the coleoptile elongation was significantly inhibited by R even in *phyB* mutants, although the extent of inhibition was less than that in the wild type. Such an incomplete effect is not observed in Arabidopsis, where etiolated *phyB* seedlings display a marked insensitivity to R with respect to almost all the aspects of seedling de-etiolation (Whitelam et al 1998). Thus, phyB plays a main role on responding to R in Arabidopsis whereas in rice, phyB does not seem to be a predominant player for the R-mediated response. Such a difference was also observed in the expression modes of light-inducible genes, mentioned below.

In Arabidopsis, most phytochrome mutants have been isolated by the forward-genetics screenings designed to identify mutants with reduced sensitivity to light. However, such screenings in Arabidopsis did not yield any *phyC* mutants, suggesting the marginal contribution of phyC to photomorphogenesis. In rice as well, phyC has minor roles in photomorphogenesis. The monogenic mutation of *PHYC* gene did not cause clear phenotypic differences in the seedling and vegetative growth. We examined the light-dependent inhibition of coleoptile and mesocotyl elongation under continuous R or FR, but no differences were observed between the wild type and *phyC* mutants.

Expression of Light-Inducible Genes in the Rice Phytochrome Mutants

In Arabidopsis, phyA is responsible for the *CAB* gene induction in the photoirreversible VLFR mode, while phyB induces *CAB* gene expression in the R/FR photoreversible LFR mode. As a result, *CAB* gene expression induced by R did not appear to be reversed by subsequent FR irradiation in the wild type and the *phyB* mutant (Hamazato et al 1997). In contrast, rice-etiolated seedlings of wild type and *phyB* mutant did show the R/FR photoreversibility on the induction of *CAB* gene expression. Unexpectedly, even *phyB/C* double mutants, in which phyA is the only active phytochrome, showed the R/FR reversibility. These results clearly indicate that phyA is involved in the *CAB* gene induction in not only VLFR mode but also R/FR reversible LFR mode. Therefore, the molecular property of phyA seems to be different between Arabidopsis and rice, and more specialized for the FR perception in Arabidopsis, while perceiving both R and FR in rice.

Phenotypes of *phyA* seedlings grown under continuous FR were indistinguishable from those of dark-grown seedlings, such as long coleoptiles, elongated mesocotyls, and loss of gravitropic response of lateral roots (Takano et al 2001). However, light-inducible genes such as *CAB* and *RBCS* were slightly induced by FR even in the *phyA* mutant, and this induction was completely diminished in the *phyA/C* double mutant. These results indicate that phyC is able to perceive FR to induce at least *CAB* and *RBCS* gene expressions, although the contribution is minor.

Flowering Times of Phytochrome Mutants in Rice

Light is a crucial factor to determine the flowering time. We grew wild-type and phytochrome mutant plants in the paddy field under the natural day-length condition, which resembled the long-day (LD) conditions, and measured the time taken to flowering. Under these conditions, wild-type plants flowered in about 100 days and so did the *phyA* mutants. However, *phyB* and *phyC* mutants flowered about two weeks earlier than the wild type. The flowering time of *phyB/C* double mutant was same as that of *phyB* or *phyC* monogenic mutant. These results suggest that both functions of phyB and phyC are necessary to suppress flowering under the long-day condition. Interestingly, *phyA/C* double mutants were very early to flower with floral initiation more than 20 days ahead of *phyC* monogenic mutants. Thus, *phyA* mutation alone does not affect the flowering time; however, in the *phyC* mutant background, *phyA* mutation has a big effect on the flowering time determination in LD conditions.

In short-day (SD) conditions, *phyC* mutants flowered at the same time as the wild type. On the other hand, *phyA* mutants showed slightly late flowering compared with Nipponbare or *phyC* mutant and the same extent of delay of flowering was observed in *phyA/C* double mutants. The *phyB* monogenic mutants were early to flower compared with the wild type or *phyC* mutant even in SD conditions, and the flowering time of *phyB/C* double mutants was the same as that of *phyB* monogenic mutants. These observations indicate that phyC is dispensable for flowering time determination in SD conditions.

We quantified the expression levels of *Hd1* and *Hd3a* during the growing processes in phytochrome mutants in order to elucidate the mechanism of accelerated flowering in *phyA/C* double mutants under LD conditions. The *Hd1* and

Hd3a are orthologues of *CO* and *FT*, respectively, two key regulatory genes for the photoperiodic control of flowering in Arabidopsis (Yano et al 2000, Kojima et al 2002). The LD signals in rice activate *Hd1* expression via OsGI function, and the promoted *Hd1* expression represses *Hd3a* expression, causing suppression of flowering in LD conditions (Hayama et al 2003). Moreover, it has been shown that phytochromes are involved in the regulation of the *Hd3a* expression (Izawa et al 2002). If early flowering of phytochrome mutants is simply due to the reduced LD signals, the reduction of *Hd1* expression and the resultant increase of *Hd3a* expression are expected. But this scenario is not likely the case because the diurnal expression of *Hd1* gene is not affected by the *se5* mutation (phytochrome chromophore deficient mutation) upon floral transition (Izawa et al 2002). Actually, *Hd1* expression patterns in *phyA/C* double mutants were not so different from the wild type. Nevertheless, *Hd3a* expression levels were far higher in the *phyA/C* double mutant compared with those of Nipponbare or each monogenic mutant. The *phyC* monogenic mutation caused a moderate phenotype of early flowering under LD conditions. However, the *phyC* mutation in combination with the *phyA* mutation led to greater promotion of flowering in LD conditions, while *phyA* mutation had little effect on the flowering time by itself. These results suggest that phyC and phyA affect the flowering time by separate pathways, as proposed in Figure 2.

In Figure 2, we applied our data to the scheme of LD signal transduction in rice proposed by Hayama et al (2003). From our results, *phyC* and *phyB* monogenic mutants showed the same early flowering and the flowering time of *phyB/C* double mutants was the same as that of *phyB* or *phyC* monogenic mutants. Furthermore, *Hd1* mutation also caused moderate early flowering (Lin et al 2000). These lines of evidence suggest that LD signals perceived by phyC in combination with phyB are incorporated upstream of *Hd1* to delay flowering. As long as this pathway is fully functional, the LD signals are enough to suppress flowering and phyA function is dispensable. Once these signals are reduced by the *phyC* mutation, however, the contribution of phyA to suppress the *Hd3a* expression is relatively increased so that the loss of phyA function causes a big promotion in the flowering time. This model seems to be supported by the observation that *se5* is epistatic to *se1* (*Hd1* mutation) (Izawa et al 2002). As yet,

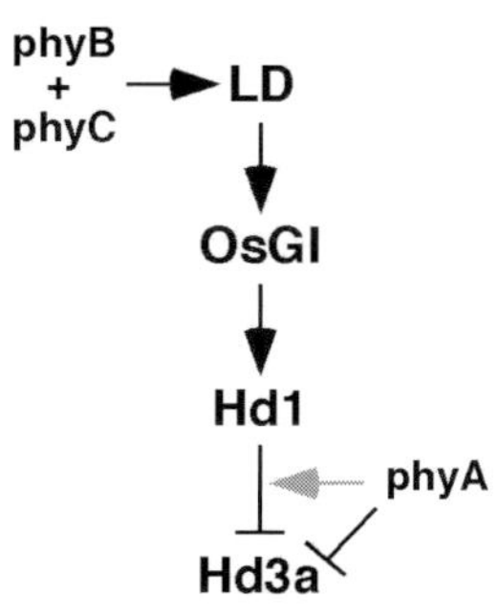

FIG. 2. Proposed model for the function of phytochromes on the long-day signaling for flowering of rice. This model was adapted from Figure 4c in Hayama et al (2003)

we have no idea about the mechanism of how phyA represses the *Hd3a* expression. Another possibility is that phyA suppresses the pathway that promotes the *Hd3a* expression extensively. Epistatic analyses between *Hd1* or *Hd3a* and phytochrome genes will clarify the relationships between them.

Acknowledgments. We thank H. Hanzawa for providing us monoclonal antibody against rye PHYA. We are grateful to Dr. M. Tahir for critical reading of the manuscript. We also thank K. Yagi, M. Sato, and Y. Iguchi for their technical assistance. This work was partly supported by the Rice Genome Project (IP-1006).

References

Basu D, Dehesh K, Schneider-Poetsch HJ, Harrington SE, McCouch SR, Quail PH (2000) Rice *PHYC* gene: structure, expression, map position and evolution. Plant Mol Biol 44: 27–42

Clack T, Mathews S, Sharrock RA (1994) The phytochrome apoprotein family in Arabidopsis is encoded by five genes: the sequences and expression of PHYD and PHYE. Plant Mol Biol 25: 413–427

Dehesh K, Tepperman J, Christensen AH, Quail PH (1991) phyB is evolutionarily conserved and constitutively expressed in rice seedling shoots. Mol Gen Genet 225: 305–313

Furuya M, Schäfer E (1997) Photoreception and signalling of induction reactions by different phytochromes. Trends Plant Sci 1: 301–307

Hamazato F, Shinomura T, Hanzawa H, Chory J, Furuya M (1997) Fluence and wavelength requirements for Arabidopsis *CAB* gene induction by different phytochromes. Plant Physiol 115: 1533–1540

Hayama R, Yokoi S, Tamaki S, Yano M, Shimamoto K (2003) Adaptation of photoperiodic control pathways produces short-day flowering in rice. Nature 422: 719–722

Hirochika H (1997) Retrotransposons of rice: their regulation and use for genome analysis. Plant Mol Biol 35: 231–240

Hirochika H (1999) Retrotransposons of rice as a tool for forward and reverse genetics. In: Shimamoto K (ed) Molecular biology of rice (Springer-Verlag, Tokyo). pp 43–58

Hirochika H, Sugimoto K, Otsuki Y, Tsugawa H, Kanda M (1996) Retrotransposons of rice involved in mutations induced by tissue culture. Proc Natl Acad Sci USA 93: 7783–7788

Izawa T, Oikawa T, Sugiyama N, Tanisaka T, Yano M, Shimamoto K (2002) Phytochrome mediates the external light signal to repress *FT* orthologs in photoperiodic flowering of rice. Genes Dev 16: 2006–2020

Kay SA, Keith B, Shinozaki K, Chua NH (1989) The sequence of the rice phytochrome gene. Nucleic Acids Res 17: 2865–2866

Kojima S, Takahashi Y, Kobayashi Y, Monna L, Sasaki T, Araki T, Yano M (2002) *Hd3a*, a rice ortholog of the *Arabidopsis FT* gene, promotes transition to flowering downstream of *Hd1* under short-day conditions. Plant Cell Physiol 43: 1096–1105

Lin HX, Yamamoto T, Sasaki T, Yano M (2000) Characterization and detection of epistatic interactions of 3 QTLs, *Hd1*, *Hd2*, and *Hd3*, controlling heading date in rice using nearly isogenic lines. Theor Appl Genet 101: 1021–1028

Sharrock RA, Quail PH (1989) Novel phytochrome sequences in *Arabidopsis thaliana*: structure, evolution and differential expression of a plant regulatory photoreceptor family. Genes Dev 3: 1745–1757

Shimamoto K (1995) The molecular biology of rice. Science 270: 1772–1773

Tahir M, Kanegae H, Takano M (1998) Phytochrome C (*PHYC*) gene in rice: Isolation and characterization of a complete coding sequence. Plant Physiol 118: 1535

Takano M, Kanegae H, Shinomura T, Miyao A, Hirochika H, Furuya M (2001) Isolation and characterization of rice phytochrome A mutants. Plant Cell 13: 521–534

Whitelam GC, Patel S, Devlin PF (1998) Phytochromes and photomorphogenesis in Arabidopsis. Philos Trans R Soc Lond B Biol Sci 353: 1445–1453

Yamamoto K, Sasaki T (1997) Large-scale EST sequencing in rice. Plant Mol Biol 35: 135–144

Yano M, Katayose Y, Ashikari M, Yamanouchi U, Monna L, Fuse T, Baba T, Yamamoto K, Umehara Y, Nagamura Y, Sasaki T (2000) *Hd1*, a major photoperiod sensitivity quantitative trait locus in rice, is closely related to the Arabidopsis flowering time gene *CONSTANS*. Plant Cell 12: 2473–2484

Part III
Cryptochrome

CHAPTER 13

Cryptochrome Overview

ANTHONY R. CASHMORE

Introduction

Cryptochromes were first discovered in Arabidopsis where a mutation conferring a deficiency in blue light signaling was shown to reside in a gene encoding a protein with similarities to photolyases (Ahmad and Cashmore 1993). The latter are flavoproteins that mediate the repair of pyrimidine dimers, generated as a result of exposure of DNA to UV-B light (Sancar 2003). This DNA repair activity of photolyases is dependent on irradiation with blue or UV-A light and results from transfer of an electron from the photolyase-bound flavin to the damaged pyrimidine dimer, which then undergoes isomerization to yield the monomer; the electron is returned to the photolyase. In these respects photolyases are photoreceptors mediating blue light-dependent redox reactions, and in view of the similarities between the Arabidopsis *cry1* gene and photolyases it was proposed that CRY1 was also a blue light photoreceptor. Cryptochromes lack the DNA repair activity of photolyases and, at least in plants, cryptochromes are characterized by a distinguishing C-terminal extension (Cashmore 2003). Cryptochromes have now been characterized for several additional plant species including tomato (Ninu et al 1999, Weller et al 2001) and rice (Matsumoto et al 2003). In both cases, as in Arabidopsis, these cryptochromes apparently play a role in blue light-mediated de-etiolation and photomorphogenesis.

Cryptochromes have been described for algae (Small et al 1995), and ferns (Kanegae and Wada 1998), and recently for mosses (Imaizumi et al 2002). In the fern *Adiantum capillus-veneris*, spore germination is regulated by blue light, and two of the five cryptochromes described for this fern are thought to be involved in this process. Two *CRY* genes have been described for the moss *Physcomitrella patens*, and disruption of these genes confers an increase in auxin sensitivity in a blue light-specific manner (Imaizumi et al 2002). In the green alga *Chlamydomonas*, there are blue light-specific responses in addition to the phototactic

Plant Science Institute, Department of Biology, University of Pennsylvania, Philadelphia, PA 19104, USA

response that is mediated by a rhodopsin-like photoreceptor residing with the eye spot (Deininger et al 1995). Whether CRY is the photoreceptor mediating these other blue light responses, such as gametogenesis (Weissig and Beck 1991), has not been determined.

The discovery of cryptochromes in animals resulted from two independent lines of investigation. Todo et al (1999) described for Drosophila a divergent class of photolyases that repair (6-4) pyrimidine dimers, a minor product of UV-B irradiation of DNA. Related (6-4) photolyase-like genes were described for humans. However, the proteins encoded by these human photolyase-like genes lack any detectable photolyase activity and it was proposed by Hsu et al that the human photolyase-like sequences were cryptochromes (see Todo 1999, Van Gelder 2002). In view of the similarities to Arabidopsis cryptochromes, it was suggested that the role of mammalian cryptochrome was that of a blue light photoreceptor mediating the entrainment of circadian behavioral rhythms (Miyamoto and Sancar 1998). Whereas mice lacking both *cry1* and *cry2* exhibit apparently normal oscillations in their behavior under LD conditions, they possess a striking arrhythmic running activity under constant darkness, indicating that cryptochrome proteins are essential for the maintenance of circadian rhythmicity (van der Horst et al 1999). A molecular basis for the requirement of cryptochromes in the functioning of the mammalian clock was provided in studies from the laboratories of Reppert and Weitz (see Reppert and Weaver 2002). The mouse CRY1 and CRY2 proteins interact with and translocate each of the three mouse PER proteins to the nucleus. Furthermore, transfection studies indicate that both CRY1 and CRY2 inhibited transcription driven by CLOCK and BMAL, proteins required for the expression of PER and CRY genes. These studies establish CRY1 and CRY2, along with the three PER proteins, as essential negative regulatory components of the mammalian circadian clock.

In *Drosophila* a mutation in a new (6-4) photolyase-like cryptochrome gene was identified through a genetic screen for new mutants that showed altered circadian properties. This cry^b mutant showed deficiencies in circadian rhythms and it was proposed that the *Drosophila CRY* gene served as a circadian photoreceptor (Stanewsky et al 1998). Related studies showed that expression of the *CRY* gene was circadian regulated and, furthermore, transgenic flies overexpressing the *CRY* gene were seen to exhibit an increase in circadian photosensitivity (Emery et al 1998). *Drosophila* uses multiple photoreceptors and at least three light input pathways for circadian entrainment: the pacemaker cells and the compound eye as well as extraocular photoreception. Double mutant *glass*; *cry* flies, which lack all eye structures as well as cryptochrome, were shown to exhibit no entrainment of circadian rhythms (Helfrich-Forster et al 2001).

Cryptochrome Photochemistry

In contrast to isolated *Escherichia coli* photolyase where the flavin is in the form of the semiquinone, Arabidopsis CRY1 is isolated with the flavin in the fully oxidized FAD form (Lin et al 1995). In anaerobic conditions, CRY1 undergoes photoreduction yielding first the semiquinone and then the fully reduced $FADH_2$. The semiquinone of Arabidopsis CRY1 is relatively stable, contrasting with *E. coli* photolyase, where photoreduction of the bound FAD yields $FADH_2$ without any detectable semiquinone intermediate. The biological significance of these differences is not known, although it has been speculated that the degree of sensitivity of Arabidopsis CRY1 to green light may reflect an in vivo role of the semiquinone that, in contrast to the other two redox forms of the flavin, absorbs in this region of the spectrum (Lin et al 1995).

As with *E. coli* photolyase, in vitro intraprotein electron transfer has been studied in Arabidopsis CRY1 (Giovani et al 2003). In this instance the reduction of FAD to the semiquinone was examined. This conversion was observed to involve electron transfer from a Trp (likely, Trp324) that in turn received an electron from one of four candidate Tyr residues. As these studies involved isolated cryptochrome it is difficult to know their relevance to the in vivo mode of action of cryptochrome. This conclusion follows from the observations reported for *E. coli* photolyase where Trp306 was shown to form an essential role as the primary electron donor for the in vitro photoconversion of the semiquinone to the fully reduced flavin, yet mutation of this same Trp has no effect on the in vivo activity of the photolyase (Li et al 1991). Indeed, in the case of *E. coli* photolyase, there is no evidence that reduction of the semiquinone is required in vivo; the flavin apparently existing in the catalytically active fully reduced form. In the case of Arabidopsis CRY1, the catalytically active form of the flavin has not been determined and therefore the requirement or otherwise for the conversion of the fully oxidized FAD to the semiquinone is similarly unknown.

Cryptochrome Action Spectra

Action spectra can be useful in determining whether a photoreceptor is involved in mediating a particular photoresponse. However, such determinations are difficult at the best of times, and particularly so in the case of cryptochromes. The reasons for this are several. Firstly, as mentioned, in addition to the primary catalytic chromophore there is a second light-harvesting chromophore. The latter will not only contribute to the action spectra, indeed it is likely to dominate it; this follows from the fact that the extinction coefficient for pterins such as MTHF is substantially greater than that of flavins (Sancar 2003). Thus any prediction of cryptochrome action spectra requires knowledge of the exact identity of the second chromophore. Whereas Arabidopsis CRY1 binds MTHF when expressed in *E. coli* (Malhotra et al 1995), the precise identity of the light harvesting chromophore in Arabidopsis is not known. An additional complication in predicting action spectra involves the redox state of the chromophores, as the absorption properties, and therefore associated action spectra, will vary substantially according to the redox state. The biologically relevant in vivo redox states of either the flavin or the light harvesting chromophore are not known, for any cryptochrome.

For all of these reasons it is not possible at the moment to make useful predictions concerning the action spectra for any process that might involve cryptochromes.

There is an additional complication involving the interpretation of action spectra. In some instances a photoreceptor may play an essential role for a certain process; a temptation in such cases is to conclude that such a photoreceptor is the only photoreceptor mediating the process and therefore the corresponding action spectrum should mimic the absorption properties of the receptor. However, there may be multiple photoreceptors serving distinct and possibly essential roles. For instance, the *cry1 cry2* double mutant of Arabidopsis has been reported to lack any blue light induced shortening of the hypocotyl and in view of this finding the action spectra of this response has been interpreted exclusively in terms of cryptochrome (Ahmad et al 2002). However, it is well established that PHYA plays a role in this blue light response (Neff and Chory 1998, Poppe et al 1998) and therefore, even if the cryptochromes play an essential role, it does not follow that the corresponding action spectrum will exclusively reflect the cryptochrome photoreceptors.

This general topic of photoreceptors serving non-overlapping roles is of great importance in reference to arguments concerning what role, if any, cryptochromes play in the entrainment of mammalian circadian rhythms. Here, mice lacking both melanopsin and opsins are totally deficient in their capacity to undergo entrainment and for this reason it has generally been concluded that mice cryptochromes (CRY1 and CRY2) play no role in this process (Hattar et al 2003, Panda et al 2003). Whereas this conclusion may well be correct, the argument is not correct, for the reasons outlined above. It is conceivable that cryptochromes, serving as photoreceptors, do play a role in entrainment of the mammalian clock, even though either melanopsin or opsins are essential for this process. The quite unexpected role that cryptochromes play in the functioning of the central oscillator unfortunately precludes any easy determination of this question of whether or not these same cryptochromes function as photoreceptors for entrainment (van der Horst et al 1999). However, whereas there has been a willingness in the field to accept that CRY1 and CRY2 play an essential role in the central oscillator, distinct from the similarly essential role played by the PERIOD proteins, it is now the general consensus that cryptochromes play no role in entrainment (Hattar et al 2003, Panda et al 2003). As noted, whereas this conclusion may conceivably be correct, the logic of the argument is incorrect. In reference to this debate it is of interest that the *Drosophila* CRY does function as a photoreceptor for entrainment (Helfrich-Forster et al 2001).

Cryptochromes and Blue Light-Dependent Inhibition of Cell Expansion

Light-mediated inhibition of hypocotyl growth can be detected within 30 s of stem irradiation with a pulse of blue light (Parks et al 1998). With continuous irradiation, stem elongation continues at increasingly reduced rates for about 30 min after light exposure, whereupon growth continues at a markedly reduced rate for several days. Surprisingly, from mutant studies it appears that this early inhibition observed for the first 30 min of exposure to blue light is mediated not by cryptochrome but by PHOT1, the blue light photoreceptor responsible for phototropism (Folta and Spalding 2001). However, after 30 min the response is largely mediated by CRY1 and CRY2 as, in contrast to the continued reduced growth rate of wild-type plants, normal growth is observed in both the *cry1* and *cry2* mutants after about 60 min of light exposure. Correlated with the rapid blue light-mediated inhibition of growth is an associated depolarization of the plasma membrane; anion channel blockers have a similar effect to that observed for the *cry1* and *cry2* mutants. From these studies it appears that blue light, acting through CRY1 and CRY2, activates an anion channel resulting in plasma membrane depolarization, which in turn inhibits cell expansion.

Related observations concerning the role of CRY1 in mediating blue light inhibition of cell expansion have come from studies with Arabidopsis protoplasts. Protoplasts isolated from hypocotyl tissue and kept under continuous red light undergo rapid and transient shrinkage over a period of 5 min subsequent to exposure to a pulse of blue light. This blue light-induced protoplast shrinkage does not occur in protoplasts prepared from the *cry1* mutant, demonstrating a role for the CRY1 photoreceptor in this process (Wang and Iino 1998). The observed responsivity to blue light requires previous exposure to red light and this response to red light is lost in protoplasts from the *phyA phyB* mutant.

CCT, the C-terminal Domain of Arabidopsis Cryptochrome, Mediates a Constitutive Light Response

In early studies it was presumed that plant cryptochromes would function in a manner similar to that of photolyases. Namely, it was assumed that cryptochromes would mediate a light-dependent redox reaction with an electron being transferred from the bound flavin to a signaling partner, the latter likely being bound by the distinguishing C-terminal domain (Cashmore et al 1999). In contrast to this line of thinking was the demonstration that transgenic plants expressing the C-terminal domain (CCT) of Arabidopsis cryptochrome mediate a constitutive light response (Yang et al 2000). This constitutive response was observed for CCT from both CRY1 and CRY2, but not with mutant CCT sequences corresponding to loss-of-function *cry1* alleles. These observations demonstrated that CCT, which lacks the flavin and therefore cannot respond to light, performs a role over and above that of simply binding the signaling partner. The signaling potential of CCT must be repressed in the dark in the native CRY molecule and the role of light is to relieve this repression, possibly via an intramolecular redox reaction involving the flavin and resulting in an alteration in the activity of CCT.

Arabidopsis Cryptochrome Negatively Regulates its Signaling Partner, COP1

The properties of plants expressing CCT are similar to the constitutive photomorphogenic phenotype of Arabidopsis *cop/det/fus* mutants. By yeast two-hybrid and other studies CCT was shown to bind to COP1, a ring finger and WD-40 repeat protein that functions like a component of an E3 ubiquitin ligase (Wang et al 2001, Yang et al 2001). COP1 binds to HY5, a bZIP transcriptional regulator that binds to the promoters of several genes that are expressed during photomorphogenesis. This binding of COP1 to HY5 in the dark facilitates proteasome-mediated degradation of HY5. In light, in a CRY1-dependent manner (in the case of blue light), this COP1-dependent and proteasome-mediated degradation of HY5 is repressed and HY5 levels are observed to rise (see Wang et al 2001). Neither in yeast nor in Arabidopsis is the binding of CRY to COP1 found to be light dependent, with these two proteins being bound constitutively in both cases (Yang et al 2001). One interpretation of these data is that there exists a CRY1:COP1:HY5 heterotrimeric complex and that the effect of light on CRY1 is to attenuate the ubiquitin ligase properties associated with COP1, the outcome being that HY5 no longer undergoes proteasome-mediated degradation. This light-dependent change in the properties of COP1 must in some way be mediated through a change in CCT. Note that in contrast to the interaction between photolyase and pyrimidine dimers, the modification of COP1 by CRY does not depend on transfer of an electron between the two signaling partners, as signaling can be achieved via CCT, which lacks the flavin co-factor.

Cryptochromes and Phosphorylation

There are several reports concerning the phosphorylation of Arabidopsis cryptochromes. Arabidopsis CRY1 was reported to be a substrate for phosphorylation by the kinase activity associated with purified oat PHYA (Ahmad et al 1998). Conversely, CRY2 was shown to undergo blue light-dependent phosphorylation in vivo (Shalitin et al 2002). This activity was not observed in red light or in far-red light, and phosphorylation occurred in several phytochrome-deficient mutants. Of interest, GUS:CCT2 was phosphorylated in a light-independent manner in transgenic seedlings. This correlation between constitutive phosphorylation and constitutive CCT signaling activity was interpreted as evidence in support of a model whereby phosphorylation is a necessary requirement for signaling. An alternative interpretation of the data would simply be that CCT exists in an active conformation, and that this active form of CCT is a substrate for phosphorylation. One possible role for phosphorylation is that it serves as a trigger for degradation, possibly via the activity of the proteasome. In keeping with this interpretation, the phosphorylated form of CRY2 is degraded more slowly in *cop1* mutants that in wild-type Arabidopsis (Shalitin et al 2002).

A similar series of observations were made for the phosphorylation of Arabidopsis CRY1 (Shalitin et al 2003). As in the case with CRY2, phosphorylation of CRY1 was blue light dependent and occurred in several phytochrome deficient mutants. Several long hypocotyl *cry1* mutants were selected and all were found to be deficient in phosphorylation, consistent with a model whereby phosphorylation is a necessary prerequisite for cryptochrome activity.

An interesting finding is that Arabidopsis CRY1 protein isolated from insect cells is phosphorylated in vitro in a blue light-dependent manner (Bouly et al 2003, Shalitin et al 2003). Similar observations were made for human CRY1 and both the Arabidopsis and human cryptochromes were shown to bind to an ATP affinity column (Bouly et al 2003). These findings suggest that the observed phosphorylation of cryptochrome may reflect autophosphorylation. However, there is no obvious kinase domain within the Arabidopsis cryptochrome sequence and therefore it is difficult to eliminate the possibility that the observed phosphorylation simply reflects a contaminating kinase.

Evolution of the Cryptochrome/Photolyase Gene Family

The plant and animal cryptochrome genes are not orthologues (Cashmore et al 1999). This conclusion is based on the fact that the animal cryptochrome sequences are more similar to the (6-4) photolyases than they are to plant cryptochromes. Indeed, the *Drosophila* and mammalian cryptochrome sequences are more similar to Arabidopsis (6-4) photolyase than they are to Arabidopsis CRY1 and CRY2. It has been argued that cryptochromes were likely of eukaryotic origin as, at the time, no prokaryotic cryptochromes had been described (Cashmore et al 1999). This latter conclusion must now be modified in light of a recent finding that there exists, both in bacteria (*Synechocystis*) and Arabidopsis, a new class of cryptochromes (Brudler et al 2003, Kleine et al 2003). Like other cryptochromes, this new family of proteins shows sequence similarity to photolyases and binds a flavin, yet lacks detectable photolyase activity. No physiological phenotype has been detected for a mutant *Synechocystis* lacking the *cry* gene, although gene expression profiling indicates that some genes are more strongly expressed in the mutant than the wild-type strain. From an X-ray crystallographic study of the *Synechocystis* CRY protein it was observed that FAD was bound in the U-shaped conformation previously found for type I photolyases (Brudler et al 2003).

An intriguing finding concerning the nuclear-encoded Arabidopsis CRY3 is that it is targeted to both chloroplasts and mitochondria. In view of this, and on the basis of sequence analysis, it was suggested that the origin of Arabidopsis CRY3 may reflect an endosymbiotic event distinct from that postulated to have given rise to Arabidopsis CRY1 and CRY2; the former sequence possibly having evolved from an endosymbiont related to cyanobacteria and the latter two CRY sequences being derived from an endosymbiont related to an ancestral α-proteobacterium (Kleine et al 2003).

The existence of a prokaryotic cryptochrome sequence allows one to re-examine an even more basic question concerning the evolution of this gene family: which came first, cryptochrome or photolyase? The apparent absence of a bacterial cryptochrome initially gave this "honor" to photolyases. In retaining this order of priority, Brudler et al (2003) have suggested that the *Synechocystis* CRY may have evolved from a photolyase as a redox-regulated transcriptional regulator. However, is it not possible that an ancestral flavoprotein first evolved into a molecule mediating a light-dependent redox reaction that did not involve photolyase activity? By "definition" this would be a cryptochrome and would be the evolutionary precursor to both the cryptochromes and photolyases that we know today.

References

Ahmad M, Cashmore AR (1993) HY4 gene of *A. thaliana* encodes a protein with characteristics of a blue-light photoreceptor. Nature 366: 162–166

Ahmad M, Jarillo JA, Smirnova O, Cashmore AR (1998) The CRY1 blue light photoreceptor of *Arabidopsis* interacts with phytochrome A in vitro. Mol Cell 1: 939–948

Ahmad M, Grancher N, Heil M, Black RC, Giovani B, Galland P, Lardemer D (2002) Action spectrum for cryptochrome-dependent hypocotyl growth inhibition in Arabidopsis. Plant Physiol 129: 774–785

Bouly JP, Giovani B, Djamei A, Mueller M, Zeugner A, Dudkin EA, Batschauer A, Ahmad M (2003) Novel ATP-binding and autophosphorylation activity associated with Arabidopsis and human cryptochrome-1. Eur J Biochem 270: 2921–2928

Brudler R, Hitomi K, Daiyasu H, Toh H, Getzoff ED (2003) Identification of a new cryptochrome class. Structure, function, and evolution. Mol Cell 11: 59–67

Cashmore AR (2003) Cryptochromes: enabling plants and animals to determine circadian time. Cell 114: 537–543

Cashmore AR, Jarillo JA, Wu YJ, Liu D (1999) Cryptochromes: Blue light receptors for plants and animals. Science 284: 760–765

Deininger W, Kroger P, Hegemann U, Lottspeich F, Hegemann P (1995) Chlamyrhodopsin represents a new type of sensory photoreceptor. EMBO J 14: 5849–5858

Emery P, So WV, Kaneko M, Hall JC, Rosbash M (1998) CRY, a *Drosophila* clock and light-regulated cryptochrome, is a major contributor to circadian rhythm resetting and photosensitivity. Cell 95: 669–679

Folta KM, Spalding EP (2001) Unexpected roles for cryptochrome 2 and phototropin revealed by high-resolution analysis of blue light-mediated hypocotyl growth inhibition. Plant J 26: 471–478

Giovani B, Byrdin M, Ahmad M, Brettel K (2003) Light-induced electron transfer in a cryptochrome blue-light photoreceptor. Nat Struct Biol 10: 489–490

Hattar S, Lucas RJ, Mrosovsky N, Thompson S, Douglas RH, Hankins MW, Lem J, Biel M, Hofmann F, Foster RG, Yau KW (2003) Melanopsin and rod-cone photoreceptive systems account for all major accessory visual functions in mice. Nature 424: 75–81

Helfrich-Forster C, Winter C, Hofbauer A, Hall JC, Stanewsky R (2001) The circadian clock of fruit flies is blind after elimination of all known photoreceptors. Neuron 30: 249–261

Imaizumi T, Kadota A, Hasebe M, Wada M (2002) Cryptochrome light signals control development to suppress auxin sensitivity in the moss *Physcomitrella patens*. Plant Cell 14: 373–386

Kanegae T, Wada M (1998) Isolation and characterization of homologues of plant blue-light photoreceptor (cryptochrome) genes from the fern *Adiantum capillus-veneris*. Mol Gen Genet 259: 345–353

Kleine T, Lockhart P, Batschauer A (2003) An Arabidopsis protein closely related to Synechocystis cryptochrome is targeted to organelles. Plant J 35: 93–103

Li YF, Heelis PF, Sancar A (1991) Active site of DNA photolyase: Tryptophan-306 is the intrinsic hydrogen atom donor essential for flavin radical photoreduction and DNA repair in vitro. Biochemistry 30: 6322–6329

Lin C, Robertson DE, Ahmad M, Raibekas AA, Schuman Jorns M, Dutton PL, Cashmore AR (1995) Association of flavin adenine dinucleotide with the Arabidopsis blue light receptor CRY1. Science 269: 968–970

Malhotra K, Sang-Tae K, Batschauer A, Dawut L, Sancar A (1995) Putative blue-light photoreceptors from *Arabidopsis thaliana* and *Sinapis alba* with a high degree of sequence homology to DNA photolyase contain the two photolyase cofactors but lack DNA repair activity. Biochemistry 34: 6892–6899

Matsumoto N, Hirano T, Iwasaki T, Yamamoto N (2003) Functional analysis and intracellular localization of rice cryptochromes. Plant Physiol 133: 1494–1503

Miyamoto Y, Sancar A (1998) Vitamin B2-based blue-light photoreceptors in the retinohypothalamic tract as the photoactive pigments fr setting the circadian clocks in mammals. Proc Natl Acad Sci USA 95: 6097–6102

Neff MM, Chory J (1998) Genetic interaction between phytochrome A, phytochrome B, and cryptochrome 1 during Arabidopsis development. Plant Physiol 118: 27–35

Ninu L, Ahmad M, Miarelli C, Cashmore AR, Giuliano G (1999) Cryptochrome 1 controls tomato development in response to blue light. Plant J 18: 551–556

Panda S, Provencio I, Tu DC, Pires SS, Rollag MD, Castrucci AM, Pletcher MT, Sato TK, Wiltshire T, Andahazy M, Kay SA, Van Gelder RN, Hogenesch JB (2003) Melanopsin is required for non-image-forming photic responses in blind mice. Science 301: 525–527

Parks BM, Cho MH, Spalding EP (1998) Two genetically separable phases of growth inhibition induced by blue light in Arabidopsis seedlings. Plant Physiol 118: 609–615

Poppe C, Sweere U, Drumm-Herrel H, Schäfer E (1998) The blue light receptor cryptochrome 1 can act independently of phytochrome A and B in *Arabidopsis thaliana*. Plant J 16: 465–471

Reppert SM, Weaver DR (2002) Coordination of circadian timing in mammals. Nature 418: 935–941

Sancar A (2003) Structure and function of DNA photolyase and cryptochrome blue-light photoreceptors. Chem Rev 103: 2203–2237

Shalitin D, Yang H, Mockler TC, Maymon M, Guo H, Whitelam GC, Lin C (2002) Regulation of Arabidopsis cryptochrome 2 by blue-light-dependent phosphorylation. Nature 417: 763–767

Shalitin D, Yu X, Maymon M, Mockler T, Lin C (2003) Blue light-dependent in vivo and in vitro phosphorylation of Arabidopsis cryptochrome 1. Plant Cell 15: 2421–2429

Small GD, Min B, Lefebvre PA (1995) Characterization of a *Chlamydomonas reinhardtii* gene encoding a protein of the DNA photolyase/blue light photoreceptor family. Plant Mol Biol 28: 443–454

Stanewsky R, Kaneko M, Emery P, Beretta B, Wager-Smith K, Kay SA, Rosbash M, Hall JC (1998) The *cry*[b] mutation identifies cryptochrome as a circadian photoreceptor in *Drosophila*. Cell 95: 681–692

Todo T (1999) Functional diversity of the DNA photolyase/blue light receptor family. Mutat Res 434: 89–97

van der Horst GT, Muijtjens M, Kobayashi K, Takano R, Kanno S, Takao M, de Wit J, Verkerk A, Eker AP, van Leenen D, Buijs R, Bootsma D, Hoeijmakers JH, Yasui A (1999) Mammalian Cry1 and Cry2 are essential for maintenance of circadian rhythms. Nature 398: 627–630

Van Gelder RN (2002) Tales from the crypt(ochromes). J Biol Rhythms 17: 110–120

Wang X, Iino M (1998) Interaction of cryptochrome 1, phytochrome, and ion fluxes in blue-light-induced shrinking of Arabidopsis hypocotyl protoplasts. Plant Physiol 117: 1265–1279

Wang H, Ma LG, Li J, Zhao HY, Deng XW (2001) Direct interaction of Arabidopsis cryptochromes with COP1 in light control development. Science 294: 154–158

Weissig H, Beck CF (1991) Action spectrum for the light-dependent step in gametic differentiation of *Chlamydomonas reinhardtii*. Plant Physiol 97: 118–121

Weller JL, Perrotta G, Schreuder ME, van Tuinen A, Koornneef M, Giuliano G, Kendrick RE (2001) Genetic dissection of blue-light sensing in tomato using mutants deficient in cryptochrome 1 and phytochromes A, B1 and B2. Plant J 25: 427–440

Yang HQ, Tang RH, Cashmore AR (2001) The signaling mechanism of Arabidopsis CRY1 involves direct interaction with COP1. Plant Cell 13: 2573–2587

Yang HQ, Wu YJ, Tang RH, Liu D, Liu Y, Cashmore AR (2000) The C termini of *Arabidopsis* cryptochromes mediate a constitutive light response. Cell 103: 815–827

Chapter 14

Early Events Triggered by Light Activation of the *Arabidopsis* CRY1 Blue-Light Photoreceptor

Jean-Pierre Bouly, Baldissera Giovani, and Margaret Ahmad

Introduction

Cryptochromes are blue-light photoreceptors involved in a variety of signaling pathways in both plants and animals, including blue light-dependent growth, development and entrainment of circadian rhythms. They are characterized by their significant amino acid sequence similarity to photolyases, particularly in the N-terminal chromophore binding and catalytic domain (Ahmad 2003). Many of the amino acids known to function in flavin binding are conserved in the cryptochromes, which apparently bind to the same chromophores as do photolyases (Lin et al 1995). In addition, most cryptochromes have poorly conserved C-terminal extensions that are thought to interact with downstream cellular signaling intermediates such as transcription factors. It has been proposed that a light-induced intramolecular redox reaction may trigger conformational changes in the protein (Yang et al 2000). In particular, it has been proposed that the C-terminal domain may in this way become accessible to molecular targets such as COP1, which are downstream elements in the signaling pathway of cryptochromes (Yang et al 2001).

Photolyases and DNA Repair

Because cryptochrome photoreceptors are so similar to photolyases, efforts to determine a signaling mechanism for cryptochromes have focused on comparison to photolyases, which are flavoproteins that catalyze the light-dependent repair of UV-damaged DNA. Structurally, photolyases are monomeric proteins with two light-absorbing chromophores, an N-terminal deazaflavin or pterin and a C-terminal flavin (FAD) (Sancar and Sancar 1984, Sancar 2003). Photolyases catalyze the repair of several types of UV-induced lesions in DNA including thymine dimers (by CPD photolyases) and type 6-4 photoproducts (by 6-4 pho-

Université de Paris VI, 4 Place Jussieu, Casier 156, 75005 Paris Cedex, France

tolyases). The repair reaction results from blue light-stimulated excitation of the flavin and subsequent electron transfer directly to the UV-damaged lesion in the DNA (Sancar 2003). Photolyases are widely dispersed and have been found in both prokaryotes and eukaryotes.

Photoactivation in Photolyases

In addition to DNA repair, photolyases undergo a light-dependent electron transfer reaction involving amino acids of the apoprotein. This reaction is known as photoactivation and is distinct from DNA repair. In *Escherichia coli* photolyase, the electron is transferred from a tryptophan residue of the apoprotein to the photoexcited flavin radical (Kim et al 1993). Kinetic resolution of rapid absorbance changes subsequent to flavin excitation have shown that this electron transfer to the flavin occurs on a time scale that suggests a chain of three tryptophan residues (Aubert et al 2000). These residues, based on the known structure of the protein, are oriented in such a way that the electron passes from an externally oriented final tryptophan (Trp 306) towards the flavin at the center of the protein (via Trp 359 and Trp 382, which is proximal to the flavin) (Aubert et al 2000). Mutations in either Trp 306 (to Phe) or Trp 382 (to Phe) result in marked reduction in efficiency of electron transfer and photoactivation in *E. coli* photolyase (Kim et al 1993, Byrdin et al 2003), supporting this pathway of electron transfer. The mechanism of photoactivation has also been studied in *Anacystis nidulans* photolyase, where both tryosine and tryptophan residues are involved in electron transfer (Aubert et al 1999a,b, Popovic et al 2002), and in the type 6-4 photolyases of *Xenopus laevis* (Weber et al 2002).

Light-Dependent Electron Transfer in Cryptochromes

In spite of their significant homology to photolyases, cryptochromes do not catalyze DNA repair, possibly because of structural constraints in their DNA binding pockets (Brudler et al 2003). However, there is conservation in all cryptochromes of tryptophan residues implicated in photoactivation in *E. coli* photolyase (Trp 306, Trp 359, and Trp 382). Preliminary studies performed with purified preparations of Arabidopsis cry1 protein showed that the flavin (FAD) could be induced to undergo a light-dependent reduction reaction in vitro (Lin et al 1995). More rapid spectroscopic techniques involving laser flash excitation demonstrated that this photoreduction involves the formation of tryptophan radicals (Giovani et al 2003). In addition, it was demonstrated that tyrosine radicals are also formed and that at least some of these tyrosines are externally oriented in the protein. Therefore, a difference from the *E. coli* photolyase pathway is suggested. Such a chain of intramolecular electron transfer to an external location occurs very rarely in flavoproteins, strongly supporting the functional relevance of this mechanism for cryptochrome activity.

Autophosphorylation Activity of Cryptochromes

In addition to light-dependent electron transfer, a light-stimulated autophosphorylation activity is associated with cryptochrome photoreceptors (Bouly et al 2003, Shalitin et al 2003). Purified preparations of cryptochromes, when incubated in the presence of radioactive ATP, became labeled, and phosphoamino acid analysis indicated that serine residues were phosphorylated (Bouly et al 2003). Because this result is unexpected in the light of complete absence of classic conserved kinase domains in the Arabidopsis cry1 amino acid sequence, several controls were performed to ensure that phosphorylation did not result from the activity of a copurified, contaminating kinase from the insect cell culture system.

The most important control was the direct demonstration that cry1 protein itself bound to ATP. This was done firstly by standard binding curves, which indicated a K_d of 19.8 μmolar for ATP binding. Furthermore, the purified cry1 proteins were run over an ATP agarose column and found to bind quantitatively to the column. The proteins could be eluted from the column with free ATP, indicating that binding is specific to ATP, and thereby providing direct support that cryptochromes have the capacity to bind ATP. Additional experiments using plant crude extracts showed that native plant cryptochrome proteins had ATP binding activity similar to the purified recombinant cry1 protein. Finally, recent evidence from crystallographic analysis of the Arabidopsis cry1 structure has directly identified the site of ATP binding adjacent to the flavin cofactor within the apoprotein (Brautigam et al 2004).

The autophosphorylation reaction was shown to be light responsive and require the presence of the flavin cofactor. Flavin antagonists such as KI and peroxide prevented light stimulation of the phosphorylation reaction, suggesting an involvement of a flavin-mediated redox reaction. Therefore, light-stimulated autophosphorylation is a reaction that is likely to be required for blue light-induced signaling by cryptochromes (Bouly et al 2003).

Conclusions

Cryptochromes arose from a gene family of photolyases, enzymes involved in the blue-light dependent repair of UV-damaged DNA. Purified preparations of Arabidopsis cryptochrome-1 undergo the light-dependent electron transfer reaction known as photoactivation that has also been found in photolyases. In addition, cryptochrome has gained a novel light-stimulated autophosphorylation activity, not found in photolyases. Together these two reactions may be involved in cryptochrome signaling.

References

Ahmad M (2003) Cryptochrome and flavoprotein blue-light photoreceptors. In: Nalwa H (ed) Handbook of photochemistry and photobiology vol. 4. American Scientific Publishers, Los Angeles, CA. pp 149–182

Aubert C, Brettel K, Eker APM, Boussac A (1999a) EPR detection of the transient tyrosyl radical in DNA photolyase from *Anacystis nidulans*. J Am Chem Soc 121: 8659–8660

Aubert C, Mathis P, Eker AMP, Brettel K (1999b) Intraprotein electron transfer between tyrosine and tryptophan in DNA photolyase from *Anacystis nidulans*. Proc Natl Acad Sci USA 96: 5423–5427

Aubert C, Vos MH, Mathis P, Eker AMP, Brettel K (2000) Intraprotein electron transfer during photoactivation of DNA photolyase. Nature 405: 586–590

Bouly JP, Giovani B, Djamei A, Mueller M, Zeugner A, Dudkin EA, Batschauer A, Ahmad M (2003) Novel ATP-binding and autophosphorylation activity associated with Arabidopsis and human cryptochrome-1. Eur J Biochem 270: 2921–2928

Brautigam CA, Smith BS, Ma Z, Palnitkar M, Tomchik DR, Machius M, Deisenhofer J (2004) Structure of the photolyase-like domain of cryptochrome-1 from *Arabidopsis thaliana*. Proc Natl Acad Sci USA 101: 12142–12147

Brudler R, Hitomi K, Dauyasu H, Toh H, Kucho KI, Ishiura M, Kanehisa M. Roberts VA, Todo T, Tainer JA, Getzoff E (2003) Identification of a new cryptochrome class: structure, function, and evolution. Mol Cell 11: 59–67

Byrdin M, Eker APM, Vos MH, Brettel K (2003) Dissection of the triple tryptophan electron transfer chain in *Escherichia coli* DNA photolyase: Trp382 is the primary donor in photoactivation. Proc Natl Acad Sci USA 100: 8676–8681

Giovani B, Byrdin M, Ahmad M, Brettel K (2003) Light-induced electron transfer in a cryptochrome blue-light photoreceptor. Nat Struct Biol 10: 489–490

Kim ST, Sancar A, Essenmacher C, Babcock GT (1993) Time resolved EPR studies with DNA photolyase: excited-state $FADH^{\cdot}$ abstract an electron from Trp-306 to generate $FADH^{-}$, the catalytically active form of the cofactor. Proc Natl Acad Sci USA 90: 8023–8027

Lin C, Robertson DE, Ahmad M, Raibekas AA, Schuman-Jorns M, Dutton PL, Cashmore A (1995) Association of flavin adenine dinucleotide with the *Arabidopsis* blue light receptor CRY1. Science 269: 968–970

Popovic DM, Zmiric A, Zaric SD, Knapp EW (2002) Energetics of radical formation in DNA photolyase. J Am Chem Soc 124: 3775–3782

Sancar A (2003) Structure and function of DNA photolyase and cryptochrome blue-light photoreceptors. Chem Rev 103: 2203–2237

Sancar A, Sancar GB (1984) *Escherichia coli* DNA photolyase is a flavoprotein. J Mol Biol 172: 223–227

Shalitin D, Yu X, Maymon M, Mockler T, Lin C (2003) Blue light-dependent in vivo and in vitro phosphorylation of *Arabidopsis* cryptochrome 1. Plant Cell 15: 2421–2429

Weber S, Kay CWM, Mogling H, Mobius K, Hitomi K, Todo T (2002) Photoactivation of the flavin cofactor in *Xenopus laevis* (6-4) photolyase: observation of a transient tyrosil radical by time resolved electron paramagnetic resonance. Proc Natl Acad Sci USA 99: 1319–1322

Yang HQ, Wu YJ, Tang RH, Liu Y, Cashmore AR (2000) The C-termini of *Arabidopsis* cryptochrome mediate a constitutive light response. Cell 103: 815–882
Yang HQ, Tang RH, Cashmore AR (2001) The signaling mechanism of *Arabidopsis* CRY1 involves direct interaction with COP1. Plant Cell 13: 2573–2587

Part IV
Phototropin

Chapter 15

Phototropin Overview

Winslow R. Briggs

Introduction

The past eleven years have seen the identification and characterization of two families of blue-light receptors, the cryptochromes (Ahmad and Cashmore 1993) and the phototropins (Huala et al 1997, Christie et al 1998). Recent studies provide provocative evidence for the existence of a third family, the ZTL/ADO family (Imaizumi et al 2003). Of these families, the phototropins have been shown to mediate a wide range of physiological responses in higher plants. Among these responses are phototropism, stomatal opening, chloroplast relocation, leaf expansion, rapid inhibition of the growth of etiolated seedlings, and very likely solar tracking (Briggs and Christie 2002). As will be discussed below, the chromophore domains of these photoreceptor proteins undergo a unique kind of flavin photochemistry and are representative of chromophore domains in at least one other family of likely plant photoreceptors, the ZTL/ADO family, as well as putative photoreceptor proteins in a number of fungi and bacteria. Here we review the early and recent history of blue-light receptors before describing some of our current findings related to their cellular distribution and function.

Early History

Although the phototropins were discovered only very recently (Huala et al 1997), studies of most of the plant responses mediated by blue light have a history that goes back almost two centuries. All of the light-activated responses mentioned above—with the exception of the rapid inhibition of growth—had already spiked the curiosity of scientists during the 19th century. In what may be the earliest report of a specific plant response to blue light, Poggioli (1817) first noted that violet light was much more effective than red light in inducing a change in the

Department of Plant Biology, Carnegie Institution of Washington, 260 Panama Street, Stanford, CA 94305, USA position of leaves of *Raphanus rusticanus*. It is not clear whether the response was true phototropism, solar tracking, or some combination of the two. However, Payer (1842) noted in experiments with cress seedlings (*Lepidium sativum*) that blue light was the most effective color in inducing what was clearly phototropism and he even distinguished between blue and violet light, completing a crude sort of action spectrum. Zantedeschi (1843) came to the same conclusion in his studies of the phototropic responses of the growing shoots of *Oxalis multiflora* and *Impatiens balsamina*. In spite of the lack of adequately calibrated light sources, both studies clearly demonstrated the special efficacy of blue light in phototropism.

In 1836 Daubeny was mainly reporting on his studies on photosynthesis. He did, however, make an important observation relating to stomatal function (although he did not mention stomata): blue light was far more effective than red light in inducing water loss from bean leaves. (Incidentally, he may have been the first worker to attempt any calibration of his light sources.) Six decades later Francis Darwin (1898, 1911) first demonstrated a direct effect of light on stomatal opening. However, it took many more decades before Hsiao and Allaway (1973), Ogawa and colleagues (Ogawa et al 1978, Ogawa 1981), and Zeiger and Field (1982) were able to demonstrate the clear relationship between blue light and stomatal opening.

The first observations on the effect of light on chloroplast movements appear to be those of Böhm (1856) who observed that bright light induced a "clumping" of chloroplasts along the cell walls in two species of *Sedum*. Shortly thereafter, Böhm (1859) noted that it was blue light that induced the clumping response (the chloroplast "avoidance response"). Eight years later Famintzen (1867) observed that at least in mosses, only blue light induced a chloroplast-accumulation response.

Comparatively, leaf-expansion studies have a somewhat shorter history. Although it was well known during the 19th century that leaf expansion required light, it was generally thought that leaf growth was simply driven by photosynthesis. It was Trumpf (1925) and Priestley (1925) who showed that very brief pulses of white light were insufficient to cause any greening but nevertheless induced significant leaf expansion. Trumpf used a series of colored solutions as light filters and concluded that the effective wavelengths were in the red region of the spectrum. Much later, Parker et al (1949) confirmed that red light was effective in inducing the expansion of the leaves of *Pisum sativum* and Liverman et al (1955) showed that the response was red, far-red reversible, a property we now know implicates a phytochrome as the photoreceptor. It was only in 1990 that Van Volkenburgh et al demonstrated the special role played by blue light, independent both of photosynthesis and of phytochrome, in inducing leaf expansion.

Unlike the blue-light responses just discussed, rapid inhibition of growth by blue light could only be described when suitably sensitive equipment for measuring growth rates became available. Meijer (1968) used a position transducer to show that both blue and red light induced an inhibition of hypocotyl elongation in etiolated *Cucumis sativus* seedlings, albeit with different kinetics: the onset of the blue-light response occurred only a few minutes after the start of illumination whereas the red-light response had a lag period of at least half an hour.

In 1817, Poggioli noted in the same study mentioned above that the leaves of *Mimosa pudica* became reoriented to changes in the direction of incident light and responded much more rapidly under violet light than under red, likely the first study of the spectral sensitivity of solar tracking—or for that matter, of any other light response in plants. More than a century was to pass before Yin (1938) confirmed that solar tracking was indeed a blue-light response.

The Involvement of the Phototropins

A phototropin was first cloned from *Arabidopsis thaliana*, characterized, and shown to be essential for phototropism in 1997 (Huala et al 1997). A second phototropin sequence from *Arabidopsis* was published only a year later (Jarillo et al 1998). Christie et al (1998) showed that the protein described by Huala et al (now designated phototropin 1 or phot1) was actually a blue-light receptor capable of undergoing light-activated autophosphorylation in a *Baculovirus*-insect cell system in the absence of any other plant proteins. As the protein was essential for phototropism, they concluded that it was the long-sought photoreceptor for that much-studied blue-light response. Subsequently Sakai et al (2001) showed that the second family member, phototropin 2 (phot2), also mediated phototropic curvature in *Arabidopsis* but only in response to much higher fluences of blue light than required by phot1. Kinoshita et al (2001) then showed that the two phototropins functioned redundantly in mediating blue light-activated stomatal opening. Three studies demonstrated the photoreceptor role for the phototropins in chloroplast relocation: both Jarillo et al (2001) and Kagawa et al (2001) reported within weeks of each other that phot2 was exclusively the photoreceptor for the chloroplast avoidance response in bright light. Sakai et al (2001) subsequently found that the accumulation of chloroplasts in dim light was mediated by both phot1 and phot2. Further studies by Sakamoto and Briggs (2002) then showed the redundant roles of phot1 and phot2 in mediating blue light-induced leaf expansion and, Folta and Spalding (2001) reported the exclusive role of phot1 as the photoreceptor for the rapid inhibition of growth. Although definitive evidence is still lacking, anecdotal evidence indicates that the phototropins may also function as photoreceptors for solar tracking. Thus it took over a century and three-quarters from Poggioli's initial study on light-induced leaf movements for the assignment of specific photoreceptors to these six responses. Although the search for the relevant photoreceptors began in 1938 (Wald and DuBuy 1938, suggesting a carotenoid as the photoreceptor for phototropism of oat coleoptiles), final identification required the application of modern molecular genetics and the availability of mutants lacking one or both of the phototropins before specific assignments could be made.

Phototropin Characterization and Photochemistry

Arabidopsis phot1 (996 amino acids) and phot2 (915 amino acids) have a unique structure: at the N-terminal end are two highly conserved domains approximately 100 amino acids long (designated LOV domains for their homology with domains from other proteins involved in signaling in response to **L**ight, **O**xygen, or **V**oltage) that both serve to bind the chromophore flavin mononucleotide (FMN). These domains form a cage of β-strands and α-helices that binds the FMN chromophore tightly (see Briggs and Christie 2002). At the C-terminal end is found a classic serine/threonine protein kinase domain (see Briggs and Christie 2002). Blue light activates the kinase function resulting in hierarchical autophosphoryation: serines on either side of the LOV1 domain becoming phosphorylated in vivo at lower fluences of blue light than other residues between the LOV domains but closer to LOV2 (Salomon et al 2003). Although no one has yet succeeded in purifying a full-length phototropin from a heterologous system, constructs encoding just one or both LOV domains work well in *Escherichia coli*. Using such heterologously produced chromopeptides, Salomon et al (2000) showed that the initial long-lived product of LOV-domain photoexcitation arose by the formation of a covalent bond between a cysteine of the peptide and the C(4a) carbon of the flavin moiety. Formation of this cysteinyl adduct causes conformational changes both in the FMN and in the peptide moiety of the LOV domain itself (Swartz et al 2002, Corchnoy et al 2003). Most recently Harper et al (2003) have identified an amphipathic α-helix just downstream from the LOV2 domain of oat phot1 that is tightly appressed to the LOV domain itself in the dark state, but released on the light-induced formation of the cysteinyl adduct. Curiously there is no such α-helix downstream from LOV1, indicating that LOV1 and LOV2 do not necessarily serve redundant functions.

Why Two LOV Domains?

Recently Christie et al (2002) examined the separate roles of LOV1 and LOV2 in mediating autophosphorylation in vitro or complementing phototropism in *phot1 phot2* double mutants of *Arabidopsis*. For phot1, LOV2 was sufficient for normal phosphorylation whereas LOV1 alone was without effect. Both photosensitivity and phosphorylation kinetics were identical to those of the wild-type photoreceptor. LOV2 was also sufficient to restore phototropism in the double mutant, although a quantitative difference in sensitivity between the mutant and wild-type photoreceptors could not be ruled out. For phot2, LOV2 was predominant in the phosphorylation reaction although in this case, LOV1 could mediate limited autophosphorylation. Preliminary studies with transgenic plants indicate that phot2 LOV2 alone is sufficient to complement a phototropic response to saturating light. Indeed, LOV1 may actually down-regulate the sensitivity of phot2 in phototropism as transgenic plants carrying phot2 LOV2 as the only functional chromophore module showed greater sensitivity than did those with both LOV domains functioning (Kaiserli et al unpublished). Further work is in progress to attempt to resolve the specific roles of the LOV domains in both phototropins.

Very recently, Kagawa et al (2004) showed that the phot2 in the fern *Adiantum capillus-veneris* in which virtually all amino acids N-terminal from LOV2 had been deleted nevertheless was sufficient to complement the chloroplast avoidance response in a *phot2* mutant. They also demonstrated that in order to mediate the chloroplast avoidance response, a sequence between amino acids 979 and 999 was also required.

Phototropin Tissue and Subcellular Distribution

Both phototropins are associated with the plasma membrane (see Sakamoto and Briggs 2002, Harada et al 2003). The tissue and plasma-membrane localization of phot1 in *Arabidopsis* is completely consistent with its role in the various physiological processes that it mediates. Using a phot1 mutant transformed to express a phot1–GFP fusion protein, Sakamoto and Briggs (2002) showed that a phot1–GFP construct could complement a *phot1* mutant, and was expressed at high enough levels for good confocal microscope imaging. For example, in the etiolated hypocotyl, phot1–GFP is found in the same cell layer as the putative auxin transport protein PIN1, consistent with its role in redirecting auxin fluxes in response to unilateral light. However, whereas PIN1 is located largely at the basal end of these cells, phot1 is found adjacent both to the apical and basal ends. The existence of a direct physical relationship between phot1 and PIN1 remains to be determined. In the flowering stem, phot1 is located both in the xylem and phloem parenchyma, in the same cells in which PIN1 (xylem parenchyma) and PIN3 (phloem parenchyma), respectively, are found, again consistent with its putative role in mediating changes in the auxin transport pattern during the development of phototropic curvature. (see Sakamoto and Briggs 2002 for pertinent references on the PIN proteins.) Given the important role of the leaf epidermis in driving leaf expansion, it comes as no surprise that phot1 is evenly distributed adjacent to the plasma membrane of the epidermal cells of the *Arabidopsis* leaf. Likewise, given the epidermal role in stem elongation, it is not surprising to find it localized adjacent to the plasma membrane of epidermal cells of the hypocotyl. Finally, phot1 is strongly expressed adjacent to the plasma membrane of guard cells, consistent with its role in mediating blue light-induced stomatal opening.

Arabidopsis is not the plant of choice for studying solar tracking. Although the leaves move in response to changes in light direction, the movement is not great and movement brought on by truly differential growth (phototropism)—as opposed to the completely reversible turgor changes that drive the tracking response—cannot be ruled out. However, the existence of phot1 at the end walls of elongated subepidermal cells along the leaf veins is consistent with a model proposed for solar tracking (see Sakamoto and Briggs 2002 for discussion). It should be noted that solar tracking does not necessarily require the presence of a pulvinus, but can occur through turgor changes along a petiole or along stem tissue.

Studies of the subcellular distribution of phot2–GFP to date are only preliminary. We have recently obtained *Arabidopsis* transformant expressing phot2–GFP behind the native phot2 promoter in the *phot1 phot2* double mutant background (Tissier et al, unpublished). Unfortunately the phot2 fusion protein is not as strongly expressed as the phot1 fusion protein, giving a very weak fluorescence signal. Although the signal can be amplified by increasing the intensity of the scanning laser, this step induces a great deal of autofluorescence. In addition, the high-intensity light gradually bleaches the GFP leading to a loss of signal. Despite these drawbacks, it appears that the distribution of phot2–GFP is not greatly different from that of phot1–GFP. We are currently analyzing transformant lines with stronger expression in the hope of obtaining a more detailed picture of phot2–GFP distribution both in light- and dark-grown seedlings at different developmental stages and in different tissue layer and cell types.

Conclusions

From Poggioli's primitive experiments on the effect of different colors of light on leaf movement in *Mimosa pudica* (1817), the journey to the cryptochrome blue-light receptors (Ahmad and Cashmore 1993) and the phototropins (Huala et al 1997, Christie et al 1998) has been a long and arduous one. Biochemistry alone was insufficient to identify either of these photoreceptor families. It was only with the advent of modern molecular genetics that the photoreceptors could finally be identified and characterized. Now, several recent studies have identified three other *Arabidopsis* proteins, ZTL/ADO, FKF1, and LKP2, as each having a classic LOV domain similar to those found in the phototropins. These three proteins are all involved one way or another in the functioning of circadian rhythms or daylength measurement in *Arabidopsis*. Imaizumi et al (2003) recently presented spectral evidence that these LOV domains undergo light-activated formation of the same cysteinyl adduct as the phototropin LOV domains and present evidence strongly implicating one of them (FKF1) in the detection of daylength in the regulation of flowering time. Thus it seems likely that these three proteins will all serve some photoreceptor function, although future work is required to test this hypothesis. If the hypothesis is verified, then studies on the phototropins have led to the identification of a third family of blue-light receptors.

References

Ahmad M, Cashmore AR (1993) *HY4* gene of *A. thaliana* encodes a protein with the characteristics of a blue-light receptor. Nature 366: 162–166

Böhm JA (1856) Beiträge zur näheren Kentniss des Chlorophylls. Sitzungber mathem-Naturwiss Classes kais. Akademie Wissenschaften 22: 479–512

Böhm JA (1859) Über den Einfluss der Sonnenstrahlen auf die Chlorophyllbildung und das Wachstum der Pflanzen überhaupt. Sitzungber mathem-Naturwiss Classes kais Akademie Wissenschaften 37: 453–476

Briggs WR, Christie JM (2002) Phototropins 1 and 2: versatile plant blue-light receptors. Trends Plant Sci 7: 204–210

Christie JM, Reymond P, Powell GK, Bernasconi P, Raibekas AA, Liscum E, Briggs WR (1998) A flavoprotein with the properties of a photoreceptor for phototropism. Science 282: 1698–1701

Christie JM, Swartz TE, Bogomolni RA, Briggs WR (2002) Phototropin LOV domains exhibit dinstinct roles in regulating photoreceptor function. Plant J 32: 205–219

Corchnoy SB, Swartz TE, Lewis JW, Szundi I, Briggs WR, Bogomolni RA (2003) Intramolecular proton transfers and structural changes during the photocycle of the LOV2 domain of phototropin 1. J Biol Chem 278: 724–731

Daubeny C (1836) On the action of light upon plants, and of plants upon the atmosphere. Phil Trans R Soc Lond 126: 149–175

Darwin F (1898) Observations on stomata. Phil Trans R Soc Lond Ser B 190: 136–154

Darwin F (1911) On a new method of estimating the aperture of stomata. Proc R Soc Lond B Biol Sci 84: 136–154

Famintzen A (1867) Die Wirkung des Lichtes und der Dunkelheit auf die Vertheilung der Chlorophyllkörner in den Blättern von Mnium sp.? Jahrb Wiss Botanik 6: 49–54

Folta KM, Spalding EP (2001) Unexpected roles for cryptochrome 2 and phototropin revealed by high-resolution analysis of blue light-mediated hypocotyl growth inhibition. Plant J 26: 471–478

Harada A, Sakai T, Okada K (2003) phot1 and phot2 mediate blue light-induced transient increases in cytosolic Ca^{+2} differently in *Arabidopsis* leaves. Proc Natl Acad Sci USA 100: 8583–8588

Harper SM, Neil LC, Gardner KH (2003) Structural basis of a phototropin light switch. Science 301: 1541–1544

Hsiao TC, Allaway, WG (1977) Acton spectra for guard cell Rb^+ uptake and stomatal opening in *Vicia faba*. Plant Physiol 51: 82–88

Huala E, Oeller PW, Liscum E, Han IS, Larsen E, Briggs WR (1997) *Arabidopsis* NPH1: A protein kinase with a putative redox-sensing domain. Science 278: 2120–2123

Imaizumi T, Tran HG, Swartz TE, Briggs WR, Kay SA (2003) FKF1 is essential for photoperiodic-specific light signalling in Arabidopsis. Nature 426: 302–306

Jarillo JA, Ahmad M, Cashmore AR (1998) NPL1 (Accession no. AF053941): a second member of the NPH serine/threonine kinase family of Arabidopsis. Plant Physiol 117: 719

Jarillo JA, Gabrys H, Capel J, Ecker JR, Cashmore AR (2001) Phototropin-related NPL1 controls chloroplast relocation induced by blue light. Nature 410: 592–594

Kagawa T, Kasahara M, Abe T, Yoshida S, Wada M (2004) Function analysis of phototropin2 using fern mutants deficient in blue light-induced chloroplast avoidance movement. Plant Cell Physiol 45: 416–426

Kagawa T, Sakai T, Suetsugu N, Oikawa K, Ishiguro S, Tabata S, Okada K, Wada M (2001) Arabidopsis NPL1: a phototropin homolog controlling the chloroplast high-light avoidance response. Science 291: 2138–2141

Kinoshita T, Doi M, Suetsugu N, Kagawa T, Wada M, Shimizaki K (2001) phot1 and phot2 mediate blue light regulation of stomatal opening. Nature 414: 656–660

Liverman JL, Johnson MP, Starr L (1955) Reversible photoreaction controlling expansion of etiolated bean-leaf disks. Science 121: 440–441

Meijer G (1968) Rapid growth inhibition of gherkin hypotocyls in blue light. Acta Bot Neerl 17: 9–14

Ogawa T (1981) Blue light response of stomata with starch-containing (*Vicia faba*) and starch-deficient (*Allium cepa*) guard cells under background illumination with red light. Plant Sci Lett 22: 103–108

Ogawa T, Ishikawa H, Shimada K, Shibata K (1978) Synergistic action of red and blue light and action spectra for malate formation in guard cells of *Vicia fava* L. Planta 142: 61–65

Parker MW, Hendricks SB, Borthwick HA, Went FW (1949) Spectral sensitivities for leaf and stem growth of etiolated pea seedlings and their similarity to action spectra for photoperiodism. Am J Bot 36: 194–204

Payer J (1842) Mémoire sur la tendance des tiges vers la lumière. Compte Rendu des Seances de L'Académie des Sciences, July 4, 1842: 1194–1196

Poggioli S (1817) Della influenza che ha il raggio magnetico sulla vegetatione delle piante. Bologna-Col Tipi di Annesio Nobili Opusc Scientif Fasc I:9–23

Priestley JH (1925) Light and growth I. The effect of brief light exposure upon etiolated plants. New Phytol 24: 271–283

Sakai T, Kagawa T, Kasahara M, Swartz TE, Christie JM, Briggs WR, Wada M, Okada K (2001) Arabidopsis nph1 and npl1: Blue light receptors that mediate both phototropism and chloroplast relocation. Proc Natl Acad Sci USA 98: 6969–6974

Sakamoto K, Briggs WR (2002) Cellular and subcellular localization of phototropin 1. Plant Cell 14: 1723–1735

Salomon M, Christie JM, Kneib E, Lempert U, Briggs WR (2000) Photochemical and mutational analysis of the FMN-binding domains of the plant blue light receptor phototropin. Biochemistry 39: 9401–9410

Salomon M, Knieb E, von Zeppelin T, Rüdiger W (2003) Mapping of low- and high-fluence autophosphorylation sites in phototropin 1. Biochemistry 42: 4217–4225

Swartz TE, Wenzel PJ, Corchnoy SB, Briggs WR, Bogomolni RA (2002) Vibration spectroscopy reveals light-induced chromophore and protein structural changes in the LOV2 domain of the plant blue-light receptor phototropin 1. Biochemistry 41: 7183–7189

Trumpf C (1924) Über den Einfluss intermittierender Belichtung auf das Etiolement der Pflanzen. Bot Arch 5: 381–410

Van Volkenburgh E, Cleland, RE, Watanabe M (1990) Light-stimulated cell expansion in bean (*Phaseolus vulgaris* L.) leaves. II. Quantity and quality of light required. Planta 182: 77–80

Wald G, DuBuy HG (1936) Pigments of the oat coleoptile. Science 84: 247

Yin HC (1938) Diaphototropic movement of the leaves of *Malva neglecta*. Am J Bot 25: 1–6

Zantedeschi M (1843) De l'influence qu'exercent sur la vegetation des plantes et la germination des graines les rayons transmis à travers des verres colorés. Compte Rendue des Seances de L'Académie des Sciences, January 2, 1843: 747–749

Zeiger E, Field C (1982) Photocontrol of the functional coupling between photosynthesis and stomatal conductance in the intact leaf. Plant Physiol 70: 370–375

Chapter 16

Proton Transfer Reactions in LOV-Domain Photochemistry

Roberto A. Bogomolni[1], Trevor E. Swartz[1,2], and Winslow R. Briggs[2]

Introduction

The phototropins (phot1 and phot2) are plant blue-light photoreceptors that mediate phototropism, chloroplast relocation, stomatal opening, leaf expansion, and rapid inhibition of hypocotyl growth (Briggs and Christie 2002). They contain two 10kDa FMN-binding LOV domains (LOV1 and LOV2) with a serine/threonine kinase at the C-terminal region. Light absorbed by the flavin chromophore results in the transient formation of a covalent flavoprotein adduct and a localized protein structural perturbation. Intramolecular propagation of this conformational change results in activation of the kinase moiety and hierarchical autophosphorylation at several sites located in the protein region between the two LOV domains and near the N-terminus. Thus, adduct formation is viewed as a light-driven molecular switch that activates the subsequent molecular events.

The LOV domains expressed in *Escherichia coli* bind FMN and are photochemically active. Time-resolved spectroscopy showed that the LOV domains undergo a photocycle, which is characterized by a series of transient intermediates (Salomon et al 2000, Swartz et al 2001, Kottke et al 2002, Losi et al 2002) with a spontaneous return to the ground-state of the protein in the dark. Some of these reactions are kinetically sensitive to proton/deuterium exchange and/or are affected by the pH of the bulk medium. We discuss here the mechanistic implications of these observations and the role of proton transfer reactions in the photochemical process and the associated molecular structural changes.

[1]Department of Chemistry and Biochemistry, University of California Santa Cruz, Santa Cruz, CA 95060, USA

[2]Department of Plant Biology, Carnegie Institution of Washington, Stanford, CA 94305, USA

Photocycle Kinetics

The photochemical cycle was first characterized in *Escherichia coli*-expressed oat phot1-LOV2 (Salomon et al 2000, Swartz et al 2001). Photoreactions of several FMN-binding LOV domains studied since were shown to be qualitatively similar. Presently, two excited states and one metastable thermal intermediate have been kinetically resolved in the LOV domain photocycle. At room temperature and slightly alkaline pH, light absorption promotes transition to the FMN singlet excited state that decays with a half-time around 2 ns into a red-absorbing species labeled $LOV2^{T}_{660}$, that has been shown to be the FMN triplet state (Swartz et al 2001, Kottke et al 2002, Losi et al 2002, Kennis et al 2003). $LOV2^{T}_{660}$ decays in microseconds into the second metastable intermediate, which absorbs maximally at 390 nm and is labeled LOV^{S}_{390} (Swartz et al 2001). The LOV^{S}_{390} intermediate thermally relaxes back to the ground state, LOV^{D}_{447} in tens of seconds.

Global kinetic analysis of laser flash-induced absorption changes of oat phot1-LOV2 (actinic wavelength 450 nm, 4 ns pulse) measured from 30 ns to 100 μs gives a single decay time constant of 2 μs in this time window, and isosbestic points indicate a single species (Swartz et al 2001). Global fitting assuming a linear kinetic scheme with two intermediates (LOV^{T}_{660} and LOV^{S}_{390}) allows for the calculation of the spectra of both species (Swartz et al 2001). Because the measured amount of the initial bleach of the ground state is half that of the amount of LOV^{S}_{390} formed, we included in the scheme a simultaneous 1:1 split decay of LOV^{T}_{660} back to the ground state LOV^{D}_{447} and forward to LOV^{S}_{390}, resulting in a calculated time constant of 4 μs for each decay direction (Swartz et al 2001). The calculated LOV^{T}_{660} spectrum closely matches that of the triplet state of FMN (Swartz et al 2001). Measurements on the femtosecond time scale show direct conversion of the FMN triplet state from the singlet excited state in about 2–3 ns via intersystem crossing (ISC) (Kennis et al 2003, Schuttrigkeit et al 2003). Further analysis of the triplet-state spectrum suggests a mixture of species in which the FMN N5 exists in an equilibrium of both protonated and unprotonated forms (Kennis et al 2003).

The LOV^{S}_{390} species that is formed from the triplet state decays spontaneously in the dark back to the ground state. The rate of return of the ground state varies from a few seconds to many minutes (Salomon et al 2000, Kasahara et al 2002, Losi et al 2002, Schwerdtfeger and Linden 2003). The photocycle of some LOV domains truncates at the long-lived intermediate, and the ground state is not regenerated (Imaizumi et al 2003). This long-lived metastable intermediate involves the formation of a protein–FMN covalent bond. Specifically an S–C bond is formed between the sulfur of cys39 and the C(4a) carbon of FMN (Salomon et al 2001, Crosson and Moffat 2002, Swartz et al 2002, Ataka et al 2003). Light-induced circular dichroism (CD) changes in the 250–500 nm range during the photocycle show the appearance of a large positive band at 270–290 nm upon formation of LOV^{S}_{390}. This band coincides with a higher energy FMN absorption band around 270 nm. The large optical activity associated with this band is consistent with the formation of an sp^3 chiral C(4a) center in the LOV^{S}_{390}

adduct (Salomon et al 2000, Corchnoy et al 2003). Both circular dichroism (CD) and difference Fourier transform infrared (FTIR) have demonstrated protein conformational perturbations associated with the photocycle (Swartz et al 2002, Ataka et al 2003, Corchnoy et al 2003, Iwata et al 2003). Numerous positive and negative FTIR bands are observed in the protein amide regions (Swartz et al 2002, Ataka et al 2003, Iwata et al 2003), and CD difference spectra show transient loss in alpha helicity (Corchnoy et al 2003). The magnitude and character of these protein conformational changes depend on the size and sequence of the specific LOV constructs used for the studies. In particular, these structural changes become more apparent in larger constructs containing additional protein segments on the C terminal side of the LOV2 domain. In a larger construct, difference 3D nuclear magnetic resonance has revealed protein perturbations of an amphipathic alpha-helix C-terminal from the LOV2 domain (Harper et al 2003), the direction in which signal propagation is expected in the full-length chromoprotein.

A Proton Transfer Reaction Initiates Triplet State Decay

The overall rate of adduct formation is five times slower in D_2O than in H_2O, suggesting that the rate-limiting step between triplet state and adduct formation is a proton (or hydrogen atom)-transfer reaction (Corchnoy et al 2003). As discussed above the decay of the triplet state (LOV^T_{660}) in wild-type oat phot1-LOV2, proceeds via a branched pathway in which 50% regenerates LOV^D_{447}, the ground state, and 50% converts to LOV^S_{390} in both D_2O and H_2O. The observed forward decay of LOV^T_{660}, to LOV^S_{390} is five times slower in D_2O (Corchnoy et al 2003). Analysis of the data shows that the forwards/backward 1:1 split ratio for its decay is unchanged in D_2O and that the backward decay to the ground state is also slowed down 5-fold. This analysis suggests that both decay reactions of the triplet (forward to the adduct state and back to the ground state) are rate limited by a single proton (or hydrogen atom)-transfer reaction. The magnitude of the deuterium kinetic effect is suggestive of a primary isotope effect (Melander and Saunders 1980), presumably associated with protonation of (or hydrogen atom transfer to) FMN's N5. In oat phot1-LOV2 adduct formation is nearly independent of bulk pH (Corchnoy et al 2003), it occurs even in the frozen state at cryogenic temperatures (Iwata et al 2003), and at room temperature at low levels of hydration (Bogomolni unpublished). This indicates that this rate-limiting step is not dependent on the presence of bulk water, and that it is a localized and almost barrier-less intramolecular proton (or hydrogen atom) transfer reaction.

Mutation of cysteine 39 to an alanine (LOV2C39A) results in a truncated photocycle as compared to the wild type. Because these proteins are missing the reactive cysteine, they do not form the LOV^S_{390} intermediate; however, after a short laser flash, these proteins do form the LOV^T_{660} intermediate. This intermediate $LOV2C39A^T_{660}$ decays directly back to the ground state ($LOV2C39A^D_{447}$) in

72 μs, one order of magnitude slower than wild type (Swartz et al 2001). The presence of a sulfur atom in the WT oat phot1-LOV2 domain may contribute to the triplet decay presumably by increasing the rate of spin flipping due to spin orbit coupling (Swartz et al 2001). The decay rate of LOV2C39A$^{T}{}_{660}$ back to the ground state, LOV2C39A$^{D}{}_{447}$, does not show a deuterium effect (Corchnoy et al 2003).

Several alternate reaction pathways for the formation of the adduct following initial activation to the triplet state have been proposed. They include (a) direct transfer of a proton from the cysteine thiol to N5 of the triplet state followed by formation of the S–C bond (Crosson and Moffat 2001), (b) excited-state proton transfer to N5 preceding triplet formation followed by reaction of a thiolate with FMN C(4a) (Kennis et al 2003), (c) involvement of a flavosemi-quinone free radical and reaction with a sulfur radical, a radical pair recombination mechanism (Bittl et al 2003, Kay et al 2003, Kottke et al 2003), and (d) proton transfer to the triplet FMN followed by reaction of a C(4a) carbo-cation with a cys39 thiolate (Swartz et al 2001). Available evidence is insufficient to establish the correct mechanism. Common to all mechanisms is the redistribution of charge around the FMN N5-C(4a) double bond that occurs in the FMN triplet state, resulting in an increase of basicity of N5 (Song 1968, Fedorov et al 2003, Neiss and Saalfrank 2003). In the ionic recombination mechanism, protonation of N5 either from the C39 thiol (Crosson and Moffat 2001, Kennis et al 2003) or from an as yet unidentified acid group (Swartz et al 2001), draws electronic density from the N5-C(4a) double bond, leaving C(4a) as a reactive carbo-cation that is attacked by the ionized sulfur of cysteine 39, resulting in formation the flavin–cysteinyl adduct.

Kennis et al (2003) proposed that proton transfer from the C39 thiol occurs in the excited state with transient thiolate formation, whereas Swartz et al (2001) have invoked the existence of a ground-state thiolate and a putative proton-acceptor residue that donates a proton to N5, initiating the split decay of the triplet state to either the adduct or back to the ground state. Vibrational-spectroscopy evidence that the sulfur of cysteine exists as a thiol has led to the argument that the proton originates directly from the cysteine SH. Although the proton could come from another group, no other acid groups have been identified in the vicinity of the chromophore. The crystal structure, however, shows two water molecules that could conduct a proton from a residue that is not directly interacting with the chromophore. The C39A mutants do produce a neutral semiquinone radical under high light intensities (see below) (Kay et al 2003). Formation of the neutral semiquinone requires donation of a hydrogen atom from an as yet unidentified group.

The oat phot1-LOV2 triplet state absorption spectrum shows no evidence for a semiquinone (Swartz et al 2001). It is possible that the triplet state is the rate limiting step, and that the rapid rate of disappearance of the semiquinone would not allow its transient accumulation and its detection in the absorption changes. Kay et al (2003) point out that protonation of the FMN triplet state is not necessarily the rate-limiting step in a radical-pair mechanism, whereas protonation of the triplet state should be a rate-limiting step in an ionic mechanism.

At physiological light intensities, the formation of a neutral flavo-semiquinone was not observed in LOV2C39A (Swartz et al 2001), but it was observed at very high light intensities (Swartz and Bogomolni unpublished). At similar high light intensities, a small fraction of wild-type oat phot1-LOV2 also forms a neutral semiquinone (Swartz and Bogomolni unpublished). At these light intensities, most of the sample forms the adduct and undergoes a fully reversible photocycle returning to the ground state in tens of seconds. The long-lived semiquinone forms via a branched reaction pathway. It returns to the ground state in tens of minutes to hours, presumably by being reoxidized by atmospheric oxygen (Swartz and Bogomolni unpublished).

It should be mentioned that although the *Chlamydomonas* phot-LOV1 photocycle contains intermediates similar to those of oat phot1-LOV2, it involves a more complex kinetic scheme. The *Chlamydomonas* phot-LOV1 triplet state decays forward with two time constants, one of 800 ns and one of 4 µs; it has been suggested that the triplet state exists as a mixture of two species with different decay times (Kottke et al 2002). However, because their spectral properties are nearly identical, it is not possible to discriminate between the two (Kottke et al 2002). In contrast to oat phot1-LOV2, *Chlamydomonas* phot-LOV1 does not show an appreciable return to the ground state from the triplet state. The isotope deuterium effect has not been reported on the *Chlamydomonas* phot-LOV1 adduct formation or decay.

Proton Transfers and/or Hydrogen-Bond Perturbations Rate-Limit Adduct Decay

The mechanism of dark recovery of the ground state LOV^{D}_{447} from LOV^{S}_{390}, requires breakage of a stable carbon sulfur-bond. Synthetic C4a-sulfur adduct model compounds have been used for structural studies (Müller 1991). A C–S covalent bond energy is typically greater than 200 kJ/mol, and is presumably stable in aqueous media. Breakage of this bond in the protein environment must involve a catalytic process provided by specific protein residues in the FMN binding pocket. In oat phot1-LOV2, the rate of adduct decay shows only a slight decrease at lower pH (apparent pK around 6.5 (Corchnoy et al 2003)), but in *Chlamydomonas* phot-LOV1 there is a marked pH dependence in which the decay rate increases several fold between pH 8 and pH 3 with an apparent pK around 5–6 (Kottke et al 2002), indicating that it is base catalyzed. Abstraction of the N(5) proton (acid–base catalysis) could be the event that initiates the back reaction. The dark recovery rate in oat phot1-LOV2 is three times slower in D_2O than in H_2O, suggesting that a proton-transfer reaction is the rate-limiting step (Swartz et al 2001).

Because the deuterium effect is only a factor of three, the rate-limiting step in the back reaction is perhaps not a primary isotope effect and could reflect breaking of a hydrogen bond rather than a proton transfer. Because the pKa of N5 is modulated by hydrogen bonding on the flavin (Yagi et al 1980), perhaps the variability of relaxation in different LOV domains is a measure of this hydrogen bonding. CD changes in the far UV indicate that the protein returns to the ground state with the same kinetics as the chromophore and with deuterium isotope effects of identical kinetic suggesting concerted events and a common rate-limiting proton-associated event (Corchnoy et al 2003). In contrast to its formation, adduct decay shows a very large activation barrier (Corchnoy et al 2003) allowing for trapping of this form at low temperatures. Adduct decay is both pH sensitive, consistent with a base-catalyzed mechanism (Kottke et al 2002), and it is blocked at low hydration levels (Bogomolni and Ciorcciari, unpublished), indicating that water molecules play a significant role in this key step of the photocycle. The X-ray structure fails to suggest a suitable catalytic basic group in the vicinity of N5. The back reaction has been measured in almost a dozen different LOV domains, with rates varying from seconds to minutes to no return of the ground state (Salomon et al 2000, Kasahara et al 2002, Kottke et al 2002, Losi et al 2002, Imaizumi et al 2003, Schwerdtfeger and Linden 2003). The sequences of these various LOV domains present no clear insight into the varying relaxation rates. Difference FTIR spectra show hydrogen-bonding perturbations associated with the two buried waters that are in close proximity to the chromophore (Iwata et al 2003), suggesting that these waters are possibly involved in the photochemistry. They may form a proton-conducting channel to a basic group not in the immediate vicinity of the chromophore, a mechanism that has been found in other proteins.

Conclusions

Both formation and decay of the flavin–cysteinyl adduct are rate limited by either localized intramolecular proton/hydrogen-atom transfer events and/or hydrogen bond perturbation. Local hydration plays a major role in adduct relaxation, but its mechanistic significance is unclear. Mechanisms invoked for either reaction must be consistent with these observations.

References

Ataka K, Hegemann P, Heberle J (2003) Vibrational spectroscopy of an algal Phot-LOV1 domain probes the molecular changes associated with blue-light reception. Biophys J 84: 466–474

Bittl R, Kay C, Weber S, Hegemann P (2003) Characterization of a flavin radical product in a C57M mutant of a LOV1 domain by electron paramagnetic resonance. Biochemistry 42: 8506–8512

Briggs WR, Christie JM (2002) Phototropins 1 and 2: versatile plant blue-light receptors. Trends Plant Sci 7: 204–210

Corchnoy SB, Swartz TE, Lewis JW, Szundi I, Briggs WR, Bogomolni RA (2003) Intramolecular proton transfers and structural changes during the photocycle of the LOV2 domain of phototropin 1. J Biol Chem 278: 724–731

Crosson S, Moffat K (2001) Structure of a flavin-binding plant photoreceptor domain: Insights into light-mediated signal transduction. Proc Natl Acad Sci USA 98: 2995–3000

Crosson S, Moffat K (2002) Photoexcited structure of a plant photoreceptor domain reveals a light-driven molecular switch. Plant Cell 14: 1067–1075

Fedorov R, Schlichting I, Hartmann E, Domratcheva T, Fuhrmann M, Hegemann P (2003) Crystal structures and molecular mechanism of a light-induced signaling switch: The phot-LOV1 domain from *Chlamydomonas reinhardtii*. Biophys J 84: 2474–2482

Harper S, Neil L, Gardner K (2003) Structural basis of a phototropin light switch. Science 301: 1541–1544

Imaizumi T, Tran H, Swartz TE, Briggs WR, Kay SA (2003) FKF1 is essential for photoperiodic-specific light signalling in Arabidopsis. Nature 426: 302–306

Iwata T, Nozaki D, Tokutomi S, Kagawa T, Wada M, Kandori H (2003) Light-induced structural changes in the LOV2 domain of *Adiantum* phytochrome3 studied by low-temperature FTIR and UV-visible spectroscopy. Biochemistry 42: 8183–8191

Kasahara M, Swartz TE, Olney MA, Onodera A, Mochizuki N, Fukuzawa H, Asamizu E, Tabata S, Kanegae H, Takano M, Christie JM, Nagatani A, Briggs WR (2002) Photochemical properties of the flavin mononucleotide-binding domains of the phototropins from Arabidopsis, rice, and *Chlamydomonas reinhardtii*. Plant Physiol 129: 762–773

Kay CW, Schleicher E, Kuppig A, Hofner H, Rüdiger W, Schleicher M, Fischer M, Bacher A, Weber S, Richter G (2003) Blue light perception in plants—Detection and characterization of a light-induced neutral flavin radical in a C450A mutant of phototropin. J Biol Chem 278: 10973–10982

Kennis J, Crosson S, Gauden M, van Stokkum I, Moffat K, van Grondelle R (2003) Primary reactions of the LOV2 domain of phototropin, a plant blue-light photoreceptor. Biochemistry 42: 3385–3392

Kottke T, Heberle J, Hehn D, Dick B, Hegemann P (2002) phot-LOV1: Photocycle of a blue-light receptor domain from the green alga *Chlamydomonas reinhardtii*. Biophys J 84: 1192–1201

Kottke T, Dick B, Fedorov R, Schlichting I, Deutzmann R, Hegemann P (2003) Irreversible photoreduction of flavin in a mutated phot-LOV1 domain. Biochemistry 42: 9854–9862

Losi A, Polverini E, Quest B, Gärtner W (2002) First evidence for phototropin-related blue-light receptors in prokaryotes. Biophys J 82: 2627–2634

Melander L, Saunders W (1980). Reaction rates of isotopic molecules. Wiley, New York.

Müller F (1991). Chemistry and biochemistry of flavoenzymes. CRC, Boca Raton.

Neiss C, Saalfrank P (2003) Ab initio quantum chemical investigation of the first steps of the photocycle of phototropin: a model study. Photochem Photobiol 77: 101–109

Salomon M, Christie JM, Knieb E, Lempert U, Briggs WR (2000) Photochemical and mutational analysis of the FMN-binding domains of the plant blue light receptor, phototropin. Biochemistry 39: 9401–9410

Salomon M, Eisenreich W, Durr H, Schleicher E, Knieb E, Massey V, Rüdiger W, Müller F, Bacher A, Richter G (2001) An optomechanical transducer in the blue light receptor phototropin from *Avena sativa*. Proc Natl Acad Sci USA 98: 12357–12361

Schuttrigkeit T, Kompa C, Salomon M, Rüdiger W, Michel-Beyerle M (2003) Primary photophysics of the FMN binding LOV2 domain of the plant blue light receptor phototropin of *Avena sativa*. Chem Phys 294: 501–508

Schwerdtfeger C, Linden H (2003) VIVID is a flavoprotein and serves as a fungal blue light photoreceptor for photoadaptation. EMBO J 22: 4846–4855

Song PS (1968) On the basicity of the excited state of flavins. Photochem Photobiol 7: 311–313

Swartz TE, Corchnoy SB, Christie JM, Lewis JW, Szundi I, Briggs WR, Bogomolni RA (2001) The photocycle of a flavin-binding domain of the blue light photoreceptor phototropin. J Biol Chem 276: 36493–36500

Swartz TE, Wenzel PJ, Corchnoy SB, Briggs WR, Bogomolni RA (2002) Vibration spectroscopy reveals light-induced chromophore and protein structural changes in the LOV2 domain of the plant blue-light receptor phototropin 1. Biochemistry 41: 7183–7189

Yagi K, Ohishi N, Nishimoto K, Choi JD, Song PS (1980) Effect of hydrogen bonding on electronic spectra and reactivity of flavins. Biochemistry 19: 1553–1557

Chapter 17

Vibrational Spectroscopy Explores the Photoreaction of LOV Domains

Joachim Heberle

Introduction

Plant phototropins (phot) exhibit a peculiar architecture such that two LOV domains (LOV1 and LOV2) are present in a tandem-like arrangement (N-terminal part of the protein), followed by a C-terminal serine/threonine kinase domain that catalyzes the light-driven autophosphorylation of phot (Huala et al 1997). The three-dimensional structure of the LOV1 domain from the green algae *Chlamydomonas reinhardtii* (Fedorov et al 2003) and the LOV2 domain from the fern *Adiantum capillus-veneris* (Crosson and Moffat 2001, 2002) have both been determined in the dark state as well as in the corresponding long-lived intermediate state.

Up to now, biophysical techniques like X-ray crystallography (Crosson and Moffat 2001, 2002, Fedorov et al 2003), nuclear magnetic resonance (Salomon et al 2001, Harper et al 2003), electron paramagnetic resonance (Kay et al 2003, Bittl et al 2003), vibrational (Swartz et al 2002, Ataka et al 2003, Iwata et al 2003), and visible spectroscopy (Salomon et al 2000, Swartz et al 2001, Kottke et al 2003, Kennis et al 2003) have only been applied to isolated LOV domains and to the tandem-array LOV1–LOV2 (Kasahara et al 2002), mostly due to the poor solubility of full length phot. In contrast, the prokaryotic LOV-protein YtvA from *Bacillus subtilis* (Akbar et al 2001) can be generated in agreeable yields as a full-length protein. Therefore, YtvA allows for biophysical studies of the transfer of the light-induced changes taking place in the photoreceptor domain to the downstream signaling domain.

The Photocycle of LOV1

The LOV domains absorb blue light with the absorbance maximum at around 450 nm and vibronic side bands at 425 and 475 nm. After light excitation, the LOV domains exert a self-contained photocycle (Kasahara et al 2002, Salomon et al

Forschungszentrum Jülich, IBI-2: Structural Biology, D-52425 Jülich, Germany

2001, Swartz et al 2001). The excited singlet state of flavin mononucleotide (FMN) decays by intersystem crossing (ISC) to the triplet state. The triplet is formed with a time constant of 3.3 ns (Kennis et al 2003) and its absorbance is strongly red-shifted (λ_{max} = 710 nm in LOV1 from *Chlamydomonas reinhardtii* and 660 nm in LOV2 from oat). This state decays in the early microsecond domain to a strongly blue-shifted intermediate (λ_{max} = 390 nm), which has been suggested to represent the signaling state (Swartz et al 2001). In LOV-390, the cysteine residue in the chromophore binding pocket is covalently linked to FMN via the C(4a) to form the thio-photoproduct (Crosson and Moffat 2002, Fedorov et al 2003, Salomon et al 2001). A radical-pair mechanism for the formation of the FMN-C(4a)–cysteinyl adduct has been proposed (Kay et al 2003). The time-constant for the subsequent transition of LOV1–390 back to the initial dark state LOV1–447 is in the range of minutes and is base-catalyzed (Kottke et al 2003).

FT-IR Difference Spectroscopy

The described chemical transitions during the light-activated turnover of the LOV domain can be investigated in detail by vibrational spectroscopy. Whereas resonance-Raman spectroscopy selectively probes the vibrational modes of the chromophore, infrared spectroscopy detects the vibrations of the entire protein. To resolve single vibrations from the manifold of vibrations of a protein, reaction-induced Fourier transform infrared (FT-IR) difference spectroscopy is employed (Vogel and Siebert 2000, Nyquist et al 2004). For light-sensitive proteins, the infrared spectrum of the dark state is referenced against the spectrum of the illuminated state. The light-induced FT-IR difference spectrum between the long-lived intermediate (LOV1–390) and the ground state (LOV1–447) is depicted in Figure 1. The convention is that negative bands correspond to the ground state whereas positive bands are due to vibrations of the intermediate state.

The S–H Vibration of the Reactive Cysteine

The reactivity of the sulfur atom of the reactive cysteine is probed by the S–H stretching vibration which appears isolated from other vibrational modes (Bare et al 1975, Moh et al 1987). The disappearance of a band at around 2570 cm^{-1} (Figure 2, left panel) has been assigned to the deprotonation of C57 in the LOV1–390 intermediate (Ataka et al 2003). Since this band has been observed also in the LOV domains from various hosts (Iwata et al 2002, 2003, Bednarz et al 2004), it is evident that the reactive cysteine is protonated prior to the reaction with FMN. This result refutes the suggestion based on fluorescence spectroscopy that the thiolate of the corresponding cysteine is the reactive species (Swartz et al 2001).

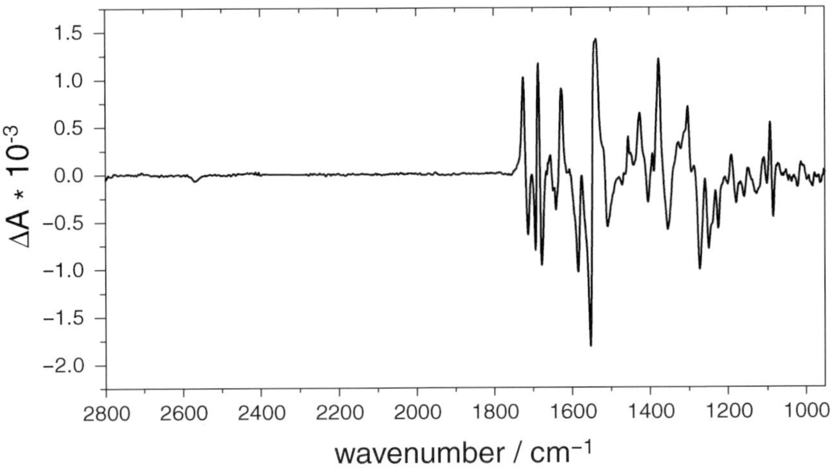

FIG. 1. Light-induced FT-IR difference spectrum of LOV1 associated with the transition from the dark state of the LOV1 domain to the long-lived photointermediate LOV1–390 upon excitation with blue light. The very slow decay time (τ = 150 s) under the applied conditions (Kottke et al 2003) leads to the accumulation of sufficient molecules in the intermediate state to be detectable by infrared difference spectroscopy

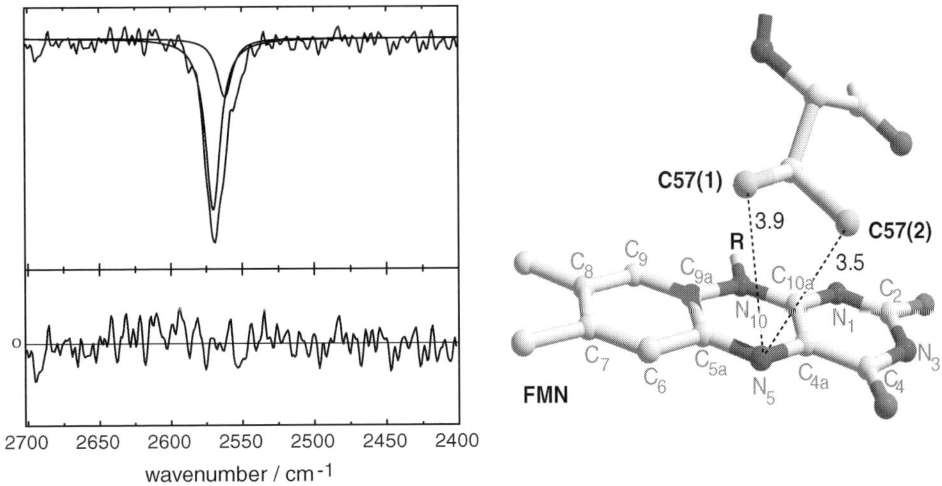

FIG. 2. **Left panel** Lorentzian fitting to the IR difference spectrum of LOV1 (*red trace*) in the S–H stretching region. Residuals of the fit are shown at the bottom. **Right panel** Crystallographic structure of the chromophore FMN in the LOV1 domain from *Chlamydomonas reinhardtii*. *R* corresponds to the phosphorylated ribityl side chain of FMN. The side chain of the reactive cysteine is shown in two different rotamer conformations (C57(1) and C57(2)). *Dashed lines* indicate the distance (in Å) between the cysteine sulfur and the N_5 of the FMN moiety. Atom coordinates have been taken from the protein data bank (entry code: 1N9L (Fedorov et al 2003))

Careful inspection of the S–H stretching band reveals that the band shape for the S–H stretching band of C57 of LOV1 is asymmetric (Bednarz et al 2004). The fit is only satisfactory when two Lorentzians are used (see residuals). The resulting minima are at 2570 cm^{-1} and 2562 cm^{-1}. The detection of two different S–H bands is in accordance with X-ray crystallographic results (Fedorov et al 2003) where two different rotamer conformations of C57 have been determined for the dark state of LOV1 (Figure 2, right panel). Since the FT-IR experiments have been performed under ambient conditions, the two rotamer conformations found by X-ray crystallography at cryogenic temperature are of functional relevance.

FMN Vibrations and the Protein Environment

The formation of the covalent bond between C57 and FMN results in vibrational changes of the chromophore. Therefore, the light-induced difference spectrum in the frequency range of 1750–950 cm^{-1} shows a manifold of difference bands (Figure 1). Most of the difference bands arise from the chromophore because FMN is the strongest dipole of the protein. However, vibrational contributions from the surrounding protein moiety can also be expected.

The comparison of the resonance-Raman (RR) spectrum with the light-induced infrared difference spectrum provides a means to assign the chromophore bands (Ataka et al 2003). The selection rules usually exclude the observation of a particular vibration by Raman scattering if it is infrared active, and vice versa. This principle holds true for small molecules but is violated when coupled vibrations are studied such as occur in large molecules. In the latter case, the same vibration is observable by both techniques, albeit with different intensity.

While the RR spectrum of the intermediate state can be conveniently observed, it is virtually impossible to record the spectrum of the dark state of LOV under fully resonant conditions. The long cycling times of blue-light photoreceptors inevitably leads to the accumulation of the intermediate state. To prevent this, pre-resonant conditions have been employed to record the vibrations of the FMN in the ground state of LOV (Ataka et al 2003). As an alternative, the mutant C57S of LOV1 can be used which is unable to form the long-lived intermediate.

The spectra reveal that many of the bands in the infrared difference spectrum (Figure 1) coincide with the Raman bands (data not shown). Consequently, they can be assigned to chromophore vibrations state of LOV1. The other vibrational bands in the IR spectrum include contributions from the protein environment, i.e., from the protein backbone and the amino acid side chains.

It is certainly not the goal of this paper to present the molecular assignment of all of the vibrational bands (the reader is referred to Swartz et al 2002 and Ataka et al 2003 for part of this discussion), which would also require further experimental (use of isotopomers) and theoretical approaches (normal mode analysis). It should rather stressed at this point that, with the assignment at hand, the chemical transformations can be monitored in great detail which follow after light-excitation. Finally, this will lead to an atomistic picture of the molecular mechanism of such a nano-machine.

Signal Transduction to the Downstream Effector

Up to this point, we focussed on the molecular description of the reactions in the light-absorbing domain. The molecular observation of the signal transfer from the LOV domain to the effector (mostly kinases) is usually hampered by the fact that the full-length protein is isolated in too low yield for biophysical studies. YtvA from *Bacillus subtilis* is superior because it is at present the only phot which is functionally expressed and soluble in full length. Thus, YtvA represents a good model system to understand the molecular mechanism by which light-induced reactions regulate intradomain communication in photoreceptor proteins (Losi et al 2003).

Figure 3 shows the spectral comparison of the IR difference spectra of the full-length YtvA (continuous trace) and the isolated LOV domain from YtvA (dashed trace). Overall, both spectra look very similar with the exception that the full-length protein exhibits stronger difference bands in the regions around $1655 \pm 35\,cm^{-1}$ and $1540 \pm 30\,cm^{-1}$. These represent conformational changes of the protein backbone which are larger in the full-length protein than in the isolated LOV domain. We conclude that the additional difference bands relate to the presence of the STAS domain in the full length YtvA and indicate that upon light excitation structural changes are transmitted from the chromophore binding part to the signaling domain.

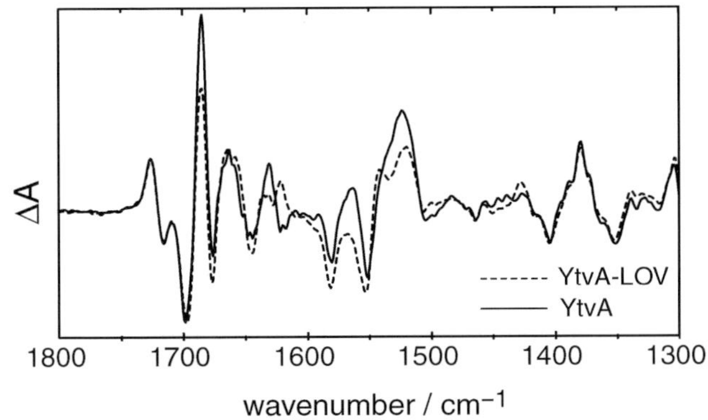

FIG. 3. IR difference spectrum of full-length YtvA from *Bacillus subtilis* (*continuous trace*). For comparison, the difference spectrum of the isolated LOV domain (YtvA-LOV, *dashed trace*) is overlaid

Summary and Conclusion

Vibrational spectroscopy represents a powerful tool to investigate the molecular mechanism of blue-light photoreceptors. It has been demonstrated, using the LOV1 domain as a model system, how the results from the combination of FT-IR and resonance-Raman spectroscopy lead to mechanistic insights not attainable by any other method. The knowledge about the three-dimensional structure is very helpful in the interpretation of the reaction mechanism, albeit not mandatory. What is still missing for a comprehensive understanding of this molecular machine is the reaction dynamics. State-of-the-art vibrational techniques that include time-resolved approaches are available, and such studies are currently underway.

It is obvious that the presented methodology can be applied to other members of the fast growing family of blue-light sensors. So, there is a bright future for vibrational spectroscopy with many unknown landscapes to explore—surprises not excluded!

References

Akbar S, Gaidenko TA, Kang CM, O'Reilly M, Devine KM, Price CW (2001) New family of regulators in the environmental signaling pathway which activates the general stress transcription factor sigma(B) of *Bacillus subtilis*. J Bacteriol 183: 1329–1338

Ataka K, Hegemann P, Heberle J (2003) Vibrational spectroscopy of an algal phot-LOV1 domain probes the molecular changes associated with blue-light reception. Biophys J 84: 466–474

Bare GH, Alben JO, Bromberg PA (1975) Sulfhydryl groups in hemoglobin. A new molecular probe at the $\alpha 1$ $\beta 1$ interface studied by Fourier transform infrared spectroscopy. Biochemistry 14: 1578–1583

Bednarz T, Losi A, Gärtner W, Hegemann P, Heberle J (2004) Functional variations among LOV domains as revealed by FT-IR difference spectroscopy. Photochem Photobiol Sci 3: 575–579

Bittl R, Kay CWM, Weber S, Hegemann P (2003) Characterization of a flavin radical product in a C57M mutant of a LOV1 domain by electron paramagnetic resonance. Biochemistry 42: 8506–8512

Crosson S, Moffat K (2001) Structure of a flavin-binding plant photoreceptor domain: insights into light-mediated signal transduction. Proc Natl Acad Sci USA 98: 2995–3000

Crosson S, Moffat K (2002) Photoexcited structure of a plant photoreceptor domain reveals a light-driven molecular switch. Plant Cell 14: 1067–1075

Fedorov R, Schlichting I, Hartmann E, Domratcheva T, Fuhrmann M, Hegemann P (2003) Crystal structures and molecular mechanism of a light-induced signaling switch: The phot-LOV1 domain from *Chlamydomonas reinhardtii*. Biophys J 84: 2474–2482

Harper SM, Neil LC, Gardner KH (2003) Structural basis of a phototropin light switch. Science 301: 1541–1544

Huala E, Oeller PW, Liscum E, Han IS, Larsen E, Briggs WR (1997) Arabidopsis NPH1: a protein kinase with a putative redox-sensing domain. Science 278: 2120–2123

Iwata T, Tokutomi S, Kandori H (2002) Photoreaction of the cysteine S–H group in the LOV2 domain of *Adiantum* phytochrome3. J Am Chem Soc 124: 11840–11841

Iwata T, Nozaki D, Tokutomi S, Kagawa T, Wada M, Kandori H (2003) Light-induced structural changes in the LOV2 domain of *Adiantum* phytochrome3 studied by low-temperature FTIR and UV-visible spectroscopy. Biochemistry 42: 8183–8191

Kasahara M, Swartz TE, Olney MA, Onodera A, Mochizuki N, Fukuzawa H, Asamizu E, Tabata S, Kanegae H, Takano M, Christie JM, Nagatani A, Briggs WR (2002) Photochemical properties of the flavin mononucleotide-binding domains of the phototropins from *Arabidopsis*, rice, and *Chlamydomonas reinhardtii*. Plant Physiol 129: 762–773

Kay CW, Schleicher E, Kuppig A, Hofner H, Rüdiger W, Schleicher M, Fischer M, Bacher A, Weber S, Richter G (2003) Blue light perception in plants. Detection and characterization of a light-induced neutral flavin radical in a C450A mutant of phototropin. J Biol Chem 278: 10973–10982

Kennis JT, Crosson S, Gauden M, van Stokkum IH, Moffat K, van Grondelle R (2003) Primary reactions of the LOV2 domain of phototropin, a plant blue-light photoreceptor. Biochemistry 42: 3385–3392

Kottke T, Heberle J, Hehn D, Dick B, Hegemann P (2003) Phot-LOV1: Photocycle of a blue-light receptor domain from the green alga *Chlamydomonas reinhardtii*. Biophys J 84: 1192–1201

Losi A, Quest B, Gärtner W (2003) Listening to the blue: the time-resolved thermodynamics of the bacterial blue-light receptor YtvA and its isolated LOV domain. Photochem Photobiol Sci 2: 759–766

Moh PP, Fiamingo FG, Alben JO (1987) Conformational sensitivity of β-93 cysteine SH to ligation of hemoglobin observed by FT-IR spectroscopy. Biochemistry 26: 6243–6249

Nyquist RM, Ataka K, Heberle J (2004) The molecular mechanism of membrane proteins probed by evanescent infrared waves. Chem Bio Chem 5: 431–436

Salomon M, Christie JM, Knieb E, Lempert U, Briggs WR (2000) Photochemical and mutational analysis of the FMN-binding domains of the plant blue light receptor, phototropin. Biochemistry 39: 9401–9410

Salomon M, Eisenreich W, Dürr H, Schleicher E, Knieb E, Massey V, Rüdiger W, Müller F, Bacher A, Richter G (2001) An optomechanical transducer in the blue light receptor phototropin from *Avena sativa*. Proc Natl Acad Sci USA 98: 12357–12361

Swartz TE, Corchnoy SB, Christie JM, Lewis JW, Szundi I, Briggs WR, Bogomolni RA (2001) The photocycle of a flavin-binding domain of the blue light photoreceptor phototropin. J Biol Chem 276: 36493–36500

Swartz TE, Wenzel PJ, Corchnoy SB, Briggs WR, Bogomolni RA (2002) Vibration spectroscopy reveals light-induced chromophore and protein structural changes in the LOV2 domain of the plant blue-light receptor phototropin 1. Biochemistry 41: 7183–7189

Vogel R, Siebert F (2000) Vibrational spectroscopy as a tool for probing protein function. Curr Opin Chem Biol 4: 518–523

Chapter 18

LOV Domain-Containing Proteins in *Arabidopsis*

Trevor E. Swartz[1,2], Winslow R. Briggs[2], and Roberto A. Bogomolni[1]

Introduction

Arabidopsis contains five photosensory proteins that use LOV domains as their light-sensing modules. These proteins are phototropin 1 (phot1), phototropin 2 (phot2), FKF1, LKP2, and ZTL (Briggs and Christie 2002). The phototropins are blue-light photoreceptors that mediate phototropism, chloroplast relocation, stomatal opening, leaf expansion, and rapid inhibition of hypocotyl growth (Huala et al 1997, Folta and Spalding 2001, Jarillo et al 2001, Kagawa et al 2001, Kinoshita et al 2001, Sakamoto and Briggs 2002). The phototropins each contain two LOV domains (LOV1 and LOV2) with a serine/threonine kinase at the C-terminal end. The other three LOV domain-containing proteins found in *Arabidopsis* FKF1, LKP2 and ZTL are involved in circadian clock or photoperiod functions (Somers et al 2000, Schultz et al 2001, Imaizumi et al 2003). These three proteins all contain a LOV domain at the N-terminal end followed by an F-box and a kelch repeat domain. The F-box is involved in targeting of a protein for degradation; the kelch-repeat domain is involved in protein-protein interactions. All *Escherichia coli*-expressed and purified *Arabidopsis* LOV domains bind an FMN chromophore and undergo photochemistry after illumination. Here we outline differences and similarities in the sequences/structures and photochemistry of these LOV domains, which presumably have evolved depending on the physiology being mediated.

Alignment of Sequences

Sequence alignment of *Arabidopsis* LOV domains is shown in Figure 1. Overall, there is high homology between all the *At*-LOV domains. In particular, the residues around the reactive cysteine (GXNCRFLQ) and those that interact with

[1] Department of Chemistry and Biochemistry, University of California Santa Cruz, Santa Cruz, CA 95060, USA
[2] Department of Plant Biology, Carnegie Institution of Washington, Stanford, CA 94305, USA

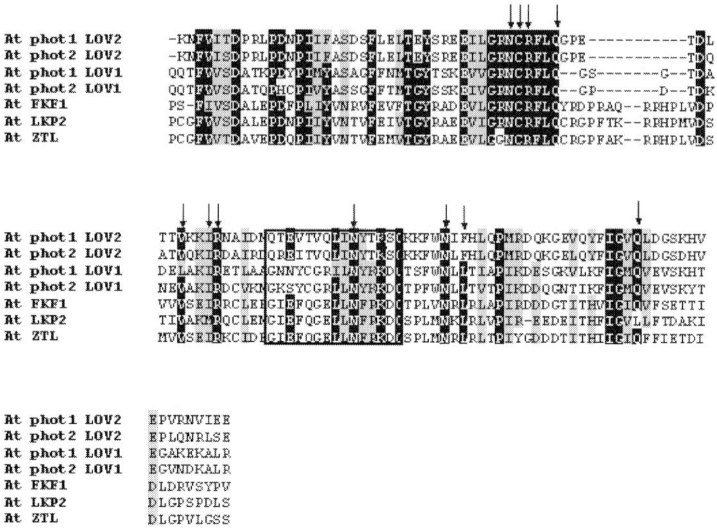

FIG. 1. Sequence alignment of *Arabidopsis*: phot1, phot2, FKF1, LKP2, and ZTL LOV domains. *Arrows* indicate residues that interact with the FMN chromophore. *Conserved area in box* corresponds to the β-sheet that lies next to the isoalloxazine ring of FMN

the FMN chromophore (arrows, Figure 1) are highly conserved (Crosson and Moffat 2001). This is characteristic of all of the LOV domains studied to date (Crosson et al 2003). There are, however, some interesting differences between the phototropin and the ZTL, FKF1, and LKP2 LOV domains. FKF1, LKP2, and ZTL contain an intervening sequence of eight residues appearing directly after GXNCRFLQ that is missing in the phototropin LOV domains. This intervening sequence is clearly seen in the predicted FKF1 structure, which is shown in Figure 2. FKF1, ZTL, and LKP2 also have a sequence of 15 residues—appearing ~30 residues downstream of the reactive cysteine (box in Figure 1)—that is completely conserved. This region is also highly conserved within subfamilies of LOV domains (within LOV1 domains and LOV2 domains). This sequence corresponds in the modeled structure (see below) to the β-sheet that lies directly behind the isoalloxazine of the FMN chromophore. Because this region is only conserved between subfamilies of LOV domains, it is possibly a site of domain-domain interaction within a given protein.

LOV-Domain Structure

The LOV-domain protein fold belongs to the superfamily of PAS-domain proteins. The X-ray crystal structure of *Adiantum* phy3-LOV2 contains 3 α-helices and five β-sheets (Crosson et al 2003). The FMN chromophore fits tightly inside

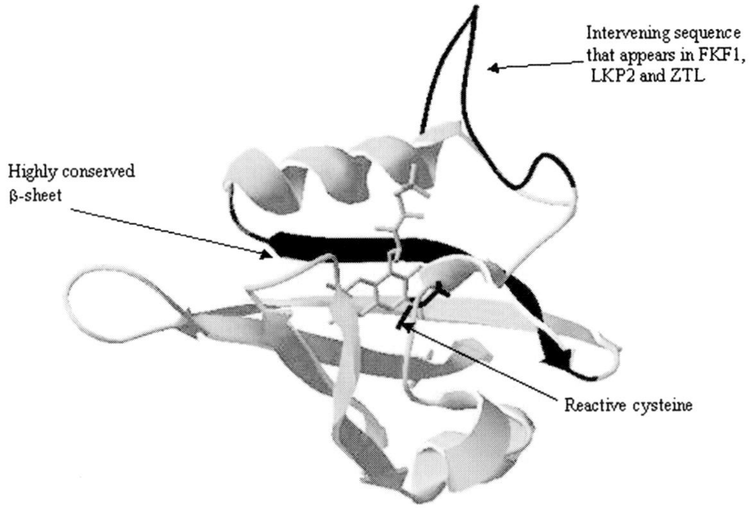

FIG. 2. X-ray structure phy3-LOV2 overlaid with predicted structure of FKF1

a protein pocket and is held in place non-covalently by both hydrogen bonding and hydrophobic interactions (Crosson et al 2003). The photo-reactive cysteine is 4.2 Å from the C(4a) carbon of FMN.

FKF1, LKP2, and ZTL expressed in *E. coli* were all found to bind an FMN chromophore (Imaizumi et al 2003). Using Swiss Model, the structure of ZTL, FKF1, and LKP2 were predicted, based on the crystal structure of phy3-LOV2. The modeled structure of FKF1 is shown in Figure 2. The structure in Figure 2 is an overlay of the phy3-LOV2 structure and the predicted FKF1 structure. The extra sequence found in the sequence alignment is the only part of the overall structure that differs between the phototropin LOV domains and the FKF1, ZTL, and LKP2 LOV domains. In addition, the modeled structures of FKF1, LKP2, and ZTL all place the reactive cysteine in close proximity to the C(4a) carbon of FMN.

LOV-Domain Photochemistry

All the LOV domains bind an FMN chromophore and exhibit an absorption profile very similar to that of FMN (Sakai et al 2001, Imaizumi et al 2003). The LOV domains have maximum absorption peaks at 450 and 370 nm with vibronic structure at 475, 425, 360, and 380 nm. The phototropin LOV domains undergo a light-activated photocycle, characterized by a series of thermally stable intermediates. The characterization of the photocycle of oat-phot1-LOV2 shows at nanoseconds after absorption of light the formation of a red-absorbing species, which resembles the triplet-state of FMN (Swartz et al 2001, Kennis et al 2003).

This intermediate then decays in microseconds to an intermediate that absorbs maximally at 390 nm. This 390 nm-absorbing intermediate, which has been labeled LOV^{S}_{390}, has been shown to involve the formation of a C—S bond between the C(4a) of FMN and the sulfur group of cysteine 39 (Salomon et al 2000, Salomon et al 2001, Crosson and Moffat 2002, Swartz et al 2002, Fedorov et al 2003). Because this intermediate is stable for many seconds, it is presumed to be the signaling state of the LOV domains (Swartz et al 2001). This presumption is supported by evidence that protein structural perturbations necessary for signaling accompany the chromophore structural changes associated with this species (Swartz et al 2002, Corchnoy et al 2003, Harper et al 2003, Iwata et al 2003).

Spontaneous breakage of this bond in the dark results in regeneration of the ground state LOV^{D}_{447} (back-reaction). The half-time of phototropin LOV domains' back-reaction varies from 5 to 29 s, depending on the phototropin LOV domain (*At*-phot1; LOV1 10 s, LOV2 29 s, and *At*-phot2; LOV1 13 s and LOV2 5 s (Kasahara et al 2002)).

LKP2, FKF1, and ZTL all bind an FMN chromophore and have an absorption similar to the phototropin LOV domains. These proteins also undergo a light-activated reaction forming a species with a maximal absorption at 390 nm. The predicted structure and similarities in the spectral properties of the ground state and intermediate state lead to the presumption that this 390 nm-absorbing species involves formation of a cysteinyl adduct. The surprising difference between the photochemistry of the phototropin LOV domains and the FKF1, LKP2, and ZTL LOV domains is that the latter do not exhibit the back-reaction from the adduct state back to the ground state (Imaizumi et al 2003). In these proteins, breakage of the carbon-sulfur bond is not catalyzed by the protein. The mechanism of the back-reaction is not understood, and there are no clear differences between the phototropin and FKF1, LKP2, ZTL LOV domains that explain the altered reaction kinetics. It has been proposed that the back reaction is initialized by removal of the N5 proton by an as yet unidentified base (Swartz et al 2001). Perhaps modulation of the pKa of either the N5 proton [via altering hydrogen bonding on the FMN chromophore (Yagi et al 1980)] or the base that abstracts the proton could alter the kinetics.

Implications of LOV-Domain Photochemistry on Plant Physiology

It is intriguing to relate the LOV-domain photochemistry kinetics to the physiology being mediated by the photoreceptor in an attempt to understand the role of the photoreceptor in these complex-signaling systems. However, one has to be cautious when projecting the function of the full-length proteins in vivo using the reaction kinetics measured for single LOV domains expressed in *E. coli*. The reaction kinetics of the LOV domains in vivo could be altered by a number of factors, including the following: (a) the native chromophore in plants may not be FMN, which the isolated LOV domains scavenge from the *E. coli*, (b) the mature/native protein may be post-translationally modified and/or phosphorylated, (c) in vivo, the photoreceptors are presumably interacting with other proteins, (d) the photocycle kinetics of the full-length proteins may differ from those of the isolated LOV domains.

The last point was demonstrated for the phototropins, in which the photocycle kinetics and quantum yields of adduct formation differed for isolated LOV domains as compared to the same LOV domains within peptides containing both LOV1 and LOV2. These measurements were accomplished by making single point mutations cys39ala, which abolishes photochemistry (Salomon et al 2000, Christie et al 2002), in a particular LOV domain within a construct containing both LOV domains. In these larger peptides, the back reaction kinetics, in which only a single LOV domain was photoactive, were vastly different as compared to isolated LOV domains (Christie et al 2002). The quantum yield of adduct formation was also altered in these peptides as compared to the single LOV domains. For isolated LOV domains, the ratio of quantum yields for LOV1: LOV2 was for phot1 ~1:10 and for phot2 ~1:2; with the peptide containing both LOV domains, the ratio was phot1 ~1:4 and phot2 ~1:2 (Christie et al 2002). Clearly, the photochemistry kinetics and quantum yield of adduct formation are altered in the longer peptides. The back-reaction half-life kinetics of the peptide containing both LOV domains did, however, closely resemble that of the full-length phototropin expressed in insect cells. The back-reaction kinetics for phot1 was 70 and 760 s and for phot2 15 s and a small very long component (Kasahara et al 2002). The faster back-reaction kinetics for phot2 would allow for a lower concentration of protein being in the adduct (signaling state) under continuous illumination as compared to phot1. This is consistent with phot2 modulating a high-light response for both phototropism and chloroplast relocation.

FKF1, LKP2, and ZTL do not show any back-reaction kinetics. These proteins are all involved in clock function. It is intriguing to postulate that proteins involved in clock function do not have a cyclic photoreaction and to consider the implications of this for a sensor feeding into the circadian oscillator. However, because of the reasons outlined earlier, further work on these proteins is necessary before assuming these proteins do not have a functional back-reaction in plants. It will be interesting to find out whether adduct formation in these proteins causes a perturbation in the F-box, resulting in the degradation of these or other proteins.

Photo-Back Reaction

A light-initiated back reaction (photo-back reaction) was first suggested for *Chlamydamonas* phot-LOV1 (Kottke et al 2002). This photo-back reaction was directly measured in oat phot1-LOV2. The reported quantum yield of the photo-back reaction for this isolated LOV domain is ~20%, with restoration of the ground state in picoseconds (Kennis et al 2004). The quantum yield in native proteins and whether the photo-back reaction is physiologically relevant are unknown at this time.

Conclusion

The demonstration and understanding of LOV-domain photochemistry were very important in implicating the phototropins, FKF1, LKP2, and ZTL, as photoreceptors. The next goal is to elucidate how these photoreceptors function in plants. For example, the quantum yield of adduct formation and photocycle kinetics of the phototropin LOV domains have evolved to effectively sense a directional light gradient across a hypocotyl in order to mediate phototropism. The next challenge will be to elucidate the behavior of these LOV domains within full-length native proteins, in order to understand how they mediate these complex responses.

References

Briggs WR, Christie JH (2002) Phototropins 1 and 2: versatile plant blue-light receptors. Trends Plant Sci 7: 204–210

Christie JM, Swartz TE, Bogomolni RA, Briggs WR (2002) Phototropin LOV domains exhibit distinct roles in regulating photoreceptor function. Plant J 32: 205–219

Corchnoy SB, Swartz TE, Lewis JW, Szundi I, Briggs WR, Bogomolni RA (2003) Intramolecular proton transfers and structural changes during the photocycle of the LOV2 domain of phototropin 1. J Biol Chem 278: 724–731

Crosson S, Moffat K (2001) Structure of a flavin-binding plant photoreceptor domain: Insights into light-mediated signal transduction. Proc Natl Acad Sci USA 98: 2995–3000

Crosson S, Moffat K (2002) Photoexcited structure of a plant photoreceptor domain reveals a light-driven molecular switch. Plant Cell 14: 1067–1075

Crosson S, Rajagopal S, Moffat K (2003) The LOV domain family: Photoresponsive signaling modules coupled to diverse output domains. Biochemistry 42: 2–10

Fedorov R, Schlichting I, Hartmann E, Domratcheva T, Fuhrmann M, Hegemann P (2003) Crystal structures and molecular mechanism of a light-induced signaling switch: The Phot-LOV1 domain from *Chlamydomonas reinhardtii*. Biophys J 84: 2474–2482

Folta KM, Spalding EP (2001) Unexpected roles for cryptochrome 2 and phototropin revealed by high-resolution analysis of blue light-mediated hypocotyl growth inhibition. Plant J 26: 471–478

Harper S, Neil L, Gardner K (2003) Structural basis of a phototropin light switch. Science 301: 1541–1544

Huala E, Oeller PW, Liscum E, Han IS, Larsen E, Briggs WR (1997) Arabidopsis NPH1: a protein kinase with a putative redox-sensing domain. Science 278: 2120–2123

Imaizumi T, Tran H, Swartz TE, Briggs WR, Kay SA (2003) FKF1 is essential for photoperiodic-specific light signalling in Arabidopsis. Nature 426: 302–306

Iwata T, Nozaki D, Tokutomi S, Kagawa T, Wada M, Kandori H (2003) Light-induced structural changes in the LOV2 domain of *Adiantum* phytochrome3 studied by low-temperature FTIR and UV-visible spectroscopy. Biochemistry 42: 8183–8191

Jarillo J, Gabrys H, Capel J, Alonso J, Ecker JR, Cashmore AR (2001) Phototropin-related NPL1 controls chloroplast relocation induced by blue light. Nature 410: 952–954
Kagawa T, Sakai T, Suetsugu N, Oikawa K, Ishiguro S, Kato T, Tabata S, Okada K, Wada M (2001) Arabidopsis NPL1: a phototropin homolog controlling the chloroplast high-light avoidance response. Science 291: 2138–2141
Kasahara M, Swartz TE, Olney MA, Onodera A, Mochizuki N, Fukuzawa H, Asamizu E, Tabata S, Kanegae H, Takano M, Christie JM, Nagatani A, Briggs WR (2002) Photochemical properties of the flavin mononucleotide-binding domains of the phototropins from Arabidopsis, rice, and *Chlamydomonas reinhardtii*. Plant Physiol 129: 762–773
Kennis J, Crosson S, Gauden M, van Stokkum I, Moffat K, van Grondelle R (2003) Primary reactions of the LOV2 domain of phototropin, a plant blue-light photoreceptor. Biochemistry 42: 3385–3392
Kennis JTM, van Stokkum IHM, Crosson S, Gauden M, Moffat K, van Grondelle R (2004) The LOV2 domain of phototropin: A reversible photochromic switch. J Am Chem Soc 126: 1412–1413
Kinoshita T, Doi M, Suetsugu N, Kagawa T, Wada M, Shimazaki K (2001) Phot1 and phot2 mediate blue light regulation of stomatal opening. Nature 414: 656–660
Kottke T, Heberle J, Hehn D, Dick B, Hegemann P (2002) Phot-LOV1: Photocycle of a blue-light receptor domain from the green alga *Chlamydomonas reinhardtii*. Biophys J 84: 1192–1201
Sakai T, Kagawa T, Kasahara M, Swartz TE, Christie JM, Briggs WR, Wada M, Okada K (2001) Arabidopsis nph1 and npl1: blue light receptors that mediate both phototropism and chloroplast relocation. Proc Natl Acad Sci USA 98: 6969–6974
Sakamoto K, Briggs WR (2002) Cellular and subcellular localization of phototropin 1. Plant Cell 14: 1723–1735
Salomon M, Christie JM, Knieb E, Lempert U, Briggs WR (2000) Photochemical and mutational analysis of the FMN-binding domains of the plant blue light receptor, phototropin. Biochemistry 39: 9401–9410
Salomon M, Eisenreich W, Durr H, Schleicher E, Knieb E, Massey V, Rüdiger W, Müller F, Bacher A, Richter G (2001) An optomechanical transducer in the blue light receptor phototropin from *Avena sativa*. Proc Natl Acad Sci USA 98: 12357–12361
Schultz TF, Kiyosue T, Yanovsky M, Wada M, Kay SA (2001) A role for LKP2 in the circadian clock of Arabidopsis. Plant Cell 13: 2659–2670
Somers DE, Schultz TF, Milnamow M, Kay SA (2000) ZEITLUPE encodes a novel clock-associated PAS protein from Arabidopsis. Cell 101: 319–329
Swartz TE, Corchnoy SB, Christie JM, Lewis JW, Szundi I, Briggs WR, Bogomolni RA (2001) The photocycle of a flavin-binding domain of the blue light photoreceptor phototropin. J Biol Chem 276: 36493–36500
Swartz TE, Wenzel PJ, Corchnoy SB, Briggs WR, Bogomolni RA (2002) Vibration spectroscopy reveals light-induced chromophore and protein structural changes in the LOV2 domain of the plant blue-light receptor phototropin 1. Biochemistry 41: 7183–7189
Yagi K, Ohishi N, Nishimoto K, Choi JD, Song PS (1980) Effect of hydrogen bonding on electronic spectra and reactivity of flavins. Biochemistry 19: 1553–1557

Chapter 19

Phototropin Phosphorylation

Wolfhart Rüdiger

Introduction

While the phenomenon of phototropism, bending of plants towards or away from a blue light source, has been investigated for more than a century, studies on the photoreceptor phototropin started less than two decades ago with the investigation of the phosphorylation of a plasma membrane-associated protein (reviewed by Short and Briggs 1994). The phosphorylation in vitro was enhanced by irradiating the membranes before addition of [γ-^{32}P]ATP and strongly reduced when the plant tissue was illuminated before membrane isolation. The latter effect, which was in fact the starting observation in this field (Gallagher et al 1988), was later explained by occupation of the potential phosphorylation sites with endogenous phosphate. Several photobiological correlations between the light-inducible phosphorylation and phototropism were established, for example similar tissue distribution, similar fluence–response curves and action spectra, and similar kinetics of recovery in the dark period after a light pulse (Short and Briggs 1994). However, the postulate of a molecular difference between irradiated and shaded cells on unilateral irradiation of phototropic sensitive plant tissue was not substantiated for phototropin phosphorylation in these early investigations.

Phototropin research became focused on areas different from phosphorylation after identification of *PHOT* genes (Huala et al 1997, Jarillo et al 1998). The progress in the knowledge of phototropin photochemistry (see Chapters 16 and 17, respectively) may in part be due to the fact that the chromophore binding LOV domains bind flavin mononucleotide also when expressed in *Escherichia coli* and are thus easily available in large amounts. By contrast, attempts to express full-length phototropin or only the kinase domain of phot1 in *E. coli* resulted only in denatured protein samples (Knieb 2002). The successful expression of the full-length *PHOT1* gene in insect cells transfected with recombinant baculovirus was the key proof for blue light-dependent autophosphorylation of

Department Biologie I (Botanik), Ludwig-Maximilians-Universität, Menzinger Strasse 67, D-80638 Munich, Germany

phototropin (Christie et al 1998); however, most of the phototropin obtained in this way was insoluble and thus not suitable for detailed biochemical investigations. In this review, work on phototropin phosphorylation carried out during the last decade, mainly in our laboratory in Munich, will be summarized.

Detection of Phosphorylation Gradients Within the Oat Coleoptile

Early investigations aimed at finding the subcellular location of the phosphorylated protein required a large amount of plant tissue and time-consuming procedures for membrane isolation. Salomon et al (1997a) developed a micromethod that allowed quantitation of the phosphorylation reaction using the homogenate prepared within 2.5 min from a single 5-mm section of an oat coleoptile. Since the homogenate contained endogenous ATP, carrier-free [γ-^{32}P]ATP was used, added exactly at the beginning of the irradiation with blue light. With this method, an exponential decrease of maximal phosphorylation from tip to base was determined (Salomon et al 1997b). Assuming that the saturating light pulse used in these experiments leads to the same degree of phototropin phosphorylation in all coleoptile sections, the authors concluded that the amount of the photoreceptor decreases exponentially from tip to base. The exponential decrease was later confirmed using immunoblots with an anti-phototropin antiserum while a linear decrease was found in basipetal sections of dicotyledonous seedlings starting from the hypocotyl hook (Knieb et al 2004).

The micromethod was also used to compare illuminated and shaded sides of oat coleoptile tips after unilateral irradiation (Salomon et al 1997a, b). The rationale behind this approach was the partial phosphorylation of phototropin in vivo with endogenous, non-labeled ATP immediately on non-saturating illumination of the tissue such that the occupied phosphorylation sites were not labeled with [γ-^{32}P]ATP in the subsequent in vitro phosphorylation reaction, as mentioned in the Introduction. The full extent of in vitro phosphorylation was achieved with the dark control while no in vitro phosphorylation was observed after saturating illumination of the coleoptiles, and the in vivo phosphorylation after non-saturating illumination was calculated as the complement to the observed in vitro phosphorylation. The experiment showed a clear gradient of in vivo phosphorylation parallel to the light gradient across the coleoptile. Thus phototropin phosphorylation meets the postulate of a molecular difference between irradiated and shaded cells on unilateral irradiation. The fluence–response curve for the difference in phosphorylation between illuminated and shaded side turned out to be a bell-shaped curve like that for the first positive curvature of phototropism; however, the phosphorylation curve is shifted to higher fluence compared with the phototropism curve. Salomon et al (1997b) hypothesized that either the average sensitivity in the investigated tissue might be lower than the sensitivity responsible for bending, given that an exponential decline from tip to base in the amount of photoreceptor (Salomon et al 1997b) and for phototropic response

(Palmer et al 1993) has been found or that, out of multiple phosphorylation sites, the rapidly reacting residues already cause phototropic curvature. This question will be discussed below.

While the light-dependent phosphorylation in the most sensitive tissues, for example the coleoptile tip, adheres to the Bunsen–Roscoe reciprocity law like the first positive curvature of phototropism (Short and Briggs 1994), this is not the case in basal tissue. Salomon et al (1997b) found that a constant fluence of $120\,\mu mol\,m^{-2}$ applied as a unilateral 30-s light pulse did not cause a phosphorylation gradient but did so when the fluence rate was reduced and the illumination period increased by a factor of 10. Such a time-dependent response is characteristic for the second positive curvature of phototropism, and this result allows one to correlate phototropin phosphorylation also with this physiological response. A more detailed investigation showed that "recovery" in the dark period after a first blue light pulse in the coleoptile base did not only restore the original phosphorylation capacity but caused an at least 3-fold amplification of phototropin phosphorylation on a second light pulse and a higher sensitivity of the plants to phototropic stimulation (Salomon et al 1997c). The original assumption that the first light pulse induced synthesis of additional phototropin turned out not to be the case: immunoanalysis showed that the amount of phototropin protein remained constant under these conditions (Knieb et al 2004). Alternative explanations like light-induced increase in kinase activity or removal of inhibitors (Salomon et al 1997c) wait still for experimental verification.

Autophosphorylation of Phototropin at Multiple Sites

As mentioned above, autophosphorylation of phototropin at several sites had already been postulated, and if some sites become phosphorylated in response to very low light while other sites react only at higher fluence, different functions can be assumed. The idea that low fluence sites trigger phototropic bending and high fluence sites trigger other functions was expressed as one possible explanation for the difference in the fluence response of phototropin autophosphorylation compared with that of phototropism (Briggs et al 1996). Early evidence for multiple phosphorylation sites was obtained by exhaustive digestion of phosphorylated phototropin with either V8 or lysyl endopeptidase yielding three major phosphopeptides; however, the relationship to blue light induction was not clear because the same pattern was found for irradiated and non-irradiated membranes and for phosphorylation early (0–2 min) and late (5–20 min) after the blue light pulse (Short et al 1994). Experimental evidence for blue light-dependent multiple phosphorylation sites and their characterization was achieved more recently (Salomon et al 2003). After in vitro phosphorylation and sodium dodecyl sulfate gel electrophoresis, the isolated phototropin was subjected to combined degradation with cyanobromide and trypsin, and the resulting peptides were analyzed by two-dimensional electrophoresis and chromatography. Ten phosphopeptide spots were reproducibly obtained; as it turned out, these spots belong to eight phosphopeptides: two phosphorylation sites gave two peptides each because of incomplete tryptic digestion. The indirect determination of in vivo phosphorylation after illumination of coleoptiles (see above) revealed that four sites were phosphorylated in vivo at low fluence, two at medium and two at high fluence. The recovery of phosphorylation in vivo in the dark period after saturating illumination occurred in the opposite order: the sites that were phosphorylated at high fluence were dephosphorylated prior to the low-fluence sites.

Salomon et al (2003) emphasized that low and high fluence sites were only found in vivo but not for in vitro phosphorylation; apparently it is not an intrinsic property of phototropin but rather depends on its interaction with the membrane or some reaction partners. NPH3, which interacts with phototropin, has been supposed to act as an adapter protein to bring together the reaction partners rather than being a reaction partner itself in the phosphorelay (Motchoulski and Liscum 1999). In an elegant combination of physiological experiments and site-directed mutagenesis, Kinoshita et al (2003) demonstrated that early phosphorylation of phototropin leads to transient binding of a 14-3-3 protein as presumptive step in the signal transduction chain of blue light-dependent opening of stomata of broad bean. The argument for "early" phosphorylation was based on several sets of experiments: the kinetics of binding of the 14-3-3 protein preceded that of maximal phosphorylation of phototropin and, on the other hand, phosphorylation of phototropin was a precondition for binding as shown by inhibition of the phosphorylation in guard cells and by in vitro phosphorylation of recombinant phototropin. Since blue light resulted in binding of the 14-3-3 protein to phototropin also in etiolated seedlings and green leaves, Kinoshita et al (2003) proposed this as general mechanism of phototropin action.

Mapping of the single phosphorylation sites of phototropin became possible when it turned out that the same sites are targets for autophosphorylation and for protein kinase A (Knieb 2002, Salomon et al 2003). Two serine residues of the N-terminal domain and 6 serine residues of the hinge region between LOV1 and LOV2 domains are the phosphorylation sites, no phosphorylation was found within the LOV domains and the kinase domain. The low-fluence sites are those that are closest to the LOV1 domain in the phototropin sequence. The assignment for all phosphorylation sites was confirmed by the respective point mutations (Salomon et al 2003). Kinoshita et al (2003) mapped the phosphorylation site for binding of the 14-3-3 protein by site-directed mutagenesis to be Ser-358 in phototropin1a and Ser-344 for phototropin 1b, both in the hinge region between LOV1 and LOV2.

Phosphorylation Affects Electrophoretic Mobility and Immunoreactivity of Phototropin

One of the early observations in phototropin research was a reduced mobility of the phototropin band in sodium dodecyl sulfate gel electrophoresis after light-induced autophosphorylation, and this was considered to be characteristic of phototropin (Short and Briggs 1994) in spite of the fact that it had been observed with pea seedlings only in vivo but not after in vitro phosphorylation (Short et al 1993) and not at all with maize seedlings (Palmer et al 1993). Later, the shift was observed with *Arabidopsis* phototropin after both in vivo and in vitro phosphorylation (Liscum and Briggs 1995), and a detailed study verified the shift corresponding to an apparent increase in size by 2–3 kDa in all investigated mono- and dicotyledonous plants which included seedlings of oat, maize, pea, mustard, cress, tomato, soybean, and *Eruca* (Knieb et al 2004). A mobility shift was also observed after phosphorylation of the recombinant N-terminal domain and was assigned by site-directed mutagenesis in this case to the phosphorylation at Ser-30 and to a lesser extent at Ser-27: the mutant S27A showed a slightly reduced shift, the mutant S30A almost no shift and the double mutant S27, 30A lacked the shift completely (Salomon et al 2003). The authors assumed that phosphorylation at Ser-27 and Ser-30 was also responsible for the mobility shift of native phototropin, although they could not exclude that phosphorylation at other sites contributed to the magnitude of the shift.

Investigation of the recombinant N-terminal domain of *Avena* phototropin allowed the explanation of another effect (Salomon et al 2003): the immunoreaction of oat phototropin with an antiserum, raised against a peptide fragment containing the N-terminus and the LOV1 domain of *Arabidopsis* phototropin (Christie et al 1998), almost disappears on autophosphorylation. The N-terminal domain of *Avena* phototropin, consisting of amino acids 1 to 127, shows a positive immunoreaction with the same antiserum. This immunoreaction disappears either on phosphorylation at S27 and S30 or by mutation of both serines to alanines (Salomon et al 2003). Of the single mutants, again the S30A mutant shows a stronger effect than the S27A mutant. It is obvious that the free hydroxy-groups of Ser-27 and Ser-30 are part of an important epitope recognized by the antiserum: when they are blocked by phosphorylation, the immunoreaction almost disappears not only in the N-terminal domain but also in full-length phototropin.

Phosphorylation Under UV-C Light

Baskin and Iino (1987) determined the action spectrum in the blue and ultraviolet for phototropism in alfalfa and detected, besides the well-known UV-A and blue light peaks, an additional peak near 280 nm. Since the fluence–response curves below 300 nm differed from those above 300 nm, the authors could not decide whether the newly detected UV-C peak and the peaks in the blue region belonged to the same photoreceptor. Therefore, it was desirable to test whether phototropin autophosphorylates under UV-C light. Several methods were used for this test (Knieb et al, unpublished). The reduction of in vitro phosphorylation with [γ-^{32}P]ATP after illumination of oat coleoptiles with UV-C light at 270 and 300 nm clearly showed in vivo phosphorylation. Since the reduction was smaller than after blue light illumination, the UV-C light is less effective for phototropin phosphorylation than blue light: this corresponds to the action spectrum for phototropic curvature in alfalfa (Baskin and Iino 1987). Further, the mobility shift of phototropin was shown after illumination with UV-C light; by contrast, no reduction in the immunoreaction was observed in the same experiments. This means that autophosphorylation after illumination with UV-C light leaves out the residues Ser-27 and Ser-30 of phototropin: it had been shown that the immunoreactivity requires the free hydroxy-group at Ser-27 and Ser-30 and disappears after phosphorylation at these sites (see above). It is not clear by which mechanism UV-C light activates phototropin differently from UV-A and blue light; the authors speculated about differential activation of single LOV domains via energy transfer from aromatic amino acids. Christie et al (2002) inactivated single LOV domains of phototropins from *Arabidopsis* by mutation of the active-site cysteine to alanine and found that the LOV2 domain was more important for light-dependent autophosphorylation and phototropic curvature than the LOV1 domain. It will now be of considerable interest to compare phototropin autophosphorylation in wild-type plants under UV-C illumination with mutants in which single LOV domains are inactivated.

Summary

Phototropin had been detected as a protein that autophosphorylates in blue light. We showed that the phosphorylation is asymmetric after unilateral illumination of oat coleoptiles, i.e., it is an early biochemical reaction that follows the light gradient necessary for phototropism. We mapped the phosphorylation sites in the phototropin molecule by site-directed mutagenesis and showed differences under low and high fluence. Differential phosphorylation was also found under UV-C and blue light illumination.

References

Baskin TI, Iino M (1987) An action spectrum in the blue and ultraviolet for phototropism in alfalfa. Photochem Photobiol 46: 127–136

Briggs WR, Liscum E, Oeller PW, Palmer JM (1996) Light as an energy source and information carrier in plant physiology. NATO ASI Series, Series A: Life Sciences 287: 159–167

Christie JM, Reymond P, Powell GK, Bernasconi P, Raibekas A, Liscum E, Briggs WR (1998) Arabidopsis NPH1: A flavoprotein with the properties of a photoreceptor for phototropism. Science 282: 1698–1701

Christie JM, Swartz TE, Bogomolni RA, Briggs WR (2002) Phototropin LOV domains exhibit distinct roles in regulating photoreceptor function. Plant J 32: 205–219

Gallagher S, Short TW, Ray PM, Pratt LH, Briggs WR (1988) Light-mediated changes in two proteins found associated with plasma membrane fractions from pea stem sections. Proc Natl Acad Sci USA 85: 8003–8007

Huala E, Oeller PW, Liscum E, Han IS, Larsen E, Briggs WR (1997) *Arabidopsis* NPH1: a protein kinase with a putative redox-sensing domain. Science 278: 2120–2123

Jarillo JA, Ahmad M, Cashmore AR (1998) NPL1 (accession no. AF053941): a second member of the NPH serine/threonine kinase family of Arabidopsis. Plant Physiol 117: 719

Kinoshita T, Emi T, Tominaga M, Sakamoto K, Shigenaga A, Doi M, Shimazaki K-I (2003) Blue-light- and phosphorylation-dependent binding of a 14–3-3 protein to phototropins in stomatal guard cells of broad bean. Plant Physiol 133: 1453–1463

Knieb E (2002) Struktur, Funktion und spektroskopische Eigenschaften der flavin-bindenden Domänen des pflanzlichen Blaulichtrezeptors Phototropin (phot1). Dissertation, University of Munich

Knieb E, Salomon M, Rüdiger W (2004) Tissue-specific and subcellular localization of phototropin determined by immuno-blotting. Planta 218: 843–851

Liscum E, Briggs WR (1995) Mutations in the nph1 locus of *Arabidopsis* disrupts the perception of phototropic stimuli. Plant Cell 7: 473–485

Motchouski A, Liscum E (1999) Arabidopsis NPH3: a NPH1 photoreceptor interacting protein essential for phototropism. Science 286: 961–964

Palmer JM, Short TW, Gallagher S, Briggs WR (1993) Blue light-induced phosphorylation of a plasma membrane-associated protein in *Zea mays* L. Plant Physiol 102: 1211–1218

Salomon M, Zacherl M, Rüdiger W (1997a) Phototropism and protein phosphorylation in higher plants. Unilateral blue light irradiation generates a directional gradient of protein phosphorylation across the oat coleoptile. Bot Acta 110: 214–216

Salomon M, Zacherl M, Rüdiger W (1997b) Asymmetric blue light-dependent phosphorylation of a 116 kDa plasma-membrane protein can be correlated with the first- and second positive phototropic curvature of oat (*Avena sativa* L) coleoptiles. Plant Physiol 115: 485–491

Salomon M, Zacherl M, Luff L, Rüdiger W (1997c) Exposure of oat seedlings to blue light results in amplified phosphorylation of the putative photoreceptor for phototropism and in higher sensitivity of the plants to phototropic stimulation. Plant Physiol 115: 493–500

Salomon M, Knieb E, von Zeppelin T, Rüdiger W (2003) Mapping of low- and high-fluence autophosphorylation sites in phototropin 1. Biochemistry 42: 4217–4225

Short TW, Briggs WR (1994) The transduction of blue light signals in higher plants. Annu Rev Plant Physiol Plant Mol Biol 45: 143–171

Short TW, Reymond P, Briggs WR (1993) A pea plasma membrane protein exhibiting blue light-induced phosphorylation retains photosensitivity following Triton solubilisation. Plant Physiol 101: 647–655

Short TW, Porst M, Palmer J, Fernbach E, Briggs WR (1994) Blue light induces phosphorylation at seryl residues on a pea (*Pisum sativum* L) plasma membrane protein. Plant Physiol 104: 1317–1324

Chapter 20

NPH3 and RPT2: Signal Transducers in Phototropin Signaling Pathways

Tatsuya Sakai

Introduction

Phototropisms of plant organs such as hypocotyls, coleoptiles, and roots are induced by unilateral blue light. These phototropisms result from differential growth, thought to be mediated by asymmetrical distribution of auxin (see Chapter 31 by Iino and Haga). Phototropins 1 and 2 (phot1 and phot2) have been identified as the photoreceptors involved in phototropism of *Arabidopsis thaliana* (see Chapter 15 by Briggs and Chapter 31 by Iino and Haga, respectively). Mutants of Arabidopsis impaired in phototropism, including *nph2*, *nph3*, *nph4*, and *rpt2*, have been analyzed. These studies are providing valuable information on the signaling pathway of phototropism (Liscum 2002). NPH3 and RPT2 have been identified as signal transducers functioning specifically for phototropism. NPH4 has been identified as a transcriptional regulator involved in the control of phototropic and gravitropic differential growth. NPH2 has not yet been analyzed in detail. This article reviews the functions and structural properties of NPH3 and RPT2.

Functions of NPH3 and RPT2 in Phototropism

The *nph3* mutant shows no phototropic response of hypocotyls and roots (Liscum and Briggs 1996). From an in vitro phosphorylation assay of the 120-kDa protein (PHOT1) in the microsomal fraction, NPH3 was thought to be a player in the signaling pathway downstream of phot1 (Motchoulski and Liscum 1999). The *NPH3* gene encodes a protein of 745 amino acid residues. It contains a BTB (*broad-complex*, *tramtrack*, and *bric à brac*)/POZ (pox virus and zinc finger) domain in the N-terminal region and a coiled-coil domain in the C-terminal region, both of which are involved in protein–protein interaction. A yeast two-hybrid study and an in vitro binding assay showed that the C-terminal region

---

RIKEN Plant Science Center, Yokohama, Kanagawa 230-0045, Japan

of NPH3 physically interacts with PHOT1. Cell fractionation studies indicated that NPH3 is plasma-membrane associated (Motchoulski and Liscum 1999), as is phot1 (Briggs and Christie 2002). The above results suggest that NPH3 functions as a scaffold protein downstream of phot1.

The *rpt2* mutant was isolated as a root phototropism mutant (Sakai et al 2000). The in vitro phosphorylation assay of the 120-kDa protein in the microsomal fraction suggested that RPT2 is also involved in the signaling pathway downstream of phot1 (Sakai et al 2000). The hypocotyl of the *rpt2* mutant shows a unique profile of phototropic response. Hypocotyl curvature of the *rpt2* mutant was induced by low-fluence-rate blue light, but the extent of curvature decreased as the fluence rate was increased (Sakai et al 2000). The *phot1 rpt2* double mutant showed a response similar to that of the *phot1* single mutant. Together, these results indicate that loss of RPT2 has no effect on the phot2-mediated phototropism. In addition, the *phot2 rpt2* double mutant showed a response similar to that of the *rpt2* single mutant and had a defect in the phototropic response at $100\,\mu mol\,m^{-2}s^{-1}$ (the *phot1* single mutant showed a normal phototropic response at $100\,\mu mol\,m^{-2}s^{-1}$) (Inada et al 2004). Thus, the mutation of *RPT2* influenced the phot1-mediated phototropism and it is suggested that RPT2 is involved in the phot1-mediated phototropism. In contrast, green light-induced phototropism, for which phot1 is necessary, occurred normally in the *rpt2* mutant, at least at $\leq 5\,\mu mol\,m^{-2}s^{-1}$ (Sakai unpublished). Thus, RPT2 was not required for the green light-induced phot1-mediated phototropism. The requirement for RPT2, therefore, is conditional in phot1-mediated phototropism. There is no evidence, however, that RPT2 is involved in the phot2-mediated phototropism.

RPT2 belongs to the same family as NPH3, and contains 593 amino acid residues with a BTB/POZ domain in the N-terminal region and a coiled-coil domain in the C-terminal region (Sakai et al 2000). A yeast two-hybrid study and a co-immunoprecipitation assay showed that RPT2 binds to phot1 in vivo (Inada et al 2004). RPT2 also binds to NPH3, at least in yeast. A cell fractionation study revealed that, like phot1 and NPH3, RPT2 is partitioned into the microsomal fraction. These results suggest that phot1, RPT2, and NPH3 constitute a plasma membrane-associated complex.

NPH3 is constitutively expressed in seedlings and undergoes some modification in the dark, as the molecular mass of the NPH3 protein from the seedlings was larger than that of an unmodified NPH3 (Motchoulski and Liscum 1999, Liscum 2002). Although one possibility of this modification is phosphorylation, further studies are necessary to elucidate whether that is phosphorylation and whether such modification/de-modification is involved in the regulation of phototropic response. In contract, the expression of RPT2 is induced by blue light irradiation (Sakai et al 2000, Inada et al 2004). Furthermore, there is no evidence on the modification of the RPT2 protein, as the molecular mass of this protein in both conditions of light and dark was the same as that of an unmodified RPT2 synthesized by an in vitro transcription and translation system (Inada et al 2004). Therefore, it appears that functions of NPH3 and RPT2 in the phototropin-signaling pathways are fundamentally different, although they show similar protein structures.

Involvement of RPT2 and NPH3 in Other Blue Light Responses

The phot1 and phot2 photoreceptors function in blue light-activated stomatal opening and chloroplast relocation, in addition to the phototropic response (see Chapters 15, 21 and 22 by Briggs, Wada, and Shimazaki). The involvements of RPT2 and NPH3 in these responses were examined with *rpt2*, *nph3*, and a series of double mutants generated with *phot1* and *phot2* (Inada et al 2004). Stomatal opening induced by blue light irradiation in the *rpt2* mutant was small, as was in *phot1* and *phot2*. Furthermore, the stomata of the *phot2 rpt2* double mutant failed to open in response to blue light as did the *phot1 phot2* double mutant, and the *phot1 rpt2* double mutant showed a similar phenotype with the *phot1* single mutant. In contrast, stomata of *nph3* opened normally, and the *phot1 nph3* and *phot2 nph3* mutants showed the same phenotypes as *phot1* and *phot2*, respectively. These results suggest that RPT2 is involved in the phot1-signaling pathway, but not in the phot2-signaling pathway, for stomatal opening, and that NPH3 does not mediate a blue light-induced stomatal opening. On the other hand, chloroplast movement was normal in the *rpt2* and *nph3* single mutants and the series of double mutants. Also, *rpt2* and *nph3* mutations did not give any phenotype additional to the *phot1* and *phot2* mutants (Inada et al 2004). These results suggest that RPT2 and NPH3 are not involved in chloroplast relocation. Thus, phot1 and phot2 choose different signaling transducers, NPH3, RPT2, and others, to induce phototropic response, stomatal opening, and chloroplast relocation.

Structure and Function of NPH3 and RPT2

NPH3 and RPT2 are BTB/POZ family proteins. The BTB/POZ domain is evolutionarily conserved in eukaryotes. A protein carrying this domain often works as an adapter/scaffold protein with another protein–protein interaction domain. In particular, this domain is owned by transcriptional regulators, actin-binding proteins with kelch domains, and certain adapter proteins of CUL3, a constituent of ubiquitin ligase complexes (Albagli et al 1995, Pintard et al 2004). Analysis of the SMART and InterProScan programs predicted that the *Arabidopsis* genome contains 79 genes encoding proteins with BTB/POZ domains and that many of these proteins possess at least one more protein–protein interaction domain, such as the ankyrin repeat, tetratricopeptide repeat (TPR), armadillo repeat, MATH domain, Zf-TAD domain, WD-40-like domain, and coiled-coil domain (Table 1). Some members in *Arabidopsis* of the BTB/POZ family protein are known as

TABLE 1. The BTB/POZ domain family in *Arabidopsis*

1. RPT2/NPH3 family[a]				
(1) Members with a NPH3 domain[b] and a coiled-coil domain				
At1g03010	At1g30440[c]	At2g14820	At2g30520/RPT2[c]	At2g47860
At3g03510	At3g08570	At3g08660	At3g22104	At3g44820
At3g49970	At3g50840	At4g37590	At5g03250	At5g10250
At5g13600	At5g17580	At5g48130[c]	At5g48800	At5g64330/NPH3
At5g66560				
(2) Members with a NPH3 domain only				
At1g67900	At1g50280[d]	At2g23050	At3g19850	At3g26490
At4g31820	At5g47800	At5g67440		
(3) Member with a coiled-coil domain only				
At3g49900				
2. Members with an ankyrin repeat domain				
At1g64280/NPR1[e]	At2g04740	At2g41370	At3g57130	At4g19660
At4g26120	At5g45110			
3. Members with a MATH (meprin and TRAF homology) domain				
At2g39760	AT3g03740	At3g06190	At3g43700	At5g19000
At5g21010				
4. Members with a Zf-TAZ domain[f]				
At1g05690/ AtBT3	At3g48360/ AtBT2	At4g37610/ AtBT5	At5g63160/ AtBT1	At5g67480/ AtBT4
5. Members with a WD-40-like domain				
At2g24240	At3g09030	At4g30940	At5g41330	
6. Members with an armadillo repeat domain				
At5g13060	At5g19330			
7. Members with a TPR (tetratricopeptide repeat) domain[g]				
At4g02680/EOL1	At3g51770/ETO1		At5g58550/EOL2	
8. Member with a pentapeptide repeat domain[h]				
At5g55000/FIP2				
9. Member with a BSD domain				
At1g21780				
10. Member with the second BTB domain and a F5/8 Type C domain				
At2g30600				
11. Member with a coiled-coil domain				
At4g08455				
12. Others (without a typical conserved domain)				
At1g01640	At1g55750	At1g55760	At1g63850	At2g05330
At2g40440	At2g40450	At2g46260	At3g01790	At3g05675
At3g29740	At3g56230	At3g61600	At4g01160	At4g04090
At4g15840	At5g48510	At5g60050		

[a] Liscum (2002). This family includes two other members, At1g52770 and At3g15570, which contain the NPH3 conserved domain and a coiled-coil domain but not a BTB domain.
[b] The NPH3 domain consists with domains II, III, and IV, described by Liscum (2002).
[c] Coiled-coil domain predicted by MARCOILS (http://www.isrec.isb-sib.ch/webmarcoil/webmarcoilC1.html).
[d] Coiled-coil domain included in the NPH3 domain.
[e] Cao et al (1997).
[f] Du and Poovaiah (2004).
[g] Wang et al (2004).
[h] Banno and Chua (2000).

transcriptional regulators (e.g. NPR1 and AtBT1–5; Cao et al 1997, Du and Poovaiah 2004). A member of this family, ETO1, is known as adapter protein of CUL3 (Wang et al 2004).

The *RPT2/NPH3* family includes 32 genes (Liscum 2002). The structure of a typical protein in this family has a BTB/POZ domain at the N-terminus (30 out of 32 members), a coiled-coil domain at the C-terminus (24 members), and the NPH3-conserved domain between the above two domains (31 members) (Table 1). The BTB/POZ domains of RPT2/NPH3 are distinct from those of other *Arabidopsis* members, and it is difficult to predict the molecular function of the RPT2/NPH3 family only by comparing protein structures. Although RPT2 and NPH3 have typical nuclear localization signals (KPRRRR and RKPRRRWR, respectively) in their C-termini, phot1 localizes at the plasma membrane and in the cytoplasm, not in the nucleus (Sakamoto and Briggs 2002). Furthermore, microarray analysis suggests that phot1 has little activity in transcriptional regulation of blue light-responsive genes (Jiao et al 2003, Ohgishi et al 2004). For the present, there is no evidence that NPH3 and/or RPT2 are involved in the regulation of transcription or protein degradation. On the other hand, in yeast expression system, the N-terminal BTB/POZ-containing region of RPT2 interacts with the N-terminal BTB/POZ-containing region of NPH3. In addition, BTB/POZ domains are known to mediate homodimerization or multimerization for certain proteins (Aravind and Koonin 1999). Therefore the BTB/POZ domain of RPT2/NPH3 might be involved in homodimerization and/or heterodimerization with some member of the RPT2/NPH3 family.

Many questions regarding the structure and function of NPH3 and RPT2 remain unsolved. What is the function of the NPH3 domains conserved in members of the RPT2/NPH3 family? What are functions of other members of this family? How do phot1, phot2, NPH3, and RPT2—all of which lack transmembrane domains—associate with cell membranes? If RPT2 works as a scaffold protein in the phot1-signaling pathway, which protein binds to its C-terminal region? The answers to these questions will provide insights into phototropin-signaling pathways.

Summary

Genetic analysis with *Arabidopsis* mutants has shed light on the molecular mechanisms of light responses of plants. Study of the phototropin-signaling pathway is now expanding from phototropism to other blue light responses, including stomatal opening and chloroplast relocation. NPH3 and RPT2 were identified as signal transducers in the phototropic response. These proteins are able to directly interact with phot1. RPT2 is also found to be involved in the phot1-signaling pathway for stomatal opening. Further genetic and biochemical analyses will reveal the overall network of phototropin signaling pathways and the involvement of RPT2 and NPH3 in this signaling network.

References

Albagli O, Dhordain P, Deweindt C, Lecocq G, Leprince D (1995) The BTB/POZ domain: a new protein-protein interaction motif common to DNA- and actin-binding proteins. Cell Growth Differ 6: 796–815

Aravind L, Koonin EV (1999) Fold prediction and evolutionary analysis of the POZ domain: structural evolutionary relationship with the potassium channel tetramerization domain. J Mol Biol 285: 1353–1361

Banno H, Chua NH (2000) Characterization of the Arabidopsis forming-like protein AFH1 and its interacting protein. Plant Cell Physiol 41: 617–626

Briggs WR, Christie JM (2002) Phototropin 1 and 2: versatile plant blue-light receptors. Trends Plant Sci 7: 204–210

Cao H, Glazebrook J, Clarke JD, Volko S, Dong X (1997) The *Arabidopsis* NPR1 gene that controls systemic acquired resistance encodes a novel protein containing ankyrin repeats. Cell 88: 57–63

Du L, Poovaiah BW (2004) A novel family of Ca^{2+}/calmodulin-binding proteins involved in transcriptional regulation: interaction with fsh/Ring3 class transcription activators. Plant Mol Biol 54: 549–569

Inada S, Ohgishi M, Mayama T, Okada K, Sakai T (2004) RPT2 is a signal transducer involved in phototropic response and stomatal opening by association with phot1. Plant Cell 16: 887–896

Jiao Y, Yang H, Ma L, Sun N, Yu H, Liu T, Gao Y, Gu H, Chen Z, Wada M, Gerstein M, Zhao H, Qu LJ, Deng XW (2003) A genome-wide analysis of blue-light regulation of *Arabidopsis* transcription factor gene expression during seedling development. Plant Physiol 133: 1480–1493

Liscum E (2002) Phototropism: mechanisms and outcomes. In: Somerville CR, Meyerowitz EM (eds) The *Arabidopsis* book. American Society of Plant Biologists, Tockville, MD. http://www.aspb.org/publications/arabidopsis

Liscum E, Briggs WR (1996) Mutations of Arabidopsis in potential transduction and response components of the phototropic signaling pathway. Plant Physiol 112: 291–296

Motchoulski A, Liscum E (1999) *Arabidopsis* NPH3: A NPH1 photoreceptor-interacting protein essential for phototropism. Science 286: 961–964

Ohgishi M, Saji K, Okada K, Sakai T (2004) Functional analysis of each blue light receptor, cry1, cry2, phot1, and phot2 using combinatorial multiple mutants in *Arabidopsis*. Proc Natl Acad Sci USA 101: 2223–2228

Pintard L, Willems A, Peter M (2004) Cullin-based ubiquitin ligases: Cul3–BTB complexes join the family. EMBO J 23: 1681–1687

Sakai T, Wada T, Ishiguro S, Okada K (2000) RPT2: A signal transducer of the phototropic response in *Arabidopsis*. Plant Cell 12: 225–236

Sakamoto K, Briggs WR (2002) Cellular and subcellular localization of phototropin 1. Plant Cell 14: 1723–1735

Wang LC, Yoshida H, Lurin C, Ecker JR (2004) Regulation of ethylene gas biosynthesis by the *Arabidopsis* ETO1 protein. Nature 428: 945–950

Chapter 21

Molecular Mechanism of Blue Light Response in Stomatal Guard Cells

Ken-ichiro Shimazaki

Introduction

Stomata open in response to blue light (BL), and facilitate gas exchanges between plants and atmosphere. The response of stomata is a typical case of blue light responses in higher plants including phototropism, chloroplast relocation, leaf expansion, and rapid inhibition of hypocotyl elongation (Briggs and Christie 2002). Stomatal opening is initiated by the perception of blue light, and finally activates the plasma membrane H^+-ATPase, which drives the opening via the cascade of ion fluxes. Important question is how does BL activate the H^+-ATPase? In this short review, I summarize recent progress on the subjects in stomatal guard cells.

Opening Mechanisms of Stomata

Stomatal pores surrounded by a pair of guard cells in the epidermis regulate gas exchanges between leaves and the atmosphere, and allow CO_2 entry for photosynthesis and transpirational stream in higher plants (Assmann and Shimazaki 1999, Schroeder et al 2001). The opening of stomata is induced by the passive water uptake into a pair of guard cells from the external medium, and the swelling of guard cells causes open stomata via the special orientation of cellulose microfibrils and uneven thickness of cell walls. The water uptake is driven by low water potential in guard cells and the low water potential is generated by the accumulation of potassium salt. The accumulation of K^+-salt in guard cells is mediated through the voltage-gated K^+ channel and is driven by an inside-negative, electrical potential across the plasma membrane. This electrical potential is created by a BL-activated H^+ pump which has been suggested to be H^+-ATPase in the plasma membrane. Creation of negative electrical potential mediated by the H^+-ATPase in response to BL is demonstrated in a single guard cell of *Vicia* intact leaves using double-barreled microelectrode (Roelfsema et al 2001).

Department of Biology, Kyushu University, Ropponmatsu 4-2-1, Fukuoka 810-8560, Japan

Activation Mechanisms of the Plasma Membrane H^+-ATPase

Recent biochemical investigation has elucidated the mechanism by which the perception of BL is transduced into activation of H^+ pump in guard cells (Kinoshita and Shimazaki 1999). The H^+-pump is demonstrated to be the plasma membrane H^+-ATPase. When guard-cell protoplasts (GCPs) were illuminated with a short pulse of BL (30s) under background red light, H^+ is extruded to the medium for more than 15min. To show that this BL-activated H^+ pump is the plasma membrane H^+-ATPase, ATP hydrolysis in response to BL was determined after immediate disruption of GCPs. The rate of H^+ pumping and that of ATP hydrolysis in response to BL revealed the similar kinetic properties, suggesting that BL-dependent H^+ pumping is mediated by the plasma membrane H^+-ATPase.

It is likely that the plasma membrane H^+-ATPase undergoes some stable modification of the protein molecule by BL, and such modified state may have a high activity. To show this, GCPs were incubated with ^{32}P and phosphorylation levels of plasma membrane H^+-ATPase were determined in response to BL (Kinoshita and Shimazaki 1999). The phosphorylation level of H^+-ATPase was increased by BL, and the changes in the phosphorylation levels were closely correlated with those of ATP hydrolytic activities. The results suggest that the activation of the H^+-ATPase is due to its phosphorylation. The phosphorylation occurred exclusively in the C-terminus of the plasma membrane H^+-ATPase on both serine and threonine residues.

A 14-3-3 protein was found to bind to the C-terminus of plasma membrane H^+-ATPase in response to BL, and the amount of bound 14-3-3 protein depended on phosphorylation levels of the H^+-ATPase. Binding of 14-3-3 protein to the H^+-ATPase in the presence of synthetic phosphopeptide designed from the C-terminus was investigated. The presence of phosphopeptide including phospho-T_{950}V peptide (P-950) prevented binding of 14-3-3 protein to the phosphorylated H^+-ATPase, but neither dephosphorylated T-950 nor other phosphopeptides had effect on the binding (Kinoshita and Shimazaki 2002). This result indicates the binding site of 14-3-3 protein localized on the penultimate T-950 in the plasma membrane H^+-ATPase (Figure 1) and is in accord with the previous data provided by genetic analysis (Svennelid et al 1999).

We then investigated whether the binding of a 14-3-3 protein is required for the activation of the H^+-ATPase. When GCPs were illuminated with BL, ATP hydrolytic activity in guard-cell extract increased twofold. However, the addition of P-950 to guard-cell extract completely eliminated the increase in the activities and inhibited 14-3-3 protein binding to the H^+-ATPase. Interestingly, phosphorylation levels of the H^+-ATPase did not change by the addition of P-950. The results indicate that activation of the H^+-ATPase by BL requires the binding of 14-3-3 protein to the phosphorylated C-terminus, and the phosphorylation is not sufficient for the H^+-ATPase activation (Kinoshita and Shimazaki 2002).

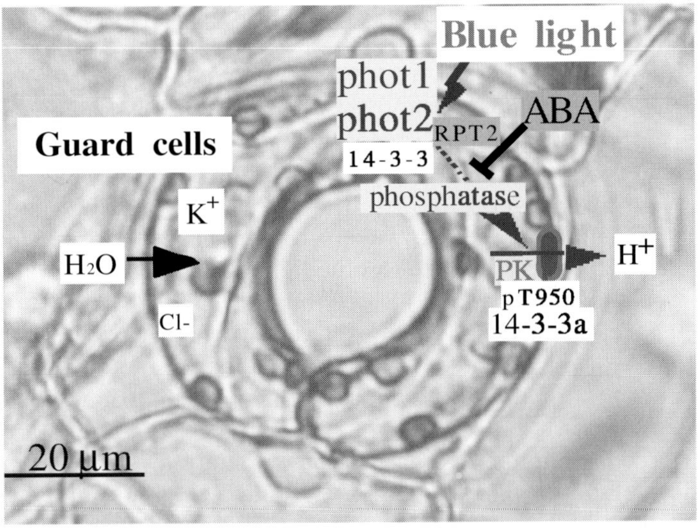

FIG. 1. A proposed model for BL signaling in guard cells. *ABA*, abscisic acid; *PK*, protein kinase; *pT950*, phosphorylated Thr-950 in the plasma membrane H^+-ATPase; *RPT2*, Root Phototropism 2

There are several isoforms of 14-3-3 protein in plants. In *Vicia* guard cells, four 14-3-3 transcripts were expressed, being with order of higher expression levels of *vf* (*Vicia faba*) 14-3-3a, *vf*14-3-3b, *vf*14-3-3c, and *vf*14-3-3d (Emi et al 2001). The isoform that bound to the plasma membrane H^+-ATPase in response to BL was identified as *vf*14-3-3a by the analysis with matrix-assisted laser-desorption ionization time of flight (MALDI-TOF) mass spectrometry (Figure 1). *Vf*14-3-3a possessed a higher binding affinity for the H^+-ATPase than *vf*14 3 3b.

Photoreceptors

Zeaxanthin, one of the carotenoids in the thylakoid membrane, has been proposed as a blue light receptor in guard cells based on several lines of evidences (Zeiger and Zhu 1998). The action spectrum for stomatal opening resembles an absorption spectrum of zeaxanthin. Since BL response of stomata is enhanced by background red light, property of the photoreceptor should explain the phenomena. The amount of zeaxanthin easily increases in guard cells in response to red light, and this increase can be interpreted to increase the sensitivity to blue light. The most direct evidence came from the analysis of *npq* mutant that lacked zeaxanthin. However, conflicting report against this was presented.

Recently, phototropins have been identified as blue light receptors for phototropism (Huala et al 1997, Briggs and Christie 2002) and for chloroplast relocation movement (Kagawa et al 2001, Sakai et al 2001). Using the recombinant proteins of the N-terminal of phototropin including LOV domains, the absorption spectra of the proteins were obtained (Salomon et al 2000). The spectra of the LOV domains of phot1 and phot2 that function as binding sites for the flavin mononucleotide chromophore (Christie et al 1998) closely matched the action spectrum for stomatal opening. However, stomata in *Arabidopsis phot1* mutant were still able to open in response to blue light. A possible explanation for this result could be that phot2 functions redundantly with phot1 in this response, because they have overlapping functions. To clarify this, we used a double mutant with a null allele of phot1 and a chloroplast avoidance mutant *phot2*. An additional mutant, *npq1–2* (non-phototropic quenching), that does not accumulate zeaxanthin was also used.

We determined stomatal responses using epidermal strips from wild-type plants, single mutants of *phot1*, *phot2*, *npq1*, and a double mutant *phot1 phot2* (Kinoshita et al 2001). The stomata opened in wild-type tissue, and in tissues from single mutants of *phot1*, *phot2*, and *npq1* in response to BL. However, stomata did not open in the *phot1 phot2* double mutant. The results suggest that phot1 and phot2 act redundantly as blue light receptors mediating stomatal opening. In support of this, stomata opened widely in the double mutant as well as in other plants when the H^+-ATPase was directly activated by fungal toxin fusicoccin. These results indicate that the simultaneous impairment of *PHOT1* and *PHOT2* genes results in loss of blue light-mediated stomatal opening, probably due to a lack of blue light perception (Figure 1). To confirm the requirement of phototropins in BL responses, *Arabidopsis phot1 phot2* double mutant was transformed with *PHOT1* cDNA. The transformant restored blue-light response of stomata in intact leaves (Doi et al 2004).

Phototropins are most likely to be blue light receptors in stomatal guard cells. Phot1 is localized evenly in the plasma membrane of guard cells of *Arabidopsis* (Sakamoto and Briggs 2002). We obtained biochemical evidence that strongly supports this notion. We used guard-cell protoplasts from *Vicia faba*, which are more suitable for biochemical analysis than those from *Arabidopsis thaliana*, and determined phosphorylation levels of *Vicia* phototropins in response to BL. Vfphots (*V*icia *f*aba phototropins) consist of two homologs of Vfphot1a and Vfphot1b (Kinoshita et al 2003). Vfphots having molecular masses around 125 kDa were phosphorylated by BL. The time required for reaching maximum phosphorylation levels in Vfphots was shorter than that for the plasma membrane H^+-ATPase in response to BL. Furthermore, specific inhibitors, including inhibitors of protein kinases and flavoproteins, inhibited phosphorylation of Vfphot and the same reagents inhibited phosphorylation of the H^+-ATPase to the same extent. These results support that Vfphots function as BL receptors and transduce light signal into activation of the plasma membrane H^+-ATPase, as was evidenced genetically using a *phot1 phot2* double mutant of *Arabidopsis*.

Signaling from Photoreceptor to the Plasma Membrane H^+-ATPase

Signal transduction from phot1 and phot2 to the plasma membrane H^+-ATPase is largely unknown. The physiological role of autophosphorylation of phototropins remains unknown. Recently, we found that 14-3-3 protein bound to Vfphot in response to BL, and the binding was dependent on Vfphot autophosphorylation. This binding of 14-3-3 protein to Vfphot occurred quickly, reaching the maximum level within 1 min and dephosphorylated within 15 min after the pulse of BL (Kinoshita et al 2003). Although the binding was dependent on phosphorylation, a time course of binding of 14-3-3 protein to Vfphot did not coincide with that of phosphorylation levels of Vfphot. This suggests that a part of phosphorylated amino acids is involved in the binding. Since Vfphot-14-3-3 protein complex formation preceded phosphorylation of the plasma membrane H^+-ATPase in guard cells, the complex might be a signaling state that transmits the light signal to the plasma membrane H^+-ATPase (Figure 1).

On the basis of the fact that the 14-3-3 protein binds to its targets in a sequence-specific and phosphorylation-dependent manner (Aitken 2002), we determined phosphorylation site of Vfphot in response to BL. Ser-358 in Vfphot1a and Ser-344 in Vfphot1b were identified as the sites of binding for 14-3-3 protein, and that RRKpS is most likely to be the binding motif for 14-3-3 protein in Vfphots (Kinoshita et al 2003). The site corresponded to one of the identified autophosphorylation sites in *Avena sativa* phototropin (Asphot) by Salomon et al (2003).

Very recently, RPT2 (**R**oot **P**hototropism **2**) was suggested as a signal transducer in both phototropism and stomatal opening in response to BL, but not in chloroplast movement (Inada et al 2004) (Figure 1). A mutant of *rpt2* showed partially decreased BL-dependent stomatal opening in the epidermis of *Arabidopsis*, and the double mutant of *phot2 rpt2* lost the response to BL. RPT2 interacted with phot1 but not phot2. The result suggests that BL signal perceived by phot1 might be transduced via RPT2 into the activation of the plasma membrane H^+-ATPase in guard cells (Figure 1). It is interesting to elucidate the role of RPT2 as a signal molecule in guard cells.

Ca^{2+} may act as a second messenger that relays the signal from phototropin to the plasma membrane H^+-ATPase in guard cells. The requirement of Ca^{2+} for the activation of the plasma membrane H^+-ATPase has been suggested using pharmacological tools (Shimazaki et al 1999). Recently, the increase in cytosolic concentration of Ca^{2+} in mesophyll cells was demonstrated in response to BL, and the increase was mediated mainly by phot1 (Harada et al 2003, Stoelze et al 2003). However, there is no such determination of cytosolic Ca^{2+} in guard cells in response to BL. It is interesting to note that oscillations of cytosolic Ca^{2+} concentration in guard cells are required for stomatal closure (Schroeder et al 2001). The opening response of stomata may require different concentration changes of Ca^{2+}.

A Positive Regulatory Function of Type 1 Protein Phosphatase

Phototropins possess serine/threonine protein kinases domain in the C-terminus, with two repeated motifs of so-called LOV domains that have non-covalent binding sites for the chromophore FMN (Briggs and Christie 2002). When phototropins are illuminated with BL, a covalent cysteinyl adduct with FMN chromophore is formed and results in the activation of protein kinase with subsequent physiological responses (Salomon et al 2000, Briggs and Christie 2002, Crosson et al 2003). It is possible that phototropins directly phosphorylate and activate the plasma membrane H^+-ATPase in response to BL. However, this case is unlikely. The protein kinase activity that directly phosphorylates the plasma membrane H^+-ATPase was less sensitive to protein kinase inhibitors such as K-252a and staurosporine, and such a property was different from that of phototropins (Svennelid et al 1999, Kinoshita et al 2002). Furthermore, we have evidence that a protein phosphatase is likely to be involved in BL signaling as a positive regulator in stomatal guard cells (Assmann and Shimazaki 1999). Protein phosphatase inhibitors suppressed both activation of the plasma membrane H^+-ATPase and stomatal opening in response to BL. More direct evidence will be required to demonstrate this (Figure 1).

Cross-Talk Between BL and Abscisic Acid Signaling

It is very important to prevent water loss via transpiration in the daytime under drought stress because stomata open in the light. Under drought, a higher concentration of abscisic acid (ABA) reaches stomatal guard cells through the transpirational stream and causes stomatal closure by activation of K^+-salt efflux mechanisms (Schroeder et al 2001). However, if the plasma membrane H^+-ATPase keeps its high activity under the light, it favors stomatal opening and consumes extra ATP in guard cells. To avoid this adverse effect on plant, ABA inhibits BL-dependent H^+ pumping and stomatal opening (Assmann and Shimazaki 1999). Abscisic acid inhibits BL-dependent H^+ pumping by 60%–70% in *Vicia* GCPs. Since ABA did not affect H^+ pumping by the H^+-ATPase in isolated microsomal membranes, ABA might affect BL signaling pathway in guard cells (Figure 1).

Summary and Conclusion

Recent investigations on stomatal response by BL have uncovered the mechanism of perception of BL and that of activation in the plasma membrane H^+-ATPase in guard cells. BL receptors are phototropins (phot1 and phot2), and the H^+-ATPase is activated via the exclusive phosphorylation of the C-terminus with subsequent binding of a 14-3-3a protein. The next question is how the BL signal is transduced into activation of the H^+-ATPase. There may be at least several signal molecules between phototropins and the plasma membrane H^+-ATPase although none of them is as yet conclusively identified. The candidates are RPT2, 14-3-3 protein, cytosolic Ca^{2+}, protein phosphatase, and protein kinase, that directly phosphorylate the H^+-ATPase (Figure 1). Moreover, there is cross-talk between BL and ABA signaling at some of these components. Further investigations will elucidate functions and localizations of these components in BL signaling pathways of guard cells in the near future.

References

Aitken A (2002) Functional specificity in 14-3-3 isoform interactions through dimer formation and phosphorylation. Chromosome location of mammalian isoforms and variants. Plant Mol Biol 50: 993–1010

Assmann SM, Shimazaki K (1999) The multisensory guard cell. Stomatal responses to blue light and abscisic acid. Plant Physiol 199: 809–815

Briggs WR, Christie JM (2002) Photropins 1 and 2: versatile plant blue-light receptors. Trends Plant Sci 7: 204–210

Christie JM, Reymond P, Powell GK, Bernasconi P, Raibekas AA, Liscum E, Briggs WR (1998) *Arabidopsis* NPH1: a flavoprotein with the properties of a photoreceptor for phototropism. Science 282: 1698–1701

Crosson S, Rajagopal S, Moffat K (2003) The LOV domain family: Photoresponsive signaling modules coupled to diverse output domains. Biochemistry 42: 2–10

Doi M, Shigenaga A, Emi T, Kinoshita T, Shimazaki K (2004) A transgene encoding a blue-light receptor, phot1, restores blue-light responses in the Arabidopsis *phot1 phot2* double mutant. J Exp Bot 55: 517–523

Emi T, Kinoshita T, Shimazaki K (2001) Specific binding of a vf14-3-3a isoform to the plasma membrane H^+-ATPase in response to blue light and fusicoccin in guard cells of broad bean. Plant Physiol 125: 1115–1125

Harada A, Sakai T, Okada K (2003) phot1 and phot2 mediate blue light-induced transient increases in cytosolic Ca^{2+} differently in *Arabidopsis* leaves. Proc Natl Acad Sci USA 100: 8543–8588

Huala E, Oeller PW, Liscum E, Han IS, Larsen E, Briggs WR (1997) *Arabidopsis* NPH1: a protein kinase with a putative redox-sensing domain. Science 278: 2120–2123

Inada S, Ohgishi M, Mayama T, Okada K, Sakai T (2004) RPT2 is a signal transducer involved in phototropic response and stomatal opening by association with phototropin 1 in *Arabidopsis thaliana*. Plant Cell 16: 887–896

Kagawa T, Sakai T, Suetsugu N, Oikawa K, Ishiguro S, Kato T, Tabata S, Okada K, Wada M (2001) Arabidopsis NPL1: A phototropin homolog controlling the chloroplast high-light avoidance response. Science 291: 2138–2141

Kinoshita T, Shimazaki K (1999) Blue light activates the plasma membrane H^+-ATPase by phosphorylation of the C-terminus in stomatal guard cells. EMBO J 18: 5548–5558

Kinoshita T, Shimazaki K (2002) Biochemical evidence for the requirement of 14-3-3 protein binding in activation of the guard-cell plasma membrane H^+-ATPase by blue light. Plant Cell Physiol 43: 1359–1365

Kinoshita T, Doi M, Suetsugu N, Kagawa T, Wada M, Shimazaki K (2001) phot1 and phot2 mediate blue light regulation of stomatal opening. Nature 414: 656–660

Kinoshita T, Emi T, Tominaga M, Sakamoto K, Shigenaga A, Doi M, Shimazaki K (2003) Blue-light- and phosphorylation-dependent binding of a 14-3-3 protein to phototropins in stomatal guard cells of broad bean. Plant Physiol 133: 1453–1463

Roelfsema MRG, Steinmeyer R, Staal M, Hedrich R (2001) Single guard cell recordings in intact plants: light-induced hyperpolarization of the plasma membrane. Plant J 26: 1–13

Sakai T, Kagawa T, Kasahara M, Swartz TE, Christie JM, Briggs WR, Wada M, Okada K (2001) Arabidospsis nph1 and npl1: Blue light receptors that mediate both phototropism and chloroplasts relocation. Proc Natl Acad Sci USA 98: 6969–6974

Sakamoto K, Briggs WR (2002) Cellular and subcellular localization of phototropin 1. Plant Cell 14: 1723–1735

Salomon M, Christie JM, Knieb E, Lempert U, Briggs WR (2000) Photochemical and mutational analysis of the FMN-binding domains of the plant blue light receptor, phototropin. Biochemistry 39: 9401–9410

Salomon M, Knieb E, von Zeppelin T, Rüdiger W (2003) Mapping of low- and high-fluence autophosphorylation sites in phototropin 1. Biochemistry 42: 4217–4225

Schroeder JI, Allen GJ, Hugouvieux V, Kwak JM, Waner D (2001) Guard cell signal transduction. Annu Rev Plant Physiol Plant Mol Biol 52: 627–658

Shimazaki K, Goh CH, Kinoshita T (1999) Involvement of intracellular Ca^{2+} in blue light-dependent proton pumping in guard cell protoplasts from *Vicia faba*. Physiol Plant 105: 554–561

Stoelze S, Kagawa T, Wada M, Hedrich R, Dietrich P (2003) Blue light activates calcium-permeable channels in Arabidopsis mesophyll cells via the phototropin signaling pathway. Proc Natl Acad Sci USA 100: 1456–1461

Svennelid F, Olsson A, Piotrowski M, Rosenquist M, Ottman C, Larsson C, Oecking C, Sommarin M (1999) Phosphorylation of Thr-948 at the C terminus of the plasma membrane H^+-ATPase creates a biding site for the regulatory 14-3-3 protein. Plant Cell 11: 2379–2391

Zeiger E, Zhu J (1998) Role of zeaxanthin in blue light photoreception and the modulation of light-CO_2 interactions in guard cells. J Exp Bot 49: 433–442

Chapter 22

Chloroplast Photorelocation Movement

Masamitsu Wada

Introduction

Chloroplast photorelocation movement is a well-known phenomenon, which has been investigated since the 19th century and studied physiologically for more than 100 years. It is only very recently that genetic and molecular approaches, such as screening for mutants and transient expression of photomovement-related genes, were introduced to this field. Continuous recording of chloroplasts under the microscope using infrared light have also revealed new aspects to their behavior. Here, we present recent results of chloroplast movement found in *Arabidopsis thaliana*, *Adiantum capillus-veneris*, and *Physcomitrella patens*. Please refer to other reviews for historical background (Haupt 1999, Haupt and Scheuerlein 1990, Wada et al 1993, 2003, Kagawa and Wada 2002, Sato et al 2003a,b, Kasahara and Wada 2004).

Photoreceptors

In most of the cases studied so far, both the chloroplast accumulation response under low fluence rate illumination and the avoidance response under high fluence rate illumination are induced by blue light. In the algae *Mougeotia* and *Mesotaenium*, the moss *Physcomitrella* and *Ceratodon*, and in *Adiantum* and several other fern species, red light was reported to be effective for movement, although this is clearly not the case for the model flowering plant *Arabidopsis*. Screening for mutants deficient in the chloroplast avoidance response was performed using *Arabidopsis* mutagenized by EMS or t-DNA by a newly developed method of partial irradiation of a leaf through a slit with strong light (Kagawa et al 2001, Oikawa et al 2003). Most of the mutants obtained were found to be defective for the phototropin2 gene (*phot2*) and deficient in the chloroplast avoidance

Department of Biological Science, Graduate School of Sciences, Tokyo Metropolitan University, Hachioji, Tokyo 192-0397, Japan; Division of Photobiology, National Institute for Basic Biology, Okazaki, Aichi 444-8585, Japan

response, but surprisingly chloroplast accumulation movement was normal in these mutants (Kagawa et al 2001). Similar results were obtained by a reverse genetic approach (Jarillo et al 2001). Analysis of the *phot1 phot2* double mutant revealed that the accumulation response is mediated by these two phototropins redundantly in *Arabidopsis* (Sakai et al 2001).

Phototropin cDNAs and their genes were cloned in the fern *Adiantum capillus-veneris* (Nozue et al 2000, Kagawa et al 2004) and were named *AcPHOT1* and *AcPHOT2* according to the similarity of the putative amino acid sequences to those of *Arabidopsis* PHOT1 and PHOT2. A phylogenic tree constructed from phototropin sequences available to date reveals that Acphot1 and Acphot2 are orthologs of Atphot1 and Atphot2, respectively. Mutant screening was also performed in fern gametophytes grown from spores mutagenized by heavy ion beam irradiation (nitrogen). Two among 11 mutant lines that did not show the chloroplast avoidance response were defective in the*AcPHOT2* gene (Kagawa et al 2004). As back-crossing of ferns is not easy, we attempted to rescue mutant cells by transient expression of the wild-type *AcPHOT2* gene, and demonstrated that *Ac*phot2 is the photoreceptor for the chloroplast avoidance response in *Adiantum* as well. Unfortunately, *phot1* mutants are not yet available, so it is not yet clear whether both phot1 and phot2 redundantly mediate the accumulation response in this fern.

Site-directed mutagenesis, and both N- and C-terminal deletions in the *Acphot2* gene, have been used in combination with the transient expression assay in *Acphot2* mutant cells to test which amino acids or portions of the molecule are critical to inducing the chloroplast avoidance response (Kagawa et al 2004). Deletion of the 5′ end, from the junction of the LOV1 and LOV2 domains, leaving only LOV2, results in a molecule that still functions to induce the chloroplast avoidance response by strong light irradiation, indicating that the LOV1 domain is not necessary for absorbing strong blue light. When amino acids in the C-terminus, outside of the Ser/Thr kinase domain (between positions 979 and 999) are deleted, function is lost, indicating the importance of this terminal region for Acphot2 function.

In *Adiantum*, the phototropism and chloroplast movement could be induced by red light as well as blue light (see reviews). Red-light aphototropic (*rap*) mutants were isolated from gametophytes of EMS-mutagenized spores and shown to be deficient in red light-induced chloroplast movement (Kadota and Wada 1999), as well as the phototropic phenotype used for screening. *Adiantum* has four phytochrome genes; two conventional phytochromes (phy1 and phy2), one chimeric phytochrome (phy3), which contains the chromophore binding domain of phytochrome in the N-terminus, and a full-length phototropin in the C-terminus (Nozue et al 1998). Phy3 was found to function as the photoreceptor of red light-induced chloroplast movement and phototropism in *Adiantum* (Kawai et al 2003). It is not yet known whether phy3 can function as a blue light receptor using the LOV domain(s) of phototropin-containing C-terminus.

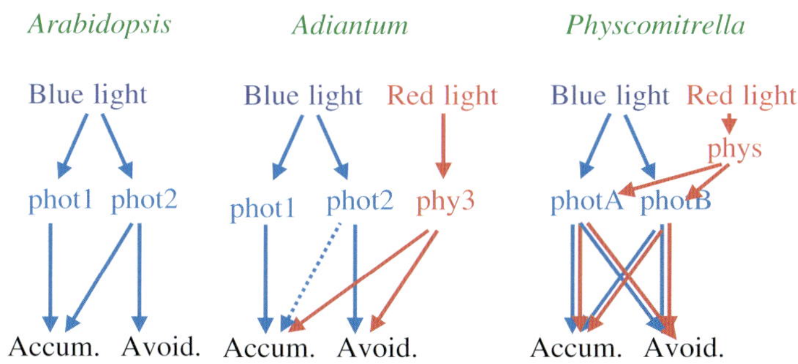

FIG. 1. Photoreceptors and signal transduction pathways for chloroplast photorelocation movement in Arabidopsis, fern *Adiantum*, and moss *Physcomitrella*. *Blue arrows* mean blue light and the pathway of blue-light signals from photoreceptors to physiological responses. *Red arrows* mean the same as those of the blue ones. An *arrow with dotted line* is a plausible case but without evidence. See text for details

In *Physcomitrella*, four phototropin genes were cloned and sequenced (Kasahara et al 2004). photA1 and photA2 share high sequence homology and photB1 and photB2 are also similar to one another. However, the phylogenic relationship showed that the four phototropins form a group independent from phot1 and phot2 of seed plant and fern phototropins, we have named the four phototropins PpphotA1 and A2 and PpphotB1 and B2. Four single, two double, and one triple mutant were created by homologous recombination. The analyses of these *Ppphot* mutants have revealed that both photA and photB function as the photoreceptors of chloroplast avoidance movement in *Physcomitrella*. Very interestingly, red light-induced chloroplast movement, which is mediated by phytochrome, was also significantly affected in these *Ppphot* mutants, indicating that the moss phototropins function downstream of the phytochrome response (Kasahara et al 2004) (Figure 1).

Components of Signal Transduction Pathways

When the central portion of dark-adapted fern gametophyte cells are irradiated continuously with either weak or strong blue light, chloroplasts around the anticlinal wall move towards the light beam. Under weak light, chloroplasts move into the beam-irradiated area, but under strong light they stop at the edge of the beam and do not directly enter into it. When the strong blue light is switched off, the chloroplasts move into the area that was formerly irradiated. These results indicate that (1) the signal of the accumulation response can traverse the distance of the cell, but the one for avoidance movement does not and stays inside the irradiated area, (2) although signals for both the avoidance and accumulation movement are generated under strong blue light, the signal for avoidance movement is dominant to that for accumulation movement during irradiation, and (3) the signal for the accumulation response remains long after irradiation, but that for the avoidance response diminishes immediately. These results suggest that the signals for both responses are different. It would, however, be unlikely that the mechanisms of chloroplast movement for both responses are different. One possibility is that there are additional factors playing a role in the strong light response, perhaps located within the chloroplasts themselves (Wada et al 2003).

Calcium is one of the candidates for a component of signal transduction in chloroplast movement, although no conclusive evidence has been shown (see reviews). Sato et al (1999, 2003a,b) showed that chloroplasts in *Adiantum* and *Physcomitrella* protonemal cells move from or towards the mechanically induced stimuli, respectively. In this system, calcium influx is needed for the movement, however the inhibition of calcium influx by La^{3+} or Gd^{3+} has no influence on the light-induced chloroplast movement (Sato et al 2001b). These results suggest the possible involvement of calcium in chloroplast movement.

Mechanisms of Movement

Chloroplast movement is believed to be mediated by an actomyosin system in general, based on the results obtained by inhibitor treatments. In *Physcomitrella*, however, both actin filaments and microtubules are used in the blue light-mediated chloroplast movement, although only microbutules are used in the red light-mediated chloroplast movement (Sato et al 2001a). In *Adiantum* gametophytes, although no direct evidence has provided unequivocally linking them, a ring-like structure made of actin could be seen when chloroplasts set their position after movement (Kadota and Wada 1992). Recently, dynamic changes in the actin filaments of living cells can be observed under a fluorescent microscope using *Arabidopsis* cells transformed with GFP-talin. Talin is a binding component of actin filaments. Thick filaments, and even thin filaments, fluorescing with GFP can be clearly seen. However, the patterns of actin filament movement and that of the distribution and movement of chloroplasts under blue light have not been correlated. While this suggests that the actin filaments observed are not involved in chloroplast movement, it is possible that very thin filaments, which cannot be seen in our technique, are involved in chloroplast movement. It is also possible that thin filaments, which can be seen during microscopy, have polymerization and depolymerization rates that occur too rapidly for our method of detecting chloroplast movement. The tentative polymerization of actin filaments underneath chloroplasts is also very much plausible, since a chloroplast can move very freely within a short distance, as observed when small parts of a chloroplast are irradiated sequentially with a microbeam of blue or red light (Yamashita et al unpublished). It would be predicted that if chloroplasts move along actin filaments already polymerized, they would only move in one direction due to actin polarity.

We selected a mutant that is deficient in chloroplast photorelocation movement called *chup1* (chloroplast unusual positioning1; Oikawa et al 2003). In this mutant, chloroplasts sediment at the bottom of mesophyll cells. The chup1 protein has an actin-binding domain and has been shown to bind to F-actin. Chup1 also has a hydrophobic region at the N-terminus, and it was shown that this domain is targeted to the chloroplast outer envelope. It is very plausible that chup1, on the chloroplast outer membrane, has a role in the polymerization of F-actin, although we do not yet have experimental evidence.

Significance of Chloroplast Movement

The chloroplast accumulation response is believed to maximize photosynthetic efficiency within the cell and the chloroplast avoidance response to minimize its photodamage. Two mutant lines deficient for the chloroplast avoidance response, i.e., *phot2* and *chup1*, were used to demonstrate that if chloroplasts could not escape from strong light they sustained damaged, cells burst, and leaves become necrotic (Kasahara et al 2002). Hence, chloroplasts should escape from strong light as soon as possible when irradiated with strong light. We tested the relationship between the speed of chloroplast avoidance movement and the intensity of light (Kagawa and Wada 2004). It was shown that the velocity of chloroplast avoidance response is fluence rate dependent under white light. The velocity is also shown to be dependent on the amount of phot2. Progeny that were obtained by crossing wild-type *Arabidopsis* and homozygous *phot2* null mutant plants are heterozygous for the phot2 allele, and thus are predicted to have an intermediate level of the phot2 protein compared to that of wild-type plants. The velocity of chloroplast avoidance movement in heterozygous cells is indeed less than that of wild-type plant cells (Kagawa and Wada 2004).

Summary and Conclusion

Recent screening of mutants deficient in chloroplast photorelocation movement in *Arabidopsis* and *Adiantum*, and the analyses of these mutants, shed light on the photoreceptors and components of signal transduction pathways in this phenomenon. Moreover, the importance of the chloroplast avoidance response for plant survival has also been clarified. Both phot1 and phot2 are the photoreceptors for the accumulation response and only phot2 controls the avoidance response. Red light-induced accumulation responses are mediated by phy3 in *Adiantum*. The chup1 protein is likely localized on the chloroplast outer membrane, and may control the actin polymerization necessary for chloroplast movement. Signal transduction pathways have yet to be clarified, but several components of these pathways are now under study. It is very possible that, with continued effort, the mechanisms of chloroplast movement can be understood within the next five years.

References

Haupt W (1999) Chloroplast movement: from phenomenology to molecular biology. Prog Bot 60: 3–36

Haupt W, Scheuerlein R (1990) Chloroplast movement. Plant Cell Environ 13: 595–614

Jarillo JA, Gabrys H, Capel J, Alonso JM, Ecker JR, Cashmore AR (2001) Phototropin-related NPL1 controls chloroplast relocation induced by blue light. Nature 410: 952–954

Kadota A, Wada M (1992) Photoinduction of formation of circular structures by microfilaments on chloroplasts during intracellular orientation in protonemal cells of the fern *Adiantum capillus-veneris*. Protoplasma 167: 97–107

Kadota A, Wada M (1999) Red light-aphototropic (*rap*) mutants lack red light-induced chloroplast relocation movement in the fern *Adiantum capillus-veneris*. Plant Cell Physiol 40: 238–247

Kagawa T, Wada M (2002) Blue light-induced chloroplast relocation. Plant Cell Physiol 43: 367–371

Kagawa T, Wada M (2004) Chloroplast avoidance movement rate is fluence dependent. Photochem Photobiol Sci 3: 592–595

Kagawa T, Sakai T, Suetsugu T, Oikawa K, Ishiguro S, Kato T, Tabata S, Okada K, Wada M (2001) Arabidopsis NPL1: a phototropin homologue controlling the chloroplast highlight avoidance response. Science 291: 2138–2141

Kagawa T, Kasahara M, Abe T, Yoshida S, Wada M (2004) Function analysis of Acphot2 using mutants deficient in blue light-induced chloroplast avoidance movement of the fern *Adiantum capillus-veneris* L. Plant Cell Physiol 45: 416–426

Kasahara M, Wada M (2004) Chloroplast avoidance movement. In: Moller SG (ed) Plastids, Annual Plant Reviews, volume 13, Blackwell, pp 267–282

Kasahara M, Kagawa T, Oikawa K, Suetsugu N, Miyao M, Wada M (2002) Chloroplast avoidance movement reduces photodamage in plants. Nature 420: 829–832

Kasahara M, Kagawa T, Sato Y, Kiyosue T, Wada M (2004) Phototropins mediate blue and red light-induced chloroplast movements in *Physcomitrella patens*. Plant Physiol 135: 1388–1397

Kawai H, Kanegae T, Christensen S, Kiyosue T, Sato Y, Imaizumi T, Kadota A, Wada M (2003) Responses of ferns to red light are mediated by an unconventional photoreceptor. Nature 421: 287–290

Nozue K, Kanegae T, Imaizumi T, Fukuda S, Okamoto H, Yeh KC, Lagarias JC, Wada M (1998) A phytochorme from the fern *Adiantum* with features of the putative photoreceptor NPH1. Proc Natl Acad Sci USA 95: 15826–15830

Nozue K, Christie J, Kiyosue T, Briggs WR, Wada M (2000) Isolation and characterization of fern phototropin (Accession No. AB037188). A putative blue-light photoreceptor for phototropism (PGR00–039). Plant Physiol 122: 1457

Oikawa K, Kasahara M, Kiyosue K, Kagawa T, Suetsugu N, Takahashi F, Kanegae T, Niwa Y, Kadota A, Wada M (2003) CHUP1 is essential for proper chloroplast positioning. Plant Cell 15: 2805–2815

Sakai T, Kagawa T, Kasahara M, Swartz TE, Christie JM, Briggs WR, Wada M, Okada K (2001) Arabidopsis nph1 and npl1: Blue-light receptors that mediate both phototropism and chloroplast relocation. Proc Natl Acad Sci USA 98: 6969–6974

Sato Y, Kadota A, Wada M (1999) Mechanically-induced avoidance response of chloroplasts in fern protonemal cells. Plant Physiol 121: 37–44

Sato Y, Wada M, Kadota A (2001a) Choice of tracks, microtubules and/or actin filaments for chloroplast photomovement is differentially controlled by phytochrome and a blue light receptor. J Cell Sci 114: 269–279
Sato Y, Wada M, Kadota A (2001b) External Ca^{2+} is essential for chloroplast movement induced by mechanical stimulation but not by light stimulation. Plant Physiol 127: 497–504
Sato Y, Kadota A, Wada M (2003a) Chloroplast movement: dissection of events downstream of photo- and mechano-perception. J Plant Res 116: 1–5
Sato Y, Wada M, Kadota A (2003b) Accumulation response of chloroplasts induced by mechanical stimulation in bryophyte cells. Planta 216: 772–777
Wada M, Grolig F, Haupt W (1993) Light-oriented chloroplast positioning. Contribution to progress in photobiology. J Photochem Photobiol B 17: 3–25
Wada M, Kagawa T, Sato Y (2003) Chloroplast movement. Annu Rev Plant Biol 54: 455–468

Part V
Other Photoreceptors

Chapter 23

Structural Dynamics of the Signal Termination Process in Rhodopsin

David L. Farrens

Introduction

This review will provide and overview of what is known, and what is not known, about the visual signal termination process in mammalian vision. The focus will be on the role of structure and dynamic changes in the primary mammalian photo-transducer rhodopsin, and the protein that attenuates rhodopsin signaling, arrestin. Although this review focuses on mammalian photoreceptor proteins, analogous mechanisms may be used in the phototransduction pathways of other organisms.

Overview of Mammalian Visual Signaling

Initiation of Visual Signal Transduction

Light-activated rhodopsin initiates a visual signal transduction cascade after light induces isomerization of its retinal chromophore from the 11-*cis* to all-*trans* configuration. The active, signaling state thus produced, called MII, amplifies the initial light signal by binding and inducing the G-protein transducin (G_T) to exchange GDP for GTP and dissociate into two subunits, $G_T\alpha$ and $G_T\beta\gamma$. The GTP-bound $G_T\alpha$ then induces a cGMP phosphodiesterase to hydrolyze cGMP, resulting in the closure of cation conductance channels and the generation of a nerve signal.

Termination of Visual Signal Transduction

Signaling by MII rhodopsin is "turned off" through two different pathways. The slower of the two pathways involves changes in the state of the activated receptor, either by conversion into an inactive retinal-bound state, called MIII, and/or

Biochemistry and Molecular Biology, Oregon Health and Science University, Portland, OR 97239-3098, USA

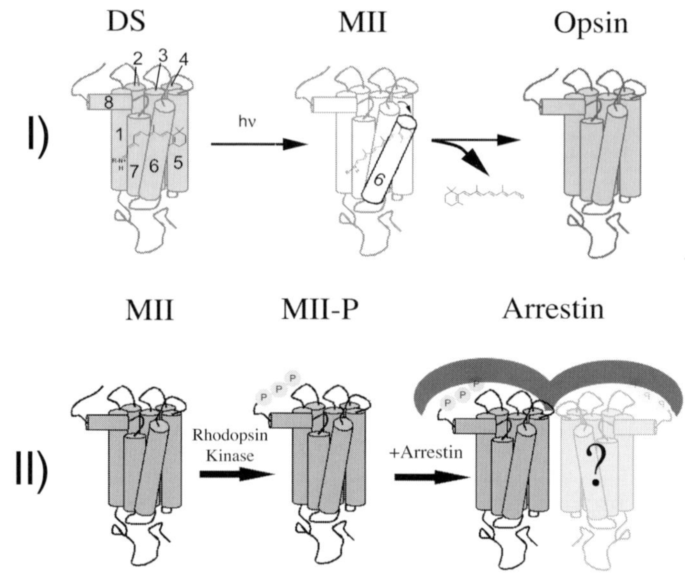

FIG. 1. Cartoon model depicting two pathways used to "shut-off" MII rhodopsin signaling. **I** Photoconversion of the 11-*cis* retinal chromophore in dark-state (*DS*) rhodopsin into all-*trans* retinal induces the formation of the active, signaling MII species. Subsequent to MII formation, the retinal Schiff-base linkage is hydrolyzed, leading to the release of all-*trans* retinal and the formation of the apoprotein, Opsin. **II** Phosphorylation of active MII rhodopsin (MII-P) by rhodopsin kinase (not shown) leads to the binding of the signal terminating protein, Arrestin. The possibility that Arrestin may bind to a dimer of rhodopsin is also suggested in the figure. A third pathway, conversion of MII to an inactive form called MIII, is not shown

hydrolysis of the Schiff-base linkage attaching retinal to rhodopsin and release of the freed retinal leaving the "inactive" apoprotein, called opsin. In a more rapid termination process, the MII form is phosphorylated on the C-terminus by a rhodopsin specific kinase. This phosphorylated rhodopsin (MII-P) is recognized by a protein called arrestin, which then binds MII-P and physically blocks any further interaction of rhodopsin with transducin. Two of these processes are shown in Figure 1 (due to space limitations, the MIII pathway is not discussed here).

Role of Structure and Dynamics in Rhodopsin Activation

The structures of the inactive forms of rhodopsin and arrestin are known to high resolution. The structure of each is discussed below, followed by a description of the dynamic changes that occur during activation and termination.

Overview of Rhodopsin Structure

The high-resolution crystal structure of rhodopsin (Palczewski et al 2000, Okada et al 2001) confirmed many previous theories—the general arrangement of the seven transmembrane helices, the relative helical alignment and orientation, and the attachment of 11-*cis* retinal chromophore at lysine-296 (Figure 2). However, several findings in the structure were unexpected, and the two most relevant to our discussion are outlined briefly below.

The rhodopsin extracellular (intradiscal) region is highly structured and surrounds the retinal chromophore. The extracellular loops of rhodopsin are sensitive to deletions and mutations, as well as numerous natural retinitis pigmentosa (disease causing) mutations (Hwa et al 2001). As shown in Figure 2, the initial 33 residues form a compact, glycosylated unit that lies across the other connecting loops. Even more striking, loop E-2 (which connects helices 4 and 5) forms a twisted β-hairpin that makes a "plug" or lid across the retinal. This "retinal plug" is postulated to shield the retinal Schiff-base linkage from solvent, contributing to its high-degree of stability.

The rhodopsin cytoplasmic face interacts with transducin and arrestin and has some unusual characteristics. Transmembrane Helix 7 extends past the membrane interface and bends along the membrane to form an eighth helix (Helix 8). This helix is amphipathic, with the hydrophobic part oriented towards the membrane, and the polar groups displayed outwards, towards the presumed interface with transducin and arrestin. Also, the end of Helix 8 is anchored to the membrane by two palmitoylation sites. The extreme cytoplasmic tail of rhodopsin appears highly mobile and disordered in rhodopsin crystal structures, in agreement with electron paramagnetic resonance (EPR) and nuclear magnetic resonance (NMR)

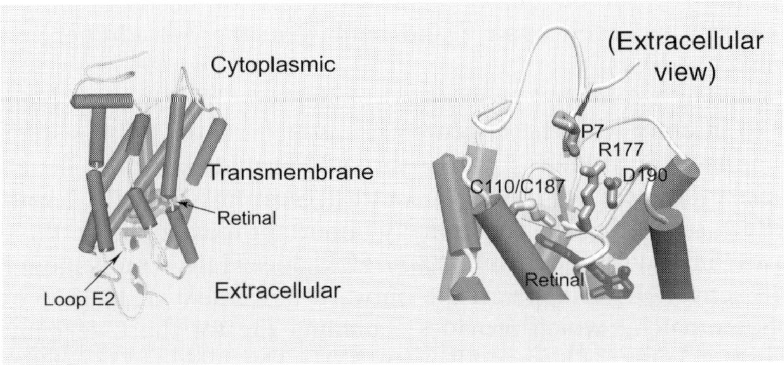

FIG. 2. **Left** Side view of rhodopsin structure, showing the arrangement of the seven transmembrane helices, the location of the retinal chromophore, and the cytoplasmic and extracellular (intradiscal) domains. **Right** Close-up view of the extracellular (intradiscal) loop region. Notice how loop E-2, the "retinal plug" forms a twisted β-hairpin structure that lies over the top of the retinal chromophore

studies, although it may adopt structure after being phosphorylated and bound by arrestin.

Structural Changes During Rhodopsin Activation

In the absence of a crystal structure of the active form of rhodopsin, dynamic structural methods have been used to assess the light-activation conversion of rhodopsin into the signaling MII form. These methods have included chemical reactivity studies and spectroscopic approaches such as fluorescence, EPR and NMR, discussed below.

The extracellular loops may move during MII formation. Evidence suggesting the extracellular (intradiscal) region of rhodopsin changes structure upon light-activation. These results include increased reactivity of a cysteine residue in loop E-2 (Ridge et al 1995), and photo-crosslinking studies using a retinal analogue, which suggest the extracellular (intradiscal) part of Helix 4 also moves (Borhan et al 2000). More work is needed to confirm and define movements in this region.

Conformational changes in the cytoplasmic face are required to convert rhodopsin into the functional MII form. The light-activated movements in the cytoplasmic face of rhodopsin have been studied using mutants that contain unique, reactive cysteines placed throughout the cytoplasmic domain. The reactivity of these cysteine residues were then studied and spectroscopic probes (spin-, fluorescent-, and NMR) were attached to these cysteines to report localized conformational changes (Dunham and Farrens 1999, Klein-Seetharaman et al 1999, Hubbell et al 2003). Together, these studies suggest that rhodopsin activation primarily involves a tilt and/or rotation of one of the helices, Helix 6 (Farrens et al 1996, Hubbell et al 2003). It appears Helix 6 movement may be a conserved activation mechanism in seven transmembrane helical receptors, as it has been observed during light activation of bacteriorhodopsin, sensory rhodopsin and even upon ligand binding in the β-2 adrenergic receptor (Ghanouni et al 2001).

Why does Helix 6 move? Helix 6 movement may be required to enable the receptor to interact with the G-protein transducin. Cross-linking studies show that linking Helix 6 to Helix 3, either through disulfide bonds or metal binding sites, blocks transducin activation. In contrast, cross-linking Helix 1 with Helix 7 has no effect, suggesting the functionally important movements in the cytoplasmic face are limited (Hubbell et al 2003). How does Helix 6 movement facilitate G-protein activation? It appears, the outward movement of Helix 6 exposes a "hydrophobic patch," which provides a binding site for the C-terminus of the transducin $G_T\alpha$ subunit (Janz and Farrens 2004). This region of transducin is well known to play a key role in transducin binding and stabilizing the MII form of rhodopsin.

Role of Structure and Dynamics in Rhodopsin Deactivation (MII Decay)

Structural Dynamics of MII Decay and Opsin Formation

Once formed, MII rhodopsin begins to decay with a half-life of minutes. During the decay process, the protein loses the retinal chromophore and the ability to catalytically activate transducin. Structural changes thought to occur during MII decay are discussed below.

The rhodopsin structure controls retinal Schiff-base hydrolysis, retinal release and MII decay. The retinal Schiff-base linkage is dramatically stabilized in the dark-state rhodopsin (the $t_{1/2}$ for hydrolysis is days). In contrast, in MII rhodopsin, the Schiff base is dramatically destabilized, leading to hydrolysis and retinal release within minutes. Model retinal Schiff-base compounds hydrolyze even more rapidly.

What stabilizes the retinal Schiff base in rhodopsin? The tight interaction of loop E-2 with retinal (shown in Figure 2) is a factor in stabilizing the retinal Schiff-base linkage, and it has been proposed that this structure forms a "plug" or lid that protects the retinal Schiff-base linkage from solvent. Another key factor appears to be the protonation state of the Schiff-base counterion. The high pK_a of the retinal Schiff-base linkage (and complementary low pK_a of the counter-ion) act to suppress thermal isomerization and spontaneous hydrolysis of the retinal chromophore in dark-state rhodopsin (Ebrey and Koutalos 2001). Interestingly, mutation of the counterion to a neutral residue actually *slows* the rate of Schiff-base hydrolysis in MII rhodopsin, in a pH-dependent fashion (Janz and Farrens 2003). Similarly, mutating several polar residues that surround the Schiff base to non-polar residues also slows MII hydrolysis and retinal release (Janz 2004). These latter results suggest that Schiff-base hydrolysis, even in MII rhodopsin, is tightly controlled by the protein structure.

Variations in the stability of the Schiff-base linkage can have important consequences in vision. For example, the light activated MII state of cone (color sensing) opsins is much less stable than dim-light rhodopsin, due to rapid hydrolysis of retinal Schiff-base linkage in their MII form. This allows cone opsins to undergo rapid regeneration during continuous illumination and extend their activity by constant recycling (Babu et al 2001). It is also becoming apparent that mutations that affect retinal Schiff-base stability are a factor in some retinal disease states.

Structure of opsin formed by MII decay and release of retinal. In vivo, after the active MII state decays by retinal Schiff-base hydrolysis and retinal release, the apoprotein opsin is reconverted into a photosensitive pigment through reconstitution with exogenous 11-*cis*-retinal provided through the retinoid cycle (McBee et al 2001). Interestingly, opsin in detergent loses its ability to be regenerated with exogenously added 11-*cis*-retinal (Sakamoto and Khorana 1995). Stabilization of the "retinal plug" may be a key factor for maintaining the ability of opsin to regenerate with retinal in detergent (Xie et al 2003). The ability of opsin to regenerate with retinal indicates MII decay does not induce a denatured form of the receptor.

How then does the structure of opsin differ from MII or dark-state rhodopsin? Fourier transform infrared studies indicate no large scale protein unfolding occurs during MII to opsin decay (Vogel and Siebert 2002), and ^{19}F NMR studies show that chemical shifts which occur in labeled rhodopsin following photobleaching gradually return to near dark state levels during MII decay (Klein-Seetharaman et al 1999). Unfortunately, specific conformational changes that occur during the decay of the MII intermediate are not well understood and further study is needed.

A model for why opsin loses the ability to activate transducin. Why do these changes in opsin result in a form of the receptor that is unable to activate transducin? Recently, it was shown that a peptide analogous to the C-terminal tail of $G_T\alpha$ binds to a cleft on the cytoplasmic face of rhodopsin that forms as a result of the outward Helix 6 movement during MII formation. The peptide binds only to the MII intermediate, and ceases to bind upon decay of the MII species. In addition, peptide binding is abolished upon addition of hydroxylamine, which rapidly cleaves the MII retinal Schiff-base linkage (Janz and Farrens 2004). There appears to be a potential link between Helix 6 movement "resetting" and retinal release during MII decay. To a first approximation, it appears that after TM Helix 6 moves away from Helix 3 during activation, it moves back towards the helical bundle during MII decay, thus no longer exposing the hydrophobic patch to interact with the C-terminal tail of $G_T\alpha$ (Janz 2004).

Role of Structure and Dynamics in Rhodopsin Deactivation (Arrestin Binding and Signal Termination)

Overview of Arrestin Structure and Binding

Arrestin attenuates light stimulus in the retina by binding to photoactivated, phosphorylated rhodopsin (MII-P) thus blocking transducin activation. Little is known about what exactly happens to trigger arrestin into a form that can bind MII rhodopsin.

Structure of arrestin. The crystal structures of arrestin show a highly conserved, bi-lobed protein structure. Biochemical studies suggest arrestin exists in slightly different conformational forms. The cleft between the lobes in arrestin contains several buried, interlobal salt bridges and this region is often referred to as the "polar core." Biochemical experiments suggest the C-terminus of arrestin lies across the polar core in arrestin's inactive conformation (Figure 3A). Unfortunately, the crystal structures give conflicting results regarding the location of the C-terminus.

Arrestin activation and binding to MII rhodopsin is thought to involve multiple steps. First, phosphorylation sensors within the N-domain and a "polar core" (several residues salt-bridged with residue R175) are triggered by interaction

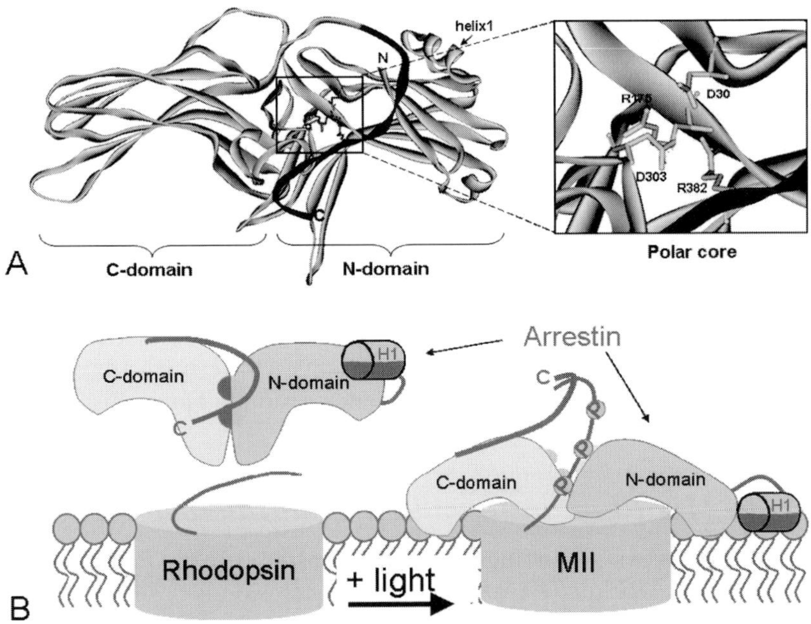

FIG. 3. **A** Structure of arrestin showing key structural features. The bi-lobed arrestin consists of the crescent shaped N- and C-domains. The flexible C-terminal tail is highlighted in black, and amphipathic Helix 1 is labeled. On the *right*, a magnification of the interface between the two domains is shown, including the important salt-bridge pairs R175-D303 and R382-D30, which are thought to link the N- and C- domains and the C-terminal tail. **B** Proposed arrestin activation mechanism (Gurevich and Gurevich 2004). On the *left*, the inactive arrestin conformation is seen to be stabilized by an intact "polar core" and the C-terminal tail. Interaction of the phosphorylated rhodopsin C-terminal tail with the positively charged arrestin N-domain disrupts the polar core, flipping arrestin into an active conformation. On the *right*, the proposed large structural changes in arrestin are shown which are proposed to involve displacement of the arrestin C-terminal tail, a "hinge opening" movement, and insertion of Helix 1 into the membrane

with the phosphorylated MII rhodopsin C-terminal tail. This causes arrestin to expose high-affinity binding sites in the N- and C-domains, which recognize the "active" loops of MII rhodopsin. Consistent with this theory, arrestin mutants R175E and R175N are constitutively active proteins that bind receptor independent of their phosphorylation status (Gurevich and Gurevich 2004).

Arrestin May Undergo Large Structural Changes During Arrestin Activation

The structure of arrestin has been proposed to undergo large-scale changes during activation and binding (Gurevich and Gurevich 2004). In the inactive state, the arrestin C-terminus is thought to lie across the polar core, and then move upon activation and rhodopsin binding. In agreement with this hypothesis, deletion of the C-terminus leads to a constitutively active arrestin. The amphipathic arrestin Helix 1 (residues 101–112) is also proposed to relocate from the protein and interact directly with the membrane surface, where it can act as a reversible membrane anchor. These proposed changes are shown in Figure 3B.

However, although a compelling model, large-scale structural changes in arrestin are not supported by all the data. For example, the two arrestin crystal structures show the C-terminus in different locations, and do not detect the extreme C-termini. Small-angle X-ray scattering studies do not find evidence for a large conformational change in the structure of activated arrestin (Shilton et al 2002), and no evidence for large-scale changes have been observed by circular dichroism and tryptophan fluorescence studies. Thus, this issue is still unresolved.

Where do MII Rhodopsin and Arrestin Interact?

The high-affinity arrestin binding site on rhodopsin probably involves rhodopsin cytoplasmic loops I, II, and III, although this conclusion is based on indirect evidence (such as lack of arrestin binding) that may also be influenced by other effects such as structural changes in the rhodopsin loops caused by the mutations used. The likely regions of contact on arrestin are also known to some extent. Phage display studies (Smith and Hargrave 2000) and peptide mapping studies (Pulvermuller et al 2000) suggest that both the N- and C- domains are involved. Taken together, the results indicate several sites on arrestin are involved in high affinity binding to rhodopsin. However, numerous questions remain about rhodopsin–arrestin interactions. Does rhodopsin really trigger large-scale changes in the arrestin structure? Where do arrestin and rhodopsin interact? Does arrestin bind to a dimer of rhodopsin? How are retinal release and arrestin binding related, i.e., what makes arrestin let go?

Summary and Conclusions

The structures of rhodopsin and arrestin only show the proteins in their inactive forms. However, studies are now building on the information provided in these structures to understand how photoactivation enables MII rhodopsin to bind and activate transducin, what causes MII rhodopsin to decay and release retinal, and how arrestin is able to recognize and bind activated rhodopsin (Ridge et al 2003). Future efforts will undoubtedly continue to focus on addressing the role of structural dynamics in visual signal termination.

References

Babu KR, Dukkipati A, Birge RR, Knox BE (2001) Regulation of phototransduction in short-wavelength cone visual pigments via the retinylidene Schiff base counterion. Biochemistry 40: 13760–13766

Borhan B, Souto ML, Imai H, Shichida Y, Nakanishi K (2000) Movement of retinal along the visual transduction path. Science 288: 2209–2212

Dunham TD, Farrens DL (1999) Conformational changes in rhodopsin. Movement of helix F detected by site-specific chemical labeling and fluorescence spectroscopy. J Biol Chem 274: 1683–1690

Ebrey T, Koutalos Y (2001) Vertebrate photoreceptors. Prog Retin Eye Res 20: 49–94

Farrens DL, Altenbach C, Yang K, Hubbell WL, Khorana HG (1996) Requirement of rigid-body motion of transmembrane helices for light activation of rhodopsin. Science 274: 768–770

Ghanouni P, Steenhuis JJ, Farrens DL, Kobilka BK (2001) Agonist-induced conformational changes in the G-protein-coupling domain of the beta 2 adrenergic receptor. Proc Natl Acad Sci USA 98: 5997–6002

Gurevich VV, Gurevich EV (2004) The molecular acrobatics of arrestin activation. Trends Pharmacol Sci 25: 105–111

Hubbell WL, Altenbach C, Hubbell CM, Khorana HG (2003) Rhodopsin structure, dynamics, and activation: a perspective from crystallography, site-directed spin labeling, sulfhydryl reactivity, and disulfide cross-linking. Adv Protein Chem 63: 243–290

Hwa J, Klein-Seetharaman J, Khorana HG (2001) Structure and function in rhodopsin: Mass spectrometric identification of the abnormal intradiscal disulfide bond in misfolded retinitis pigmentosa mutants. Proc Natl Acad Sci USA 98: 4872–4876

Janz JM (2004) Structural dynamics of rhodopsin: relationships between retinal Schiff base integrity and receptor signaling states. Oregon Health and Science University, Portland, OR

Janz JM, Farrens DL (2003) Assessing structural elements that influence Schiff base stability: mutants E113Q and D190N destabilize rhodopsin through different mechanisms. Vis Res 43: 2991–3002

Janz JM, Farrens DL (2004) Rhodopsin activation exposes a key hydrophobic binding site for the transducin alpha-subunit C terminus. J Biol Chem 279: 29767–29773

Klein-Seetharaman J, Getmanova EV, Loewen MC, Reeves PJ, Khorana HG (1999) NMR spectroscopy in studies of light-induced structural changes in mammalian rhodopsin: applicability of solution (19)F NMR. Proc Natl Acad Sci USA 96: 13744–13749

McBee JK, Palczewski K, Baehr W, Pepperberg DR (2001) Confronting complexity: the interlink of phototransduction and retinoid metabolism in the vertebrate retina. Prog Retin Eye Res 20: 469–529

Okada T, Ernst OP, Palczewski K, Hofmann KP (2001) Activation of rhodopsin: new insights from structural and biochemical studies. Trends Biochem Sci 26: 318–324

Palczewski K, Kumasaka T, Hori T, Behnke CA, Motoshima H, Fox BA, Le Trong I, Teller DC, Okada T, Stenkamp RE, Yamamoto M, Miyano M (2000) Crystal structure of rhodopsin: A G protein-coupled receptor. Science 289: 739–745

Pulvermuller A, Schroder K, Fischer T, Hofmann KP (2000) Interactions of metarhodopsin II. Arrestin peptides compete with arrestin and transducin. J Biol Chem 275: 37679–37685

Ridge KD, Lu Z, Liu X, Khorana HG (1995) Structure and function in rhodopsin. Separation and characterization of the correctly folded and misfolded opsins produced on

expression of an opsin mutant gene containing only the native intradiscal cysteine codons. Biochemistry 34: 3261–3267
Ridge KD, Abdulaev NG, Sousa M, Palczewski K (2003) Phototransduction: crystal clear. Trends Biochem Sci 28: 479–487
Sakamoto T, Khorana HG (1995) Structure and function in rhodopsin: the fate of opsin formed upon the decay of light-activated metarhodopsin II in vitro. Proc Natl Acad Sci USA 92: 249–253
Shilton BH, McDowell JH, Smith WC, Hargrave PA (2002) The solution structure and activation of visual arrestin studied by small-angle X-ray scattering. Eur J Biochem 269: 3801–3809
Smith WC, Hargrave PA (2000) Mapping interaction sites between rhodopsin and arrestin by phage display and synthetic peptides. Methods Enzymol 315: 437–455
Vogel R, Siebert F (2002) Conformation and stability of alpha-helical membrane proteins. 2. Influence of pH and salts on stability and unfolding of rhodopsin. Biochemistry 41: 3536–3545
Xie G, Gross AK, Oprian DD (2003) An opsin mutant with increased thermal stability. Biochemistry 42: 1995–2001

Chapter 24

Participation of Internal Water Molecules and Clusters in the Unidirectional Light-Induced Proton Transfer in Bacteriorhodopsin

Akio Maeda[1,3], Sergei P. Balashov[2], and Thomas G. Ebrey[3]

Introduction

Proton pumping is a fundamental biological process for generating free energy in the form of a transmembrane electrochemical potential for hydrogen ions; this energy is then utilized in cellular processes such as ATP synthesis, rotation of a flagellar motor, and ion transport across the plasma membrane. In the photosynthetic reaction center, light is used to induce charge separation and generate reducing and oxidizing equivalents. The subsequent electron flow through a chain of electron transporting proteins is coupled to the proton pumping, and so the pump makes indirect use of the light energy. In contrast a single small (26 kDa) protein, bacteriorhodopsin (BR), not only absorbs the light energy and does the photochemistry, it also is the proton pump. Recently similar pigments have been found in many marine eubacteria where they provide a significant portion of the photosynthetic yield of the oceans.

Bacteriorhodopsin Is an Advantageous Target for Studying the Pumping Mechanism

Bacteriorhodopsin carries out unidirectional proton transport across the membrane utilizing light energy absorbed by its all-trans retinal chromophore. The retinal, which lies near the middle of the membrane, is bound to Lys216 of the apoprotein through a protonated Schiff base linkage (Figure 1). Ionized Asp85

[1] Department of Biochemistry, University of Illinois at Urbana/Champaign, Urbana, IL 61801, USA
[2] Department of Physiology and Biophysics, University of California, Irvine, Irvine, CA 92697, USA
[3] Department of Biology, University of Washington, Seattle, WA 98195, USA

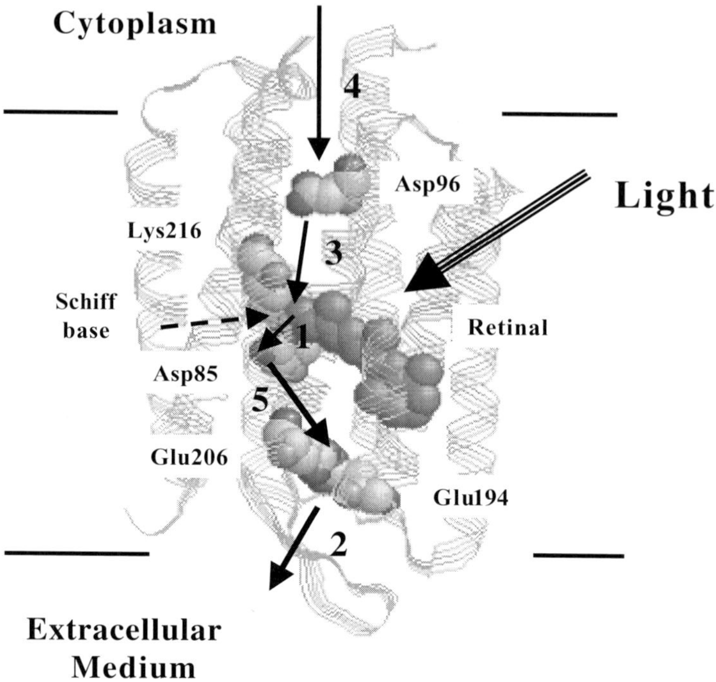

FIG. 1. Proton transfer steps between the protonatable residues of the light-activated bacteriorhodopsin on the basis of the coordinate of protein data band 1c3w by Luecke et al (1999)

is part of a counterion to the positive charge of the Schiff base and is the initial proton acceptor. This site is directly involved in the initiation of proton transport. The latter occurs in a cyclic multistep process which includes several intermediates, BR → K → L → M → N → O → BR. The initial event is the light-induced trans-cis isomerization of the retinal chromophore. This causes a change in the shape of the chromophore, including movement of the Schiff base, which carries positive electric charge, and a change in orientation of the N—H bond of the Schiff base. These structural changes lead to changes in electrostatic and H-bonding interactions of the Schiff base and eventually to proton transfer from the Schiff base to the carboxyl group of Asp85 buried inside the protein. Bacteriorhodopsin is an advantageous system for attacking the molecular mechanisms of proton pumping since it is a small and simple system. Unidirectional proton pumping must follow from the spatial arrangement and proton affinities of such protein groups as carboxylic acids, the chromophore's Schiff base, and water; they can be followed by several techniques, especially Fourier transform infrared (FTIR) spectroscopy and X-ray crystallography.

FTIR Spectroscopy Is a Powerful Tool to Study the Mechanism of Light Energy Transduction and Proton Transport

The photocycle intermediates (Figure 1) have been characterized by time-resolved spectroscopy (UV/VIS, resonance Raman, and IR) at room temperature and steady-state spectroscopy at cryogenic temperatures where the intermediates can be trapped in quasi-stable states. Fourier transform infrared (FTIR) spectroscopy has been successfully used to determine the changes in protonation state of carboxylic acids participating in the proton transfer (reviewed by Maeda 1995), the changes in orientation and hydrogen bonding of the Schiff base N—H (Maeda 1995, Kandori 2004), and other light-induced changes near the active site of BR. In particular, internal water molecules in BR have been identified and changes in their hydrogen bonding followed by monitoring their OH (or OD in D_2O) vibrations. The results of these spectroscopic studies now can be compared with structures obtained by recent X-ray diffraction studies (references cited in Lanyi and Schobert 2003, Kouyama et al 2004).

Steps of Proton Transport

After the all-trans-to-13-cis isomerization of the chromophore during the primary light reaction (BR-to-K transition), the chromophore's Schiff base undergoes deprotonation in the L-to-M transition, with Asp85 being the proton acceptor (Figure 1; Step 1). This step is accompanied by proton release from the proton release complex (PRC) at the extracellular surface of BR, which is composed of Glu194, Glu204, and waters (Step 2). Next the Schiff base is reprotonated in the M-to-N transition by a proton donated from an initially protonated Asp96 (Step 3). A proton is then taken up from the cytoplasmic side of the membrane to reprotonate Asp96 (Step 4) and finally Asp85 is returned to its initial state by donating its proton to the PRC (Step 5), completing the photocycle and leaving all residues in their initial state and one proton transported from the cytoplasmic to the extracellular medium. The protonation of Asp85 on the extracellular side along with the deprotonation of the Schiff base which occur in forming M (Step 1) and the deprotonation of Asp96 to reprotonated the Schiff base from the cytoplasmic side in forming N (Step 3) are crucial steps in inducing the unidirectional proton transfer (reviewed in Balashov 2000, Heberle 2000, Herzfeld and Tounge 2000, Balashov and Ebrey 2001, Maeda 2001, Kandori 2004, Lanyi and Schobert 2003).

Role of Internal Hydrogen Bonded Water

Discovery of discrete water bands which undergo changes in hydrogen bonding during the photocycle reactions was proof of water molecules being part of the photocycle and provided a tool for their investigations. Studies of mutants helped to assign vibrational bands to certain locations of water molecules in the extracellular domain, close to Asp85 and in the cytoplasmic domain, close to the proton donor Asp96–Thr46 complex (Yamazaki et al 1996). The subsequent X-ray structure of initial state of BR confirmed the presence of water molecules at these sites and resolved additional water molecules (total of about ten) some of which are likely to be a part of the extracellular proton translocation pathway (Luecke et al 1999). One of the waters was found right in the active site, between the Schiff base, Asp85, and Asp212. Together with other two molecules interacting with Asp85 and Asp212 they stabilize the ionized states of these residues and apparently help to keep the pKa of the Schiff base high (above 12) and pKa of the carboxyls low (below 3) in the initial state. Four water molecules interact with the residues comprising the proton release site, Arg82, Glu194, and Glu204. Hydrogen bonding involves substantial energy, and it can directly affect the pKa of the groups. Water molecules can accept and donate protons and serve as mobile proton carriers; they can create networks for long-range proton transfer. This makes the study of changes of hydrogen bonding of internal waters essential for understanding the energetics and mechanism of proton translocation.

The X-ray structures of several intermediates (K, L, M, N) show relocation and changes of hydrogen bonding of internal water as had been also documented with FTIR. However, the exact picture still needs to be refined, particularly in respect to the early (K and L) intermediates where several substantially different X-ray structures were obtained (reviewed by Lanyi 2004). FTIR spectroscopy should be able to reveal local polarization changes around the active site, small changes that may be hard to detect by X-ray crystallography. In our recent studies we addressed several aspects of the role of water in proton translocation.

One study was to identify bands of water molecules and protein residues most closely interacting with the chromophore and particularly the Schiff base. Since light causes photo reisomerization of the chromophore from its temperature stabilized state of every intermediate (K, L, M and N), one can "perturb" the chromophore in K, L, M, N, and follow the changes in the interacting residues (Maeda et al 1999, 2000). Another approach is to identify chromophore and protein vibrations which are coupled with water (Maeda et al 2002). Finally, we looked at the role of specific residues and water molecules in formation of the intermediates. Some of the findings are briefly discussed below.

Role of Water in Formation and Stabilization of L

The L intermediate of bacteriorhodopsin's photocycle is the state before the first proton transfer from the Schiff base to Asp85. L may be a transition-like state, which stabilizes the energy of the absorbed quantum and forms a structure favorable for proton transfer to Asp85 while keeping the Schiff base in its protonated state. The free energy level of L is close to that of M (Varo and Lanyi 1991) so small movements of the side chains and relocation of small molecules, water, are expected to be the driving force in this transition.

The difference FTIR spectrum for the formation of the L intermediate exhibits many intense vibration bands (Figure 2), which were assigned by isotope labeling and mutant studies. Among them water vibrations seen in the 3700–3400 cm^{-1} range undergo a shift to a lower frequencies and large increase in intensity compared to the initial BR state. Studies by use of mutants have shown that a cluster of polarized water molecules appears in L (Maeda et al 1999, 2002, 2003b). L formation involves water rearrangement on the cytoplasmic side of the Schiff base along with the movement of the side chains of Trp182 and Leu93 (Maeda et al 2003a), and the peptide bonds of Lys216 and Gly220 (Maeda et al 2003b). This water cluster is surrounded by the Schiff base, Leu93, Val49, Thr46, Asp96 and Trp182, and apparently participates in shifting the L-to-M equilibrium and also K-to-L equilibrium to L through their interaction with the Schiff base. Thus the protonated state of the Schiff base in L is stabilized by its interaction with water molecules. Independent evidence for this strong interaction comes

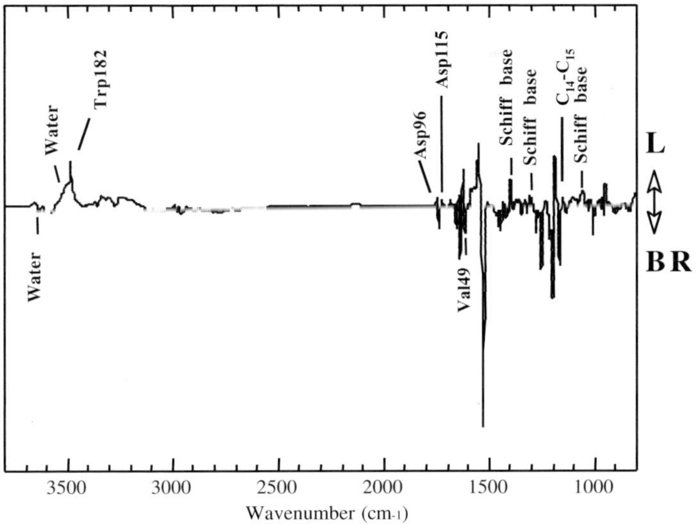

Fig. 2. The L-minus-bacteriorhodopsin Fourier transform infrared (FTIR) difference spectrum. Vibration bands assigned by isotope labeling and mutants were labeled (see reviews by Maeda 1995, 2001)

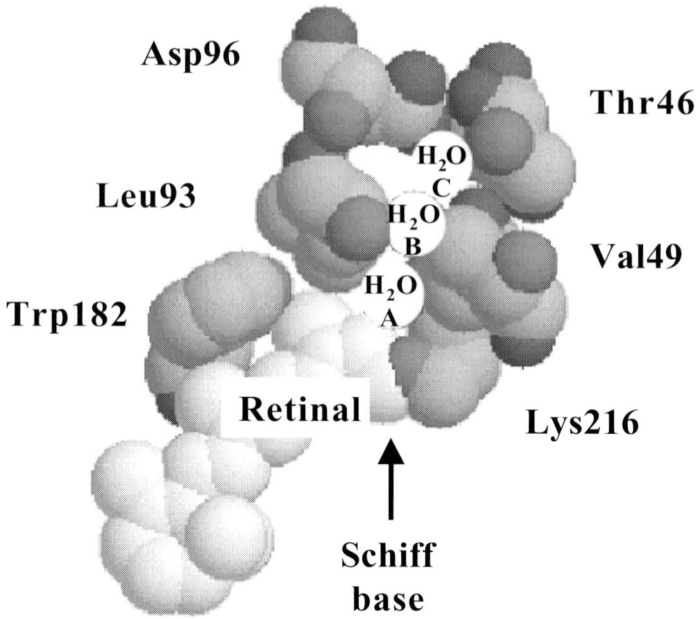

FIG. 3. The cavity that could accommodate the water cluster proposed by our FTIR studies (based on PDB1ucq of Kouyama et al 2004)

from coupling of some of the chromophore vibrations with water vibrations in L (Maeda et al 2002). This is in agreement with the L structures by Lanyi and Schobert (2003) and Kouyama et al (2004). Several other features of the L intermediate revealed by our FTIR studies (changes in Leu93 and Trp182 and a water hydrogen bonded network in the cytoplasmic domain in L) are also present in the X-ray structure of L by Kouyama et al (2004). Figure 3 depicts the proposed water cluster for L.

Water Relocation Coupled to pKa Changes of the Schiff Base and Asp85 in the L-to-M Transition

One of the factors to cause proton transfer from the Schiff base to Asp85 is a decrease in proton affinity of the Schiff base, which could result from the disappearance of the water cluster interaction with the Schiff base on its cytoplasmic side. Some of these water molecules are relocated to the region around Phe219 in M (Yamazaki et al 1998, Maeda et al 2000). The X-ray studies also show the presence of water molecules in the region surrounded by Phe219, Thr46 and Asp96 (Luecke et al 1999) and none of water molecules are detected close to the Schiff base. This is, however, not sufficient for unidirectional proton transfer. The proton affinity of Asp85 should be increased in parallel. The thermodynamic barrier thus created between the Schiff base and Asp85 prevents reverse flow of proton. It was generally thought that exclusion of water molecules from Asp85 would cause it to increase its pKa. However, our FTIR results suggest the interaction of Asp85 with a water molecule in M (Maeda et al 2000). X-ray crystallographic structures also show that Asp85 is hydrated in M (Luecke et al 1999). Rearrangement of the water cluster on the extracellular side of the Schiff base could be responsible for retaining the protonated state of Asp85, so that the proton is stabilized by interaction with water oxygen.

Proton Transfer from the Schiff Base to Asp85 in the L-to-M Transition

This proton transfer was proposed to be mediated in L by the approach of the hydrogen of the Schiff base to Asp85 with the orientation of the N—H of the Schiff base towards Asp85 (Lanyi and Schobert 2003). Resonance Raman studies indicated that the chromophore is a strained 13-cis, 14-trans, 15 anti in L; in a relaxed state of this conformation (as in N) the N—H would point towards cytoplasmic side. Recent theoretical studies have suggested that the proton transfer from the Schiff base to Asp85 might occur through several pathways including those when the N—H is oriented towards the cytoplasmic side (Bondar et al 2004). The exact way the water molecule interacts with the Schiff base in L and how it participates in the proton transfer are still open questions.

Unidirectionality of Proton Transfer in the L-to-M-to-N Transition

The next step is proton transfer from Asp96 to the Schiff base. To do this the Schiff base has to increase its pK along with an accompanying decrease in the pK of Asp96. The latter change has been proposed to be caused by the hydration of Asp96. The pK of the Schiff base in N was estimated to be ~8.2 in D96N (Brown and Lanyi 1996). Thus, the proton affinity of the Schiff base in N is less than that of Asp85, and reverse flow of proton does not occur. One of the important differences between L and N is that the chromophore of L is distorted (Pfefferle et al 1991). The abolition of the electrostatic interaction between the Schiff base and Asp85 and H-bonding between the Schiff base with water when M is formed may relax this distortion, leading to the increase in the pK of the Schiff base. A string of waters in a direction from the Schiff base to Asp96 was found in the structure of a long-lived N for theV49A mutant (Schobert et al

2003). The waters apparently assist in internal proton transfer for the long distance (about 12 Å) from Asp96 to the Schiff base. It is likely that a similar cluster or chain is transiently formed in N of wild type where Asp96 is unprotonated. Future FTIR studies will test this.

In conclusion, both FTIR and X-ray studies indicated direct involvement of waters in every step of proton transport. They are essential factors in causing pK changes and so proton translocation in BR.

References

Balashov SP (2000) Protonation reactions and their coupling in bacteriorhodopsin. Biochim Biophys Acta 1460: 75–94

Balashov S, Ebrey T (2001) Trapping and spectroscopic identification of the photointermediates of bacteriorhodopsin at low temperature. Photochem Photobiol 73: 453–462

Bondar A, Elstner M, Suhai S, Smith J, Fischer S (2004) Mechanism of primary proton transfer in bacteriorhodopsin. Structure 12: 1281–1288

Brown L, Lanyi J (1996) Determination of the transiently lowered pK of the retinal Schiff base during the photocycle of bacteriorhodopsin. Proc Natl Acad Sci USA 96: 1731–1734

Heberle J (2000) Bacteriorhodopsin: the functional details of a molecular machine are being resolved. Biophys Chem 85: 229–248

Herzfeld J, Tounge B (2000) NMR probes of vectoriality in the proton-motive photocycle of bacteriorhodopsin: evidence for an "electrostatic steering" mechanism. Biochim Biophys Acta 1460: 95–105

Kandori H (2004) Hydration switch model for the proton transfer in the Schiff base region of bacteriorhodopsin. Biochim Biophys Acta 1658: 72–79

Kouyama T, Nishikawa T, Tokuhisa T, Okamura H (2004) Crystal structure of the L intermediate of Bacteriorhodopsin: evidence for vertical translocation of a water molecule during the proton pumping cycle J Mol Biol 335: 531–546

Lanyi JK (2004) What is the real crystallographic structure of L photointermediate of bacteriorhodopsin? Biochim Biophys Acta 1658: 14–22

Lanyi J, Schobert B (2003) Mechanism of proton transport in bacteriorhodopsin from crystallographic structures of the K, L, M1, M2, and M2′ intermediates of the photocycle. J Mol Biol 321: 727–737

Luecke H, Schobert B, Richter HT, Cartailler JP, Lanyi JK (1999) Structure of bacteriorhodopsin at 1.55 Å resolution. J Mol Biol 291: 899–911

Maeda A (1995) Application of FTIR spectroscopy to the structural study on the function of bacteriorhodopsin. Isr J Chem 35: 387–400

Maeda A (2001) Internal water molecules as mobile polar groups for light-induced proton translocation in bacteriorhodopsin and rhodopsin as studied by difference FTIR spectroscopy. Biochemistry (Moscow) 66: 1256–1268

Maeda A, Tomson FL, Gennis RB, Ebrey TG, Balashov SP (1999) Chromophore-protein-water interactions in the L intermediate of bacteriorhodopsin: FTIR study of the photoreaction of L at 80 K. Biochemistry 38: 8800–8807

Maeda A, Tomson F, Gennis RB, Kandori H, Ebrey TG, Balashov SP (2000) Relocation of internal bound water in bacteriorhodopsin during the photoreaction of M at low temperature: an FTIR study. Biochemistry 39: 10154–10162

Maeda A, Balashov SP, Lugtenburg J, Verhoeven M, Herzfeld J, Belenky M, Gennis RB, Tomson FL, Ebrey TG (2002) Interaction of internal water molecules with the Schiff base in the L intermediate of the bacteriorhodopsin photocycle. Biochemistry 41: 3803–3809

Maeda A, Tomson F, Gennis R, Balashov S, Ebrey T (2003a) Water molecule rearrangements around Leu93 and Trp182 in the formation of the L intermediate in bacteriorhodopsin's photocycle. Biochemistry 42: 2535–2541

Maeda A, Herzfeld J, Belenky M, Needleman R, Gennis R, Balashov S, Ebrey T (2003b) Water-mediated hydrogen-bonded network on the cytoplasmic side of the Schiff base of the L intermediate of bacteriorhodopsin. Biochemistry 42: 14122–14129

Pfefferle M, Maeda A, Sasaki J, Yoshizawa T (1991) Application of FTIR spectroscopy to the structural study of the function of bacteriorhodopsin. Biochemistry 30: 6548–6556

Schobert B, Brown L, Lanyi J (2003) Crystallographic structures of the M and N intermediates of bacteriorhodopsin: Assembly of a hydrogen-bonded chain of water molecules between Asp-96 and the retinal Schiff base. J Mol Biol 330: 553–570

Varo G, Lanyi JK (1991) Thermodynamics and energy coupling in the bacteriorhodopsin photocycle. Biochemistry 30: 5016–5022

Yamazaki Y, Tuzi S, Saito H, Kandori H, Needleman R, Lanyi J, Maeda A (1996) Hydrogen bonds of water and C=O groups coordinate long-range structural changes in the L photointermediate of bacteriorhodopsin. Biochemistry 35: 4063–4068

Yamazaki Y, Kandori H, Needleman R, Lanyi J, Maeda A (1998) Interaction of the protonated Schiff base with the peptide backbone of valine 49 and the intervening water molecule in the N intermediate. Biochemistry 37: 1559–1564

Chapter 25

Phototaxis Photoreceptor in *Euglena gracilis*

Donat-Peter Häder[1], Maria Ntefidou[1], Mineo Iseki[2,3], and Masakatsu Watanabe[2,4]

Introduction

Many motile microorganisms, such as flagellates and ciliates, optimize their position in the water column by motile responses controlled by external stimuli. The photosynthetic, unicellular flagellate *Euglena gracilis* primarily uses light (Lebert and Häder 2000, Lebert 2001) and gravity (Häder et al 2003) for this purpose.

In addition to photokinesis, which is a steady-state dependence of the swimming velocity on the irradiance (Wolken and Shin 1958), the cells show photophobic responses upon a sudden step-down or step-up in the actinic light intensity (Doughty and Diehn 1984). At low light intensities (below $10\,W\,m^{-2}$) the cells show a directed movement toward the light source (positive phototaxis) and swim away from the light source at higher intensities (negative phototaxis) (Häder and Lebert 1985). In the water column, these antagonistic reactions to light and gravity control the vertical position of the cells in the water column (Häder and Griebenow 1988).

In contrast to many other light-responsive, eukaryotic, unicellular algae, neither is the stigma of *Euglena gracilis* organized in a quarter wavelength stack (Kreimer 1994), nor is it the photoreceptor for phototactic orientation. The receptor is believed to be located in the paraflagellar body, also called the paraxonemal body (PAB) (Andersen et al 1991). This organelle is located on the emerging flagellum, inside an invagination of the front end, at the position where the short flagellum merges. The photoreceptor pigments are oriented dichroically. In polarized light *Euglena* cells orient at an angle of about 30° clockwise to the electrical vector (Häder 1987). The dichroic orientation is further supported by the

[1] Chair of Plant Ecophysiology, Friedrich-Alexander University, Staudtstrasse 5, D-91058 Erlangen, Germany
[2] National Institute for Basic Biology, Okazaki, Aichi 444-8585, Japan
[3] PRESTO, Japan Science and Technology Agency, Kawaguchi, Saitama 332-0012, Japan
[4] Department of Photoscience, School of Advanced Sciences, Graduate University for Advanced Studies, Shonan Village, Hayama, Kanagawa 240-0193, Japan

quasi-crystalline structure of the PAB seen in electron microscopy (Piccinni and Mammi 1978).

All published action spectra for phototaxis in *E. gracilis* show two major maxima in the UV-A and blue regions of the spectrum (see review by Lebert 2001). No sensitivity could be found above 520 nm. Based on these results as well as spectroscopic and biochemical data, most authors assume the photoreceptor molecules to be flavins and pterins (Brodhun and Häder 1990). Biochemical analysis of the PAB revealed four major proteins with molecular masses between 27 and 33 kDa carrying pterins and flavins as shown by their fluorescence spectra (Brodhun and Häder 1990, Häder and Brodhun 1991). In contrast, Gualtieri (1993) claimed that the phototaxis photoreceptor in *E. gracilis* is a rhodopsin.

Photoreceptor for Step-Up Photophobic Responses in *Euglena*

The controversy about the nature of the photoreceptor was ultimately solved by the molecular genetic analysis published in a highly recognized paper by Iseki et al (2002). The authors succeeded in isolating and identifying a flavoprotein photoactivated adenylyl cyclase (PAC). PAC is a member of a novel blue-light receptor family consisting of two PACα and two PACβ subunits with molecular weights of 105 (1019 amino acids) and 90 kDa (859 amino acids), respectively. Each subunit contains a tandem repeat of a FAD-binding domain and an adjacent adenylyl cyclase catalytic domain (Iseki et al 2002). The two genes have been sequenced and the overall similarity between PACα and PACβ is 72% at the nucleotide level. The location of the photoreceptor proteins in the PAB has been demonstrated by indirect immunofluorescence staining using polyclonal antibodies. Sequence alignment of the two flavin binding domains with genes from several bacteria and cyanobacteria revealed the presence of a BLUF (blue light receptor using flavins) domain.

Fluorescence excitation spectra of the photoreceptor pigments resembled the action spectrum for photophobic responses in *E. gracilis*. Excitation of the photoreceptor protein in vitro by UV/blue light (peaks at 370 and 450 nm) resulted in enhanced cAMP production by the cyclase. This secondary messenger is thought to control the flagellar beat pattern and trigger step-up photophobic responses.

Introducing double-stranded mRNA (dsRNA) of PACα or PACβ or both into the cells by electroporation resulted in RNA interference (RNAi) and completely blocked the step-up photophobic response. In addition, after RNAi the PAB could not be seen in the cells by Nomarski interference microscopy or autofluorescence, and also Northern blots did not show the gene products of PACα and PACβ in RNAi-treated cells. But RNAi of PACα or PACβ did not impair the step-down photophobic response (Iseki et al 2002), indicating that the PAC gene products are not the photoreceptor for the latter reaction. This is consistent with the different action spectrum of the step-up responses (Matsunaga et al 1998).

Mutant Analysis

The colorless relative of *Euglena*, *Astasia longa*, also shows step-up (but not step-down) photophobic responses. PCR and subsequent sequencing indicated that *Astasia* possesses the PAC genes that are 95% homologous to those of *Euglena* (Ntefidou et al 2003a). In addition, several colorless mutants of *E. gracilis* have been sequenced. The mutants fall into two groups, one being closely related with *E. gracilis* and the other with *A. longa* (M Ntefidou, personal communication). Despite the differences, RNAi using dsRNA of PACα or PACβ or both inhibited step-up photophobic responses in all strains but not step-down responses in those strains that show this response (Table 1). This result clearly indicates that the step-down photophobic response is mediated by a different photoreceptor than the step-up response.

In *Astasia* no PAB could be found using autofluorescence of the organelle (Lebert and Häder 1997). Likewise, this method did not reveal the presence of PABs in the strains 1F, 9F and st⁻, while small PABs could be detected in some cells of the strain FB. These findings raise the question where the photoreceptor for photophobic responses is located. Using confocal microscopy and indirect immunofluorescence with polyclonal antibodies against PACα showed that at least *Astasia* and the stigmaless strain lack the PAB while this method indicated the presence of PABs in the other strains. However, this technique clearly demonstrated that, in most strains, the PACα gene product occurs along the entire length of the flagellum. This indicate that, in contrast to the previous assumption, the photoreceptor for photophobic reactions is not located in the PAB (this is at least proven for the PAB-lacking strains), but on the flagellum.

TABLE 1. Occurrence of step-up and step-down photophobic responses in *Euglena gracilis* (E.g.) and *Astasia longa* (A.l.) and its inhibition by RNAi

DsRNA	E.g. wt	E.g. dark grown	E.g. 1F	E.g. 9F	E.g. FB	E.g. st⁻	A.l.
Step-up phobic responses							
Control	+	+	+	+	+	+	+
PACα	–	–	–	–	–	–	–
PACβ	–	–	–	–	–	–	–
PACα and PACβ	–	–	–	–	–	–	–
Step-down phobic responses							
Control	+	+	–	–	–	+	–
PACα	+	+	–	–	–	+	–
PACβ	+	+	–	–	–	+	–
PACα and PACβ	+	+	–	–	–	+	–

Wt, wild type; st⁻, stigmaless.

PAC Involvement in Phototaxis

As indicated above, wild-type (green and dark grown) *Euglena* cells show positive or negative phototaxis depending on the irradiance. None of the other strains show normal phototaxis, but some perform a diaphototaxis (swimming perpendicular to the incident light beam) at high fluence rates (Lebert and Häder 1997). Since the action spectrum for phototaxis is similar to that for photophobic responses (Häder and Reinecke 1991), it might be possible that the same or similar photoreceptor might be responsible for the phototaxis.

To tackle this question, RNAi was used again and phototactic orientation was determined at both high and low fluence rates using an automatic, real-time tracking system (Häder and Lebert 2000). In fact, both positive and negative phototaxis was eliminated after RNAi using PACα, PACβ or both as templates (Ntefidou et al 2003b). Neither the swimming velocity nor the form factor of the cells were affected by this treatment. It is interesting to note that the inhibition of light-dependent orientation lasted more than two months and at least double as long as the impairment of step-up photophobic responses by the same treatment. The latter result could mean that phototaxis is mediated by a slightly different photoreceptor protein from the same family than that involved in photophobic responses, or that different concentrations of the gene product are needed for the different reactions. In any case it is obvious that the inhibitory dsRNA is carried to the daughter cells after cell division over many generations and only dilutes out after a considerable time. Since only cells that possess a PAB show phototactic orientation, it could be speculated that the PAC photoreceptor responsible for phototaxis may be located in the PAB rather than on the flagellum. But why do some of the mutants, which do possess a PAB, not show phototactic orientation? One interesting observation is that the mutant 1F of *E. gracilis* does not show flavin fluorescence of PAB preparations (Häder and Lebert 1998). This may indicate that while the PAB is produced it does not contain a functional PAC with a bound FAD.

The Role of Pterins in Phototaxis

Biochemical and spectroscopic data indicate that pterins may be involved in phototaxis of *E. gracilis* (Brodhun and Häder 1990, 1993, 1995, Häder and Brodhun 1991, Lebert and Häder 1997). The hypothesis was that pterins operate as antenna pigments that transfer the energy absorbed to flavins, which in turn perform the primary photochemical reaction. Pterin biosynthesis starts from GTP and is mediated by a key enzyme, GTP cyclohydrolase I. In contrast, the flavin pathway is controlled by GTP cyclohydrolase II. GTP cyclohydrolase I has been found in many organisms including bacteria, fungi and mammals (Auerbach et al 2000). In *Escherichia coli* it is a homodecamer with 247 kDa. The genetic sequence and the protein structure are known. It has dimensions of 65 × 100 Å and contains 10 active sites with one zinc ion per active site (Auerbach et al 2000). The enzyme was also found in *E. gracilis* and has been sequenced. We are grateful to J. Maier (Tübingen) who donated us the gene in form of a plasmid.

We used the sequence to produce dsRNA and to insert it into *E. gracilis* by electroporation. The result was a complete inhibition of both positive and negative phototaxis by RNAi, which again lasted for many generations. As with RNAi of PAC, neither swimming velocity nor cell form was affected. But, very surprisingly, neither step-up nor step-down photophobic reactions were impaired. These results can be interpreted by a number of options. First, the energy absorbed by only flavins is not sufficient to drive phototaxis but to mediate photophobic responses. The PAC photoreceptor for step-up (but not step-down) photophobic responses could be different from that responsible for phototaxis and the latter needs the cooperation with pterins. The final option is that pterins are needed for the correct assembly of the PAB. If true, the latter assumption would support the hypothesis that the PAC photoreceptor responsible for phototaxis is different and located in a different place from that involved in step-up photophobic responses.

Summary and Conclusions

Photoactivated adenylyl cyclase (PAC) is a tetrameric 390 kDa protein which controls step-up photophobic (but not step-down) responses in *E. gracilis* wild-type and phototaxis mutant strains as well as in the colorless relative *A. longa*. This novel class of photoreceptor pigments was also found to mediate positive and negative phototaxis. While the original hypothesis that PAC is located in the paraxonemal body (PAB) seems to hold for phototaxis, it cannot be true for photophobic responses in several mutants and *Astasia* that lack a PAB. Instead the location of PAC along the entire length of the flagellum was shown by confocal immunofluorescence microscopy. This indicates that there is a whole family of related PAC proteins in *Euglena* and its relatives. This is confirmed by a sequence analysis of PAC genes in the mutants that revealed considerable divergence between the strains. It remains to be explored whether or not step-down photophobic responses, the action spectrum of which is different from that for step-up responses, is mediated by still another PAC photoreceptor. While the involvement of pterins in phototaxis was assumed and is now proven by RNAi, it was a surprise that the presence of pterins is not mandatory for step-up photophobic responses.

References

Andersen RA, Barr DJS, Lynn DH, Melkonian M, Moestrup O, Sleigh MA (1991) Terminology and nomenclature of the cytoskeletal elements associated with the flagellar/ciliary apparatus in protists. Protoplasma 164: 1–8

Auerbach G, Herrmann A, Bracher A, Bader G, Gutlich M, Fischer M, Neukamm M, Garrido-Franco M, Richardson J, Nar H, Huber R, Bacher A (2000) Zinc plays a key role in human and bacterial GTP cyclohydrolase I. Proc Natl Acad Sci USA 97: 13567–13572

Brodhun B, Häder DP (1990) Photoreceptor proteins and pigments in the paraflagellar body of the flagellate *Euglena gracilis*. Photochem Photobiol 52: 865–871

Brodhun B, Häder DP (1993) UV-induced damage of photoreceptor proteins in the paraflagellar body of *Euglena gracilis*. Photochem Photobiol 58: 270–274

Brodhun B, Häder DP (1995) A novel procedure to isolate the chromoproteins in the paraflagellar body of the flagellate *Euglena gracilis*. J Photochem Photobiol B Biol 28: 39–45

Doughty MJ, Diehn B (1984) Anion sensitivity of motility and step-down photophobic responses of *Euglena gracilis*. Arch Microbiol 138: 329–332

Gualtieri P (1993) *Euglena gracilis*: is the photoreception enigma solved? J Photochem Photobiol 19: 3–14

Häder DP (1987) Polarotaxis, gravitaxis and vertical phototaxis in the green flagellate, *Euglena gracilis*. Arch Microbiol 147: 179–183

Häder DP, Brodhun B (1991) Effects of ultraviolet radiation on the photoreceptor proteins and pigments in the paraflagellar body of the flagellate, *Euglena gracilis*. J Plant Physiol 137: 641–646

Häder DP, Griebenow K (1988) Orientation of the green flagellate, *Euglena gracilis*, in a vertical column of water. FEMS Microbiol Ecol 53: 159–167

Häder DP, Lebert M (1985) Real time computer-controlled tracking of motile microorganisms. Photochem Photobiol 42: 509–514

Häder DP, Lebert M (1998) The photoreceptor for phototaxis in the photosynthetic flagellate *Euglena gracilis*. Photochem Photobiol 68: 260–265

Häder DP, Lebert M (2000) Real-time tracking of microorganisms. In: Häder DP (ed) Image analysis: methods and applications. CRC, Boca Raton, pp 393–422

Häder DP, Reinecke E (1991) Phototactic and polarotactic responses of the photosynthetic flagellate, *Euglena gracilis*. Acta Protozool 30: 13–18

Häder DP, Lebert M, Richter P, Ntefidou M (2003) Gravitaxis and graviperception in flagellates. Adv Space Res 31: 2181–2186

Iseki M, Matsunaga S, Murakami A, Ohno K, Shiga K, Yoshida C, Sugai M, Takahashi T, Hori T, Watanabe M (2002) A blue-light-activated adenylyl cyclase mediates photoavoidance in *Euglena gracilis*. Nature 415: 1047–1051

Kreimer G (1994) Cell biology of phototaxis in flagellate algae. Int Rev Cytol 148: 229–309

Lebert M (2001) Phototaxis of *Euglena gracilis*—flavins and pterins. In: Häder DP, Lebert M (eds) Photomovement 1. Elsevier, Amsterdam, pp 297–341

Lebert M, Häder DP (1997) Behavioral mutants of *Euglena gracilis*: functional and spectroscopic characterization. J Plant Physiol 151: 188–195

Lebert M, Häder DP (2000) Photoperception and phototaxis in flagellated algae. Res Adv Photochem Photobiol 1: 201–226

Matsunaga S, Hori T, Takahashi T, Kubota M, Watanabe M, Okamoto K, Masuda K, Sugai M (1998) Discovery of signaling effect of UV-B/C light in the extended UV-A/blue-type action spectra for step-down and step-up photophobic responses in the unicellular flagellate alga *Euglena gracilis*. Protoplasma 201: 45–52

Ntefidou M, Iseki M, Richter P, Streb C, Lebert M, Watanabe M, Häder DP (2003a) RNA interference of genes involved in photomovement in *Astasia longa* and *Euglena gracilis* mutants. Recent Res Dev Biochem 4: 925–930

Ntefidou M, Iseki M, Watanabe M, Lebert M, Häder DP (2003b) Photoactivated adenylyl cyclase controls phototaxis in the flagellate *Euglena gracilis*. Plant Physiol 133: 1517–1521

Piccinni E, Mammi M (1978) Motor apparatus of *Euglena gracilis*: ultrastructure of the basal portion of the flagellum and the paraflagellar body. Boll Zool 45: 405–414

Wolken JJ, Shin E (1958) Photomotion in *Euglena gracilis*. I. Photokinesis. II. Phototaxis. J Protozool 5: 39–46

Chapter 26

A Flavin Mononucleotide-Binding Aquaporin in the Plant Plasma Membrane: A Candidate for Photoreceptor?

Rainer Hertel

Introduction

In plants and fungi some of the blue light photoreceptors must reside in or at the plasma membrane, and most, if not all, should work with "bound" flavins. A riboflavin binding protein was characterized, initially as sites of reversible association in membrane material from several higher plants (Hertel et al 1980) and from the lower fungus *Phycomyces* (Dohrmann 1983, Flores et al 1999). We showed that it was a PIP1-type aquaporin in a higher plant (Lorenz et al 2003). The sheer amount of this binding protein is impressive: it constitutes more than 1% of the total plasma membrane. Comparing flavoproteins, e.g., of *Phycomyces*, in the non-mitochondrial membranes, about five times more of our protein was present than the sum of all other flavoproteins. A similar predominance is seen in plasma membranes of higher plants. (Whether the flavin binding proteins, described by Dederichs et al 1999 and by Neumann and Hertel 1994 in flagellar preparations of *Chlamydomonas* and *Euglena*, respectively, are homologous, remains to be investigated.)

Why search for another possible blue light receptor beyond the phototropins? These proteins were originally found by their blue light-mediated phosphorylation (Short and Briggs 1990, Hager and Brich 1993). They are photochemically active flavoproteins (Salomon et al 2000), and their essential photoreceptor role, e.g., in phototropism, has been documented by mutant evidence (Christie and Briggs 2001). Several arguments however justify a further search. (a) The quantity estimate, at least for *Phycomyces* (see below), argues against an exclusive role of phototropins. (b) Defects in a "forward" and limiting photoreceptor should show some shift of the fluence response curve to the right. Analysis of partial defects in phototropins is still incomplete. (c) Furthermore, effects of two successive blue light stimuli in *Phycomyces* (Löser and Schäfer 1986) and in *Avena* (Meyer 1969) strongly suggest the involvement, in the same intensity

Institut Biologie III, Albert-Ludwigs-Universitaet, Schaenzlestrasse 1, D-79104 Freiburg, Germany

range, of two different flavin-type photoreceptors (see below). One receptor should be phototropin, the other still to be identified.

The role of cryptochromes in light signaling is undisputed. However, their role as receptors remains to be established.

A Specific Aquaporin, a Major Integral Protein of Plant Plasma Membrane, Binds Flavin Mononucleotide (FMN)

Solubilization, purification, and partial sequencing of the riboflavin binding protein from etiolated *Cucurbita* hypocotyls led to a PIP1-type plasma membrane aquaporin (Lorenz et al 2003). This surprising conclusion was confirmed by binding tests, shown in Table 1, with extracts from transgenic material, from both tobacco and *Escherichia coli.*

Concerning subcellular localization, the flavin binding activity had previously been assigned to the plant plasma membrane by density gradient separation of vesicles (Hertel et al 1980) and by 2-phase-PEG partitioning (Lorenz et al 2003). Of course, the identification as a PIP1-type aquaporin fully confirmed the binding activity as an integral protein of the plasmalemma.

The 40–50 N-terminal amino acids are unique for the PIP1 type flavo-aquaporin while other aquaporins, e.g., PIP2, lack this domain. We hypothesize that the PIP1 N-terminus interacts with the flavin. As seen in Table 2, this domain is well conserved from mosses to higher plants.

The physiological involvement of the PIP1 protein is under investigation. Uehlein et al (2003) report a role as a membrane CO_2 pore in tobacco with a significant effect on a blue light response (their Figure 4).

TABLE 1. ^{3}H-Riboflavin binding (saturable [bound/free] × 1000) using membrane material from cells with modified aquaporin NtAQP1 (PIP1) expression

	Specific riboflavin binding
Tobacco "normal"	18 ± 2.5
Tobacco overexpression	49 ± 15
Tobacco antisense	2.1 ± 5.4
Escherichia coli NtAQP1-GST	17.0
Escherichia coli control GST	2.5

Data, except "normal," from Lorenz et al 2003; transgenic material kindly provided by Beate Otto and Ralf Kaldenhoff; see Siefritz et al 2002. Membrane vesicles from leaves of three lines of *Nicotiana tabacum* (average ± SEM, $n = 5$) and from two lines of *Escherichia coli* (two independent tests), one expressing NtAQP1 from tobacco.

TABLE 2. Specific, PIP1-type aquaporin N-terminal domains

N. tabacum	maenkeedvklgankfretqplgtaaqt—dkdykepppaplfepgelssw.
A. thaliana	megkeedvrvgankfperqpigtsaqs—dkdykepppaplfepgelasw.
Z. mays	megkeedvrlgankfserqpigtaaqgagagdddkdykepppaplfepgelksw.
P. patens	mqqdkdddvalgankygtrsalgtha—pvpekdyrepsvtpffdggelrlw.

Total polypeptide lengths: 270–290 aa. Examples from different plants: *Nicotiana tabacum* (NtAQP1), *Arabidopsis thaliana* (PIP1b CAB37860), *Zea mays* (ZmPIP1–4 AF326488.1); from the moss *Physcomitrella patens* (in EST PP1103:BJ180316).

A Stable Flavin Adduct is Formed After Illumination, and in planta the Ligand is FMN

During purification of the flavin binding protein, we observed that reduced flavin, in the presence of dithionite, formed a very stable adduct of riboflavin or FMN with the protein within a few minutes, i.e. the flavin can be retained on a PEI-coated glass filters (Lorenz et al 2003). When such adducts are diluted and kept in presence of dithionite, the association remains even after 1 h. If however, diluted into buffer without dithionite under oxidizing conditions, free flavin is released from the protein with a half life of approximately 30 min. Flavin release from the binding protein could also be documented in a fluorimetric assay which allowed a quantitation of the binding sites occupied (Lorenz et al 2003).

A similar or identical association could also be induced with oxidized flavin photochemically by blue light at 450 nm, but not at 550 nm, in the presence of an electron donor like EDTA or ferredoxin (Lorenz et al 2003). The amount of flavin bound was the same as with dithionite reduction.

We determined the absorption spectrum of the light-induced adduct by measuring (a) the spectrum before release, (b) the absorption after 60 min release, and (c) a standard free flavin spectrum adjusted to the amount of flavin released; after computing a–b+c, we found a peak at 414 nm for the adduct (T Kunkel, A Lorenz, R Hertel unpublished) which could be a flavin–N_5-C-S-met-protein adduct as described for a LOV1–C57M mutant by Kottke et al (2003, their Figures 2B and 7) rather than the flavin–C_{4a}-S-protein described for wild-type phototropins (Salomon et at 2000).

The light-induced adduct formation described up to now was "artificial" because electron donors like 10 mM EDTA had to be added and the quantum efficiency was relatively low (2%–5%). Using FMN in place of riboflavin, preliminary data from our laboratory indicate adduct formation in the absence of high electron donor concentrations. However, the best evidence for an in planta photochemical adduct formation consists in an extraction of flavin-aquaporin adduct from *Cucurbita* hypocotyl tissue (U Dohrmann, A Lorenz, R Hertel, unpublished). Following methods of Lorenz et al (2003) we quickly prepared microsomal membranes, solubilized them and removed free flavins over a NAP-column, all in <20 min. Then we allowed any possible adducts to resolve during

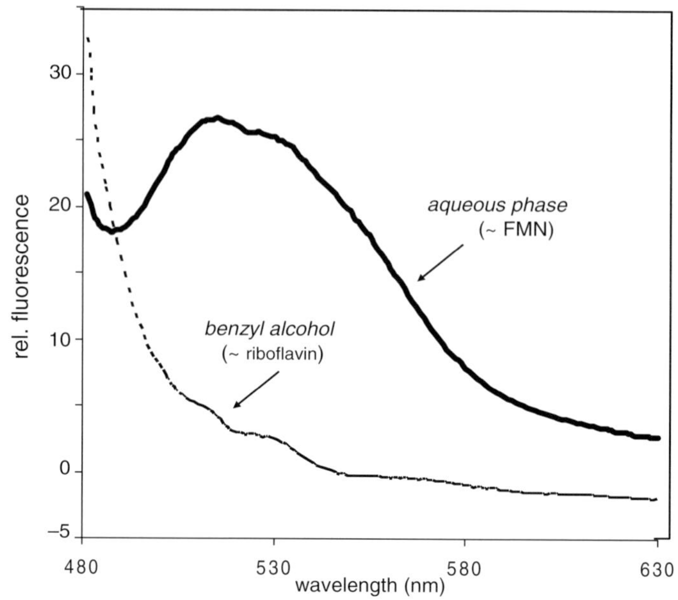

FIG. 1. FMN appears to be the natural ligand of PIP1-type flavin-binding aquaporin (Dohrmann et al, unpublished). Flavin fluorescence was measured after release from freshly extracted and solubilized *Cucurbita* microsomal proteins. FAD, FMN, and riboflavin were differentiated by benzyl alcohol/water partitioning and by measuring fluorescence yield at pH 2 and at pH 7 (see Udenfriend 1962)

a further 60 min. A large amount of flavin fluorescence was set free, about 0.1–0.2 nmol g^{-1} fresh weight. This flavin was then subjected to a selective identification test according to Udenfriend (1962): the aqueous material is partitioned with benzylalcohol; riboflavin is found in the organic phase while FMN and flavin adenine dinucleotide (FAD) remain in the buffer; pH-dependent fluorescence allows to distinguish between FMN and FAD. Surprisingly, as shown in Figure 1, essentially all flavin coming from the freshly extracted *Cucurbita* protein was found in the aqueous phase, i.e., it was neither riboflavin nor FAD but FMN as confirmed by fluorescence at different pH values.

A Similar Flavo-Protein (70 kDa) May Exist in *Phycomyces*

The molecular identity of the abundant membrane-associated flavin binding sites in *Phycomyces blakesleeanus* (Mucorales, Phycomyceta) is still under study. However, circumstantial evidence suggests a protein very similar to that described for higher plants.

Firstly, the characteristics like localization or affinity patterns, e.g., for riboflavin and FMN, correspond completely (Flores et al 1999). Adduct formation under reducing conditions is implied by the stability of the flavin-protein association (see filter test in Flores et al 1999, Polaino Orts et al unpublished). At this point it should be remembered that the photobiological characteristics of tropisms in higher plants and in *Phycomyces* are strikingly similar (Galland 1990), suggesting homology.

Secondly, FMN agarose-affinity columns as used for the purification of the plant flavo-aquaporin (Lorenz et al 2003) retain a 70kDa protein and not a 27/50kDa polypeptide (Polaino Orts et al unpublished).

Thirdly, a possible PIP1-homolog exists in yeast (Yfl054cp) with a predicted length of 646aa (~70kDa) and not ~270aa. It has a full aquaporin sequence at the C-terminus, an unknown N-terminal stretch, and in the middle a region homologous (>20%) with the critical N-terminus of plant PIP1 (Table 2).

Fourthly, Pollock et al (1985) could detect an abundant 70kDa membrane protein which tightly retained radioactive flavin even on SDS gels. An apparent discrepancy can now be settled: while Dohrmann (1983) and Pollock et al (1985) reported a large amount of flavin covalently linked to the *Phycomyces* membrane, we found only very small concentrations of tightly associated flavin (Flores et al 1999). It is likely that during our four membrane washings with centrifugations in between, the adduct could resolve while Dohrmann (1983) as well as Pollock et al (1985) used quick or denaturing preparations.

How Much Photoreceptor Do We Need in *Phycomyces*?

The receptors for phototropism in *P. blakesleeanus* are dichroic and must be fixed in or at plasmalemma (Steinhardt et al 1989). They are protein-bound flavins (evidence reviewed by Flores et al 1999).

Max Delbrück (in Bergman et al 1969) estimated the minimal concentration of flavin photoreceptor imposed by the extremely low phototropism threshold of *Phycomyces* sporangiophores. His argument rests on two plausible assumptions, a flavin extinction coefficient and a statistically significant difference between light quanta absorbed at far minus near flank.

At threshold, sporangiophores bend towards a total flux of 10^9 quanta cm^{-2} blue light. The flavin extinction coefficient of 20–40000 converts to $1.5 \times 10^{-16} cm^2$ molecular capture cross section. Consequently, at threshold there are 1.5×10^{-7} excitations/receptor molecule. To significantly distinguish proximal from distal flank in the lens focal band at least some 100 excitations are required, i.e., the establishment of an asymmetry needs a total of about $1–3 \times 10^9$ receptor molecules/growing zone; this can be expressed as a minimal receptor concentration of $0.3 nmol g^{-1}$ fresh weight. (Note that amplification factors should not change the threshold-statistical argument.)

The concentration of flavins bound covalently or tightly to sporangiophore membranes, washed for ca. 1h, was far below the minimum postulated by Delbrück. Only the concentration of the flavin binding sites was significantly higher, on the order of 1 nmol g^{-1} (Flores et al 1999). Therefore we proposed the working hypothesis that the photoreceptor consists of the described membrane protein and a flavin chromophore, freely exchanged with a cytosolic pool.

Phycomyces contains considerable amounts of free riboflavin and FMN (Dohrmann 1983). The cytoplasmic concentrations exceed the minimal concentration for photoreceptor action, and their affinity constants argue that the suspected PIP1 binding sites must be occupied under physiological conditions. This argument applies to higher plants, too.

Physiological Data Suggest Photochromicity and/or Antagonism

With any photoreceptor mechanism the following possible components have to be considered: (localized) adaptation ("modification") of the receptor, (non-linear) amplification, photochromicity, and antagonism. The first two processes will not be discussed here. Photoreversibility and antagonism of two signal chains are certainly very different processes, but both have an important physiological function.

Prime examples for photochromicity are the red/far-red reversibility of phytochrome and of many invertebrate opsins. A "phytochrome-blue-response" was suggested by Karl M. Hartmann. Modifying his idea, I proposed a flavin-photochromic mechanism and supported it with a list of photophysiological evidence (Hertel 1980). The short-lived flavin semiquinone was discussed as the active, photoreversible form; but now the long-lived FMN adducts appear to be the more plausible candidates. Recently Kennis et al (2003) demonstrated an in vitro photochromic mechanism for FMN-phototropin.

For the PIP1–FMN adduct described here, I suggest the following hypothetical mechanism: oxidized FMN and PIP1 bind loosely (PIP1 . . . FMN); then blue light induces a covalent bond (adduct PIP1met-N5–FMN_{414nm}) which delivers the signal; this adduct can further be modified (adaptation); finally it will return to the starting form by slowly releasing FMN in the dark, or by a photochromic reversion.

Considering antagonistic signaling, such chains may function when it comes to respond over a large dynamic range, relatively independent of fluence rates. In color vision, a center surround opponency of cones (e.g., De Valois et al 1966) serves to distinguish hues. In phototropism, spatial differences may be sensed by opposing receptors. Strong evidence for a two-flavin-antagonistic mechanism is provided simply by the existence of pulse-induced negative phototropism (see Iino 2001). Furthermore, Meyer (1969) and Löser and Schäfer (1986) showed in *Avena* and in *Phycomyces*, respectively, that a second blue pulse would act inhibitory, following after a first, non-saturating blue pulse at certain intervals.

Figure 2 presents a rough outline of an antagonistic blue light signaling process. Within such a hypothesis one may ask whether phototropin acts stimulatory or inhibitory. Salomon et al (1997) showed that after unilateral in vivo blue irradi-

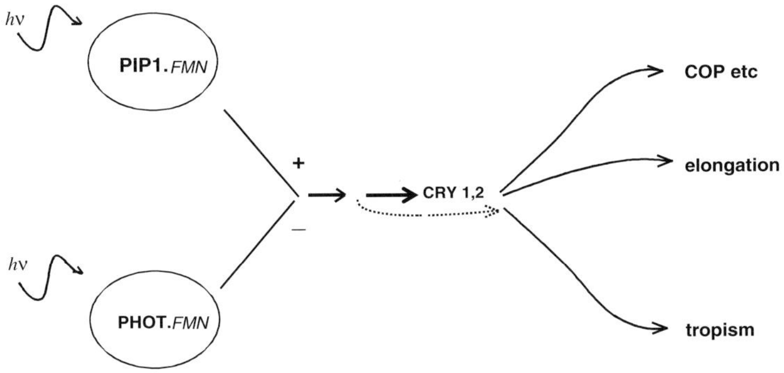

FIG. 2. A hypothesis to organize plant blue light receptors and early transduction with antagonistic signaling from PIP1–FMN adducts and PHOT–FMN adducts. Cryptochromes are depicted as essential elements in the light control but not as photoreceptors. *COP* is a complex of transcription factors

ation, with a fluence of $1\,\mu mol\,m^{-2}$, the level of phototropin phosphorylation was significantly higher at the irradiated than at the shaded side of an *Avena* coleoptile tip. At this fluence, however, the curvature response is negative (Iino 2001, his Figure 3B). Interpreting the phosphorylation levels—at face value—as reflecting excitation of phototropin, the results suggest an antagonistic role for phototropins.

Acknowledgments. I thank Astrid Lorenz for invaluable collaboration, Monier Tadros for his most generous help in peptide sequencing, Ralf Kaldenhoff and his group for an excellent cooperation, and Angelika Schäfer for enduring assistance. Help and advice by Ulrike Dohrmann, Tim Kunkel, Stefan Rensing, and Eberhard Schäfer are gratefully acknowledged, as are the unpublished data by Dohrmann, Kunkel, Lorenz, and Polaino Orts. Our Institut Biologie III provided support.

References

Bergman K, Burke PW, Cerdá-Olmedo E, David CN, Delbrück M, Foster KW, Goodell WE, Heisenberg M, Meissner G, Zalokar M, Dennison DS, Shropshire W (1969) Phycomyces. Bacteriol Rev 33: 99–157

Christie JM, Briggs WR (2001) Blue light sensing in higher plants. J Biol Chem 276: 11457–11460

Dederichs A, Schäfer A, Hertel R, van den Ende H (1999) Characterization of flavin-binding sites in *Chlamydomonas flagella*. Plant Biol 1: 315–320

De Valois RL, Abramov I, Jacobs GH (1966) Analysis of response patterns of LGN cells. J Opt Soc Am 56: 966–977

Dohrmann U (1983) In vitro riboflavin binding and endogenous flavins in *Phycomyces blakesleeanus*. Planta 159: 357–365

Flores R, Dederichs A, Cerdá-Olmedo E, Hertel R (1999) Flavin-binding sites in *Phycomyces*. Plant Biol 1: 645–655

Galland P (1990) Phototropism of the *Phycomyces* sporangiophore: a comparison with higher plants. Photochem Photobiol 52: 233–248

Hager A, Brich M (1993) Blue-light-induced phosphorylation of a plasma-membrane protein from phototropically sensitive tips of maize coleoptiles. Planta 189: 567–576

Hertel R (1980) Phototropism of lower plants. In: Lenci F, Colombetti G (eds) Photoreception and sensory transduction in aneural organisms, NATO ASI-series A 33. Plenum, New York, pp 89–105

Hertel R, Jesaitis AJ, Dohrmann U, Briggs WR (1980) In vitro binding of riboflavin in subcellular particles from maize coleoptiles and *Cucurbita* hypocotyls. Planta 147: 312–319

Iino M (2001) Phototropism in higher plants. In: Häder D-P, Lebert M (eds) Photomovement. Elsevier Science, Amsterdam, pp 659–811

Kennis JTM, Crosson S, Gauden M, van Stokkum IHM, Moffat K, van Grondelle R (2003) Ultrafast spectroscopy of the LOV2 domain of phototropin: The light-driven switch works both ways. Biophys J 84: 272a

Kottke T, Dick B, Fedorov R, Schlichting I, Deutzmann R, Hegemann P (2003) Irreversible photoreduction of flavin in a mutated Phot-LOV1 domain. Biochemistry 42: 9854–9862

Löser G, Schäfer E (1986) Are there several photoreceptors involved in phototropism of *Phycomyces blakesleeanus*? Kinetic studies of dichromatic irradiation. Photochem Photobiol 43: 195–204

Lorenz A, Kaldenhoff R, Hertel R (2003) A major integral protein of the plant plasma membrane binds flavin. Protoplasma 221: 19–30

Meyer AM (1969) Versuche zur 1. positiven und zur negativen phototropischen Krümmung der Avenakoleoptile. Lichtperception und Absorptionsgradient. Z Pflanzenphysiol 60: 418–433

Neumann R, Hertel R (1994) Riboflavin binding protein: purification from flagella of *Euglena*. Photochem Photobiol 60: 76–83

Pollock JA, Lipson ED, Sullivan DT (1985) Analysis of microsomal flavoproteins from *Phycomyces* sporangiophores: candidates for the blue-light photoreceptor. Planta 163: 506–516

Salomon M, Zacherl M, Ruediger W (1997) Phototropism and protein phosphorylation in higher plants: unilateral blue light irradiation generates a directional gradient of protein phosphorylation across the oat coleoptile. Bot Acta 110: 214–216

Salomon M, Christie JM, Knieb E, Lempert U, Briggs WR (2000) Photochemical and mutational analysis of the FMN-binding domains of the plant blue light receptor, phototropin. Biochemistry 39: 9401–9410

Short TW, Briggs WR (1990) Characterization of a rapid, blue-light-mediated change in detectable phosphorylation of a plasma membrane protein from etiolated pea (*Pisum sativum* L.) seedlings. Plant Physiol 92: 179–185

Siefritz F, Tyree MT, Lovisolo C, Schubert A, Kaldenhoff R (2002) PIP1 plasma membrane proteins in *Nicotiona tabacum*: from cellular effects to the function in plants. Plant Cell 14: 869–876

Steinhardt AR, Popescu T, Fukshansky L (1989) Is the dichroic photoreceptor for *Phycomyces* phototropism located at the plasma membrane or at the tonoplast? Photochem Photobiol 49: 79–87

Udenfriend S (1962) Fluorescence assay in biology and medicine. Academic, New York

Uehlein N, Lovisolo C, Siefritz F, Kaldenhoff R (2003) The tobacco aquaporin NtAQP1 is a membrane CO_2 pore with physiological functions. Nature 425: 734–737

Chapter 27

Green Light Effects on Plant Growth and Development

Kevin M. Folta

Evolution tends not to ignore a conditional signal. There are many instances where a seed, seedling or plant finds itself in an environment disproportionately rich in green light (500–550 nm). Is this condition meaningful to the plant, and are there mechanisms to sense these wavebands to inform physiological decisions? Cryptochromes and phytochromes readily absorb green light to initiate photomorphogenic responses. Still, the classical and contemporary literature present sporadic evidence that green light irradiation has specific influence that is not conveniently attributed to known light sensors (Frechilla et al 2000, Kim et al 2004a, Klein 1992). Recent reports corroborate early evidence that green light has specific, frequently antagonistic functions in directing light responses (Eisinger et al 2003, Folta 2004, Talbott et al 2003).

Green light responses can be grouped into two distinct categories: those that antagonize normal light-mediated responses and those that function to forward normal developmental processes. Evidence of the former is presented in the 1957 text *Experimental Control of Plant Growth*. In this book Fritz Went describes experimental evidence that green wavebands actually retard seedling growth (pp 280–284). Tomato seedlings grown under red and blue filtered light had a greater dry mass (~200%) than those grown under a comparable (or greater) PPF of red and blue with supplemental green. As fluence rate increased red + blue + green light-grown seedlings exhibited a plateau in dry mass. The interpretation was that green light actively opposed the effects of red and blue (Went 1957).

The theme of green light antagonism continued into the next decade. Early in the development of tissue culture techniques it was determined that fluorescent light repressed culture growth. A study of the relationship between light source and crown gall callus growth showed a strong relationship between fluence rate and growth inhibition (Klein 1964). A crude action spectrum showed that growth was inhibited most by UV (360 ± 40 nm) and green light (550 ± 30 nm), which limited mass to less than 50% the mass of dark, blue, red, or far-red grown tissue.

Plant Molecular and Cellular Biology Program and Horticultural Sciences Department, University of Florida, Gainesville, FL 32611, USA

The growth inhibition reverted quickly when the green light stimulus was removed. Green light-induced growth repression was shown to be fluence-rate dependent, again emphasizing that green wavebands specifically and potently repressed plant tissue growth (Klein 1964).

Sporadic reports reiterated a theme of green light-dependent repression of normal light-driven processes. Klein (1979) showed that green light inhibited root gravitropism with peak inhibition at 550 nm, and green light inhibition could be reversed with an orange/red light treatment. Artificial lighting regimes with greater percentage of 500–580 nm light show inhibitory effects on specific facets of seedling physiology, such as lower chlorophyll content and less leaf expansion (Dougher and Bugbee 2001, Klein et al 1965). The fact that these wavebands negatively influence chlorophyll accumulation and leaf expansion, attributes positively affected by red and/or blue light, suggests that red and blue sensors are not directing the antagonistic response to green light.

Some green light effects occur in the same direction as normal light-mediated responses. Green light induces phototropic curvature in Arabidopsis and lettuce seedlings with characteristics distinct from blue light-mediated phototropism. This finding suggested a separate green light-sensing pigment (Steinitz et al 1985), although this hypothesis became less attractive with the observation that both blue and green phototropic responses are absent in the *phot1* mutant (Liscum and Briggs 1995). Field studies showed that green or yellow light reflected from colored mulch increased the aroma compounds and phenolic content of sweet basil (Loughrin and Kasperbauer 2001, 2003), and the addition of green wavebands to red and blue light positively affects the long-term growth of lettuce (Kim et al 2004a).

There is a substantial body of evidence from independent laboratories that implicates green light in the control of stomatal aperture. The phytochromes and phototropins participate in light control of guard cell turgor in response to red and blue light (Talbott et al 2002). Careful studies of stomatal opening in detached epidermal cells demonstrate that specific blue-, UV-A- and UV-B-induced stomatal responses can be reversed by green light. A blue light pulse leads to an increase in stomatal aperture (Eisinger et al 2003, Frechilla et al 2000). If the blue pulse is immediately followed by a green light pulse with twice the fluence of the blue light pulse, or if blue and green treatments are delivered simultaneously, opening does not occur. Opening can be restored with a subsequent pulse of blue light. Reversal of blue light-induced stomatal opening is fluence dependent. The action spectrum for reversal of stomatal opening has a peak at 540 nm with geometry reminiscent of blue light-induced opening, red shifted 90 nm (Frechilla et al 2000). These findings suggest the existence of a blue-green sensitive toggle akin to phytochrome. Stomatal opening is stimulated by UV-B acting through a yet unidentified photoreceptor, and the response to UV-B can also be reversed by green light (Eisinger et al 2003). The absence of the response in *npq1* mutants (containing a lesion zeaxanthin de-epoxidase and therefore fail to produce zeaxanthin) suggests that blue and green light induced changes in stomatal aperture may be regulated through *cis- trans-* isomerization of bulk zeaxanthin (Frechilla et al 1999, 2000). In this scenario, isomerization of zeaxanthin in the chloroplast sets a condition where the guard cell may respond to blue light activation of cytosolic/membrane-associated components (Eisinger et al 2003, Frechilla et al 1999, 2000). The effects on stomatal aperture persist in the whole plant as lettuce grown with supplemental green light exhibits proportionately lower stomatal conductance (Kim et al 2004b).

Narrow-bandwidth green light also influences early development in the etiolated seedling. As a general rule, stem elongation is most rapid in darkness and growth rate is suppressed by light (Parks et al 2001). Green light signals present an exception to the rule. Analysis of action spectra for stem growth inhibition shows that green light treatment causes hypocotyls of dark-grown seedlings to be slightly longer than dark-grown seedlings, in a phytochrome independent manner (Goto et al 1993). A high-resolution study of hypocotyl growth kinetics shows that a short, single green light pulse can increase the elongation rate in stems of dark grown Arabidopsis seedlings (Folta 2004). The green light response persists in *phy*, *cry*, and *phot* mutant backgrounds. Green light-induced growth promotion persists in a background of dim red light, further suggesting that phytochrome is not mediating the green light response. Here again, the effect of narrow band green light contradicts the usual role of light in advancing plant development.

The central portion of the visible spectrum has not been widely considered in shaping plant decisions. Its effects have been masked by the strong developmental influence of the phytochromes, cryptochrome, and phototropins. Through implementation of today's genetic tools and high-resolution measurement techniques, it is clear that green signals guide plant responses. Emerging evidence meshes well with classical observations to describe the effects of sensory systems that subtly shape specific facets of plant physiology. Now that the conditions, temporal parameters, and tissues that facilitate green light responses have been defined, modern tools such as microarrays can be used to probe the subtle transcriptome changes coincident with green light sensing, adding greater resolution to the understanding of how diverse light qualities contribute to photomorphogenic responses.

References

Dougher TAO, Bugbee B (2001) Evidence for yellow light suppression of lettuce growth. Photochem Photobiol 73: 208–212

Eisinger WR, Bogomolni RA, Taiz L (2003) Interactions between a blue-green reversible photoreceptor and a separate UV-B receptor in stomatal guard cells. Am J Bot 90: 1560–1566

Folta KM (2004) Green light stimulates early stem elongation, antagonizing light-mediated growth inhibition. Plant Physiol 135: 1407–1416

Frechilla S, Zhu J, Talbott LD, Zeiger E (1999) Stomata from npq1, a zeaxanthin-less Arabidopsis mutant, lack a specific response to blue light. Plant Cell Physiol 40: 949–954

Frechilla S, Talbott LD, Bogomolni RA, Zeiger E (2000) Reversal of blue light-stimulated stomatal opening by green light. Plant Cell Physiol 41: 171–176

Goto N, Yamamoto KT, Watanabe M (1993) Action spectra for inhibition of hypocotyl growth of wild-type plants and of the *Hy2* long-hypocotyl mutant of *Arabidopsis-thaliana* L. Photochem Photobiol 57: 867–871

Kim HH, Goins G, Wheeler R, Sager J (2004a) Green light supplementation for enhanced lettuce growth under red and blue light-emitting diodes. Hortscience 39: 1617–1622

Kim HH, Goins G, Wheeler R, Sager J (2004b) Stomatal conductance of lettuce grown under or exposed to different light quality. Ann Bot 94: 691–697

Klein RM (1964) Repression of tissue culture growth by visible and near visible radiation. Plant Physiol 39: 546–539

Klein RM (1979) Reversible effects of green and orange-red radiation on plant cell elongation. Plant Physiol 63: 114–116

Klein RM (1992) Effects of green light on biological systems. Biol Rev Camb Philos Soc 67: 199–284

Klein RM, Edsall PC, Gentile AC (1965) Effects of near ultraviolet and green radiations on plant growth. Plant Physiol 40: 903–906

Liscum E, Briggs WR (1995) Mutations in the NPH1 locus of Arabidopsis disrupt the perception of phototropic stimuli. Plant Cell 7: 473–485

Loughrin JH, Kasperbauer MJ (2001) Light reflected from colored mulches affects aroma and phenol content of sweet basil (*Ocimum basilicum* L.) leaves. J Agric Food Chem 49: 1331–1335

Loughrin JH, Kasperbauer MJ (2003) Aroma content of fresh basil (*Ocimum basilicum* L.) leaves is affected by light reflected from colored mulches. J Agric Food Chem 51: 2272–2276

Parks BM, Folta KM, Spalding EP (2001) Photocontrol of stem growth. Curr Opin Plant Biol 4: 436–440

Steinitz B, Ren ZL, Poff KL (1985) Blue and green light-induced phototropism in *Arabidopsis thaliana* and *Lactuca-sativa* L seedlings. Plant Physiol 77: 248–251

Talbott LD, Zhu J, Han SW, Zeiger E (2002) Phytochrome and blue light-mediated stomatal opening in the orchid, *Paphiopedilum*. Plant Cell Physiol 43: 639–646

Talbott LD, Shmayevich IJ, Chung Y, Hammad JW, Zeiger E (2003) Blue light and phytochrome-mediated stomatal opening in the *npq1* and *phot1 phot2* mutants of Arabidopsis. Plant Physiol 133: 1522–1529

Went FW (1957) The experimental control of plant growth. Chronica Botanica, Waltham, MA, 343 pp

Chapter 28

Photoreceptors in Avian Magnetoreception

Thorsten Ritz

Introduction

The geomagnetic field is an important source of directional information and a physiological magnetic compass has been demonstrated in numerous species of migratory birds as well as other animals (Wiltschko and Wiltschko 1995a). Recently, a growing body of behavioral evidence indicates that the avian magnetic compass involves blue-green photoreceptors that can undergo redox chemistry (radical-pair reactions). We will review the current knowledge of the wavelength dependence of magnetic compass responses. We present the current model that the magnetic sensitivity is based on the influence of magnetic fields on singlet–triplet transitions in photoreceptors and review recent evidence supporting such a chemical sensing mechanism. Having identified the mechanism of the magnetic compass, the field of magnetoreception now calls for genetic and molecular approaches to identify the molecular basis of photoreceptor-based magnetoreception. We discuss potential receptor candidates and implications on plant morphogenesis.

Wavelength Dependence of Magnetic Compass Responses

Tests for magnetic compass orientation follow a standard test design: one bird at a time is placed in a circular funnel cage lined with coated paper on which the birds leave scratch marks as they flutter around. Analyzing scratch marks after a fixed period of time (usually between 60 and 90 min) reveals that the mean headings of birds coincide with the expected migratory direction. Change of magnetic field conditions results in a predictable change of mean direction, thus demonstrating the ability of birds to detect directional information from the geomagnetic field (Wiltschko and Wiltschko 1972). To test light-dependent effects, birds

Department of Physics and Astronomy, Frederick Reines Hall, University of California, Irvine, CA 92697-4575, USA

are exposed to diffuse light from above either produced by LEDs or from other light sources that are subsequently filtered to produce light in a desired wavelength region.

It needs to be appreciated that at least 20–30 tests are necessary in each condition for statistical significance and that data acquisition can only be performed when birds exhibit migratory behavior. Hence, spectral information on magnetic compass responses is much more scarce than, e.g., on plant light responses. Moreover, different behavioral setups and testing procedures can well result in seemingly contradictory results. Despite these caveats, a pattern of light-dependent magnetic orientation responses emerges that, albeit complex, appears to be conserved across different species of migratory birds.

Table 1 summarizes orientation experiments testing light-dependent magnetoreception in birds (Wiltschko et al 1993, Wiltschko and Wiltschko 1995b, 1999, 2001, Rappl et al 2000). For each species of migratory birds, a plus indicates orientation in the expected migratory direction, and a minus indicates disorientation. Symbols are placed at the peak wavelength of the incident light. All experiments listed above have intensities of approximately 6–8.7 × 10^{15} quanta $s^{-1}m^{-2}$. The experiments indicate that light from the blue-green part of the visual spectrum is required for magnetoreception in birds with a sharp transition to disorientation around 570 nm. Recent measurements have focused on this transition in European robins, finding strong orientation of robins under 560 and 565 nm light, but disorientation under 567.5 nm light (Muheim et al 2002).

The question arises whether disorientation under yellow and red light is caused by the photoreceptor failing to respond to this wavelength or by the antagonistic effect of a secondary photoreceptor. An increase in the intensity of yellow light (peak: 590 nm, FWHM: 37 nm) from 7 × 10^{15} quanta $s^{-1}m^{-2}$ to 43 × 10^{15} quanta $s^{-1}m^{-2}$ would be expected to produce sufficient excitation in the green region to activate the blue-green receptor, but birds continued to be disoriented under this condition (Wiltschko and Wiltschko 2001). Moreover, under a combination of green and yellow light with a quantal flux of each, birds showed a novel response, in which birds moved into a fixed direction that did not change between spring and fall (Wiltschko et al 2004a). These findings indicate involvement of a second receptor activated by long wavelength light. Antagonistic interaction between a short wavelength receptor of high sensitivity and a long-wavelength receptor of

TABLE 1. Summary of magnetic compass orientation experiments under different wavelengths of ambient light

	UV violet	*blue*		*green*	*yellow*	*red*	*IR*
Species	400	450	500	550	600	650	700 nm
Silvereyes, *Zosterops lateralis*		+		+	⊖	⊖	
European Robin, *Erithacus rubecula*	+	+	+	+	⊖	⊖	⊖
Garden Warbler, *Sylvia borin*		+		+	⊖	⊖	
Carrier Pigeon, *Columba livia*				+		⊖	⊖

lower sensitivity has also been implicated in light-dependent compass responses of newts (Deutschlander et al 1999, Ritz et al 2002).

The mechanism by which the postulated long-wavelength receptor interacts with the receptor(s) normally providing compass orientation is not yet understood. Similar fixed-axis responses have been found for high intensity turquoise light (Wiltschko and Wiltschko 2001); in many studies, retesting in a second migratory season is necessary to confirm that observed unimodal responses are indeed normal orientation behavior and not fixed-axis responses. Finally, in one study European robins have been shown to orient in the expected migratory direction under red light (peak: 645 nm) after 1 h red pre-exposure (Wiltschko et al 2004b). In another study, robins showed a unimodal response about 60° shifted clockwise from the expected migratory direction under low-intensity red light; it has not yet been clarified whether this response was normal migratory orientation or a fixed-axis response.

Radical-Pair Mechanism Can Explain Magnetic Field Effects on Photoreceptors

Excitation of a photoreceptor pigment can result in a conformational change or in a change of redox state through electron transfer. While the photochemistry of retinal in opsins has been optimized to produce a large conformational change, light can initiate electron transfer reactions in other pigments, most notably in flavins and chlorophylls.

Following an electron transfer from a donor to an acceptor, an intermediate radical pair is created in which both molecular moieties possess one unpaired electron each (cf. Figure 1). The orientation of the electron spins will change due

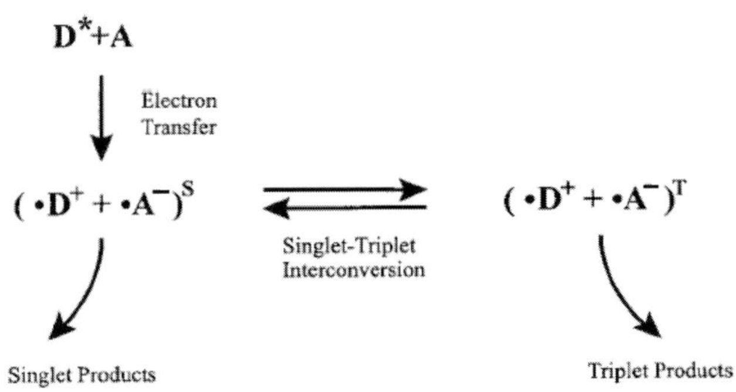

FIG. 1. Reaction scheme for a radical pair reaction with magnetic field-dependent reaction products. The radical pair is generated by a light-induced electron transfer from a donor molecule to an acceptor molecule *A* (here generated in a singlet state). An external magnetic field affects interconversion between singlet and triplet states of the radical pair

to the influence of the nuclear spins (hyperfine coupling) as well as due to the effect of external magnetic fields (Zeeman effect), resulting in the interconversion between singlet and triplet states. The radical pair intermediate will decay into either singlet or triplet products. These products are chemically different and the ratio between singlet and triplet products can be modulated by the strength and direction of an external magnetic field.

Schulten et al (1978) originally suggested that the sensitivity of such radical-pair reactions to external magnetic fields could form the basis for a physiological compass. Theory predicts that for sufficiently long-lived radical pairs (1 μs or longer), an earth-strength magnetic field can produce detectable effects on the relative yield of singlet and triplet products (Ritz et al 2000). Experimentally, the existence of a radical pair can be detected through associated signals in electron spin resonance (ESR) spectroscopy. Transient EPR spectra allow a determination of the lifetime of a spin-correlated radical pair. Indirect evidence for the existence of a radical pair comes from the detection of weak magnetic field effects on singlet or triplet yields that conform to the predictions from theory. The most common technique to detect magnetic field effects is to measure optical effects in laser flash photolysis in a solution containing two partners capable of forming radical pair intermediates through light-induced electron transfer. Effects of earth-strength magnetic fields have been detected in vitro, in good agreement with theoretical predictions (Batchelor et al 1993).

It needs to be emphasized that detection of a weak magnetic field by its influence on spin chemistry is a remarkable process from a physical point of view because the energy of magnetic interaction per particle involved in a radical pair process is much smaller than the average thermal energy per particle under physiological conditions. The reason that the magnetic signal is not buried in thermal noise is that radical pair reactions proceed from a highly ordered state (all singlet or all triplet) and on a time scale which is too fast for thermal noise to average out the magnetic signal. To date, the radical pair mechanism is the only established mechanism by which magnetic fields of less then 10 Gauss can influence chemical reactions.

Evidence for Functionally Relevant Radical-Pair Reactions in Biological Systems

Despite the potential of magnetic fields to influence the chemistry of physiological reactions, in vivo studies investigating magnetic field effects on biological function are spurious. Magnetic field effects have been reported on the singlet-triplet interconversion in the photosynthetic electron transfer cascade (see references in Schulten 1982) as well as on horseradish peroxidase. Part of the reluctance to conduct such tests may be attributed to the lack of a convincing model that could explain weak magnetic field effects on biological systems. Although the radical pair mechanism is by now a well-established mechanism, the tests for the detection of radical pairs are experimentally involved and are usually performed by physical chemists, not plant physiologists or animal behavioralists.

The radical pair model provides the framework to design diagnostic tests to investigate involvement of a radical pair mechanism (Ritz et al 2000). Recently, feasibility of such a diagnostic approach was demonstrated in the context of the avian magnetic compass (Ritz et al 2004). Application of oscillating magnetic fields at distinct resonance frequencies in the low radiofrequency range (1–100 MHz) in addition to the geomagnetic field, should disrupt a physiological compass based on radical-pair reactions because such oscillating fields will mask the effects of an earth-strength static field on radical-pair reactions. On the other hand, such fields will not affect magnetic compasses based on other mechanisms, such as magnetite-based compasses, provided that they are weaker than the geomagnetic field. Moreover, in a radical-pair based compass system, the alignment of the oscillating field with respect to the static field will determine whether oscillating fields lead to disruption or not, whereas a non-specific disturbing effect of oscillating fields should occur regardless of angle between static and oscillating fields.

The magnetic compass orientation behavior of birds in oscillating magnetic fields can thus be a good test to identify the underlying biophysical mechanism. European robins were tested for magnetic orientation in the geomagnetic field only (control condition) and in conditions in which an additional weak oscillating field was applied. In the control condition, the robins exhibited seasonally appropriate northerly orientation. In the presence of an additional broadband (0.1–10 MHz) or single-frequency 7.0 MHz oscillating field presented at 24° and 48° relative to the geomagnetic field, the birds were disorientated. All fields were between 50 and 300 times weaker than the geomagnetic field, excluding direct effects on a magnetite-based system. In contrast, when the 7.0 MHz oscillating field was parallel to the geomagnetic field, the birds oriented in the migratory direction and their response was indistinguishable from the control condition. These results indicate that a magnetically sensitive radical-pair process exists in European robins that are linked to the physiology of magnetic compass orientation (Ritz et al 2004).

Summary and Conclusions

Behavioral evidence in birds points towards an involvement of photoreceptors in magnetoreception. The radical pair mechanism can provide a compelling explanation as to how magnetic fields can be detected. A recently established diagnostic tool, i.e., the use of oscillating fields, can indicate involvement of radical pair reactions through simple physiological or behavioral measurements. Application of this diagnostic tool revealed involvement of a radical pair mechanism as the mechanism underlying magnetic compass orientation. The identification of the mechanism now puts the spotlight on the eye of birds to search for the photoreceptors involved in magnetoreception. Of the many various photoreceptors known to absorb in the blue-green region, only receptors that link light absorption to redox chemistry can function in a radical-pair mechanism. Cryptochromes as well as phototropins would be promising candidates because of their involvement in the electron transfer processes (Ahmed 2004, Kay et al 2003) and cryptochromes have, in fact, been found in the eyes of birds.

With regard to plant morphogenesis, the existence of a magnetically sensitive biological reaction in birds that is based on blue-green photoreceptors undergoing redox chemistry raises the question whether light-dependent reactions in plants that involve likely the same photopigments may not also show a dependence on weak magnetic fields. To the best of our knowledge, no conclusive studies exist on this question. The use of combined oscillating and static fields as a diagnostic device could also be applied to investigate involvement of a radical-pair reaction in plant morphogenesis, provided that these reactions are functionally relevant.

References

Batchelor S, Kay C, Mclauchlan K, Shkrob I (1993) Time-resolved and modulation methods in the study of the effects of magnetic fields on the yields of free radical reactions. J Phys Chem 97: 13250–13528

Deutschlander M, Phillips J, Borland S (1999) The case for light-dependent magnetic orientation in animals. J Exp Biol 202: 891–908

Kay CWM, Schleicher E, Kuppig A, Hofner H, Rüdiger W, Schleicher M, Fischer M, Bacher A, Weber S, Richter G (2003) Blue light perception in plants—Detection and characterization of a light-induced neutral flavin radical in a C450A mutant of phototropin. J Biol Chem 278: 10973–10982

Muheim R, Bäckman J, Akesson S (2002) Magnetic compass orientation in European robins is dependent on both wavelength and intensity of light. J Exp Biol 205: 3845–3856

Rappl R, Wiltschko R, Weindler P, Berthold P, Wiltschko W (2000) Orientation behavior of garden warblers, *Sylvia borin*, under monochromatic light of different wavelengths. Auk 117: 256–260

Ritz T, Adem S, Schulten K (2000) A model for photoreceptor-based magneto-reception in birds. Biophys J 78: 707–718

Ritz T, Dommer D, Phillips J (2002) Shedding light on vertebrate magnetoreception. Neuron 34: 503–506

Ritz T, Thalau P, Phillips J, Wiltschko R, Wiltschko W (2004) Resonance effects indicate radical pair mechanism for avian magnetic compass. Nature 429: 177–180

Schulten K (1982) Magnetic field effects in chemistry and biology. In: Treusch J (ed) Festkörperprobleme, vol. 22. Vieweg, Braunschweig, pp 61–83

Schulten K, Swenberg C, Weller A (1978) A biomagnetic sensory mechanism based on magnetic field modulated coherent electron spin motion. Z Phys Chem NF111: 1–5

Wiltschko W, Wiltschko R (1972) Magnetic compass of European robins. Science 176: 62–64

Wiltschko R, Wiltschko W (1995a) Magnetic orientation in animals. Springer, Berlin

Wiltschko W, Wiltschko R (1995b) Migratory orientation of European robins is affected by the wavelength of light as well as by a magnetic pulse. J Comp Physiol A 177: 363–369

Wiltschko W, Wiltschko R (1999) The effect of yellow and blue light on magnetic compass orientation in European robins, *Erithacus rubecula*. J Comp Physiol 184: 295–299

Wiltschko W, Wiltschko R (2001). Light-dependent magnetoreception in birds: the behavior of European robins, *Erithacus rubecula*. J Exp Biol 204: 3295–3302

Wiltschko W, Munro U, Ford H, Wiltschko R (1993) Red light disrupts magnetic orientation of migratory birds. Nature 364: 525–527

Wiltschko W, Gesson M, Stapput K, Wiltschko R (2004a) Light-dependent magnetoreception in birds: interaction of at least two different receptors. Naturwissenschaften 91: 130–134

Wiltschko W, Möller A, Gesson M, Noll C, Wiltschko R (2004b) Light-dependent magnetoreception in migratory birds: previous exposure to red light alters the orientation response under red light. J Exp Biol 207: 1193–1202

Part VI
Signal Transduction

Chapter 29

Light Control of Plant Development: A Role of the Ubiquitin/Proteasome-Mediated Proteolysis

Yuki Yanagawa, Suhua Feng, and Xing Wang Deng

Introduction

The relationship between light environment and plant development is most dramatically illustrated by seedling development in *Arabidopsis*. *Arabidopsis* seedlings exhibit two different phenotypes: photomorphogenesis in the light and skotomorphogenesis (or etiolation) in darkness, respectively. Light-grown seedlings have short hypocotyls, and open and expanded cotyledons, while skotomorphogenic seedlings have long hypocotyls and apical hooks (closed and unexpanded cotyledons) (for review, see von Arnim and Deng 1996). Recently, a major regulatory switch for the light control of seedling development was shown to be the ubiquitin/proteasome-mediated degradation of photomorphogenesis-promoting transcription factors (Osterlund et al 2000, Seo et al 2003, Saijo et al 2003). The pleiotropic *COP* (*constitutive photomorphogenic*)/*DET* (*de-etiolated*)/*FUS* (*fusca*) genes were initially identified to encode negative regulators of photomorphogenesis by genetic analysis, and recently they have been shown to play important roles in the protein ubiquitination processes. This review briefly summarizes our current understanding of the role of the COP/DET/FUS proteins and the ubiquitin/proteasome-mediated proteolysis in the light control of *Arabidopsis* seedling development.

Ubiquitin/Proteasome Pathway

Proteolysis plays an essential role in the development of all eukaryotes. Accumulating evidence indicates that the activities of many cellular proteins are regulated through the ubiquitin/proteasome-mediated proteolysis.

The ubiquitin/proteasome pathway is evolutionarily conserved. Ubiquitin, a 76-amino-acid globular protein, is covalently attached to the target protein by an

Department of Molecular, Cellular, and Developmental Biology, Yale University, New Haven, CT 06520-8104, USA

isopeptide linkage between the C-terminus of ubiquitin and the lysine E-amino groups of the target protein. Three different enzymes are required in this process, including a ubiquitin-activating enzyme (E1), a ubiquitin-conjugating enzyme (E2), and a ubiquitin ligase (E3). Subsequently, polyubiquitin chain is formed, in which C-terminus of each ubiquitin unit is linked to the K-48 residue of the previous ubiquitin (Callis and Vierstra 2002, Vierstra 2003). It is also known that K-63 lysine residue of ubiquitin can also be used to form a noncanonical polyubiquitin chain, which are involved in the regulation of DNA repair, transcription, and translation, but not in protein degradation (Aguilar and Wendland 2003).

The 26S proteasome (multicatalytic endopeptidase complex, EC 3.4.99.46) is a highly conserved large multi-subunit protein complex. It catalyzes the degradation of the K48 polyubiquitinated proteins in an ATP-dependent manner. The 26S proteasome is composed of a 20S catalytic core and a pair of symmetrically assembled regulatory particle named as 19S/22S/PA700 (Tanaka and Chiba 1998). The 19S regulatory particle can be further divided into two subcomplexes known as the "base" and the "lid," corresponding to the portions of the regulatory particle at the proximal and distal from the 20S catalytic core, respectively (Glickman et al 1998). The base is composed of six ATPases (Rpt1–6), and three non-ATPase subunits (Rpn1, Rpn2 and Rpn10); the lid contains eight distinct non-ATPase subunits, named Rpn3, 5, 6, 7, 8, 9, 11, and 12 (Glickman et al 1998).

The *Arabidopsis thaliana* genome has 2 genes for E1, 46 genes for E2 or E2-like proteins, and more than 1200 genes for the components of E3 (Sullivan et al 2003, Vierstra 2003). The ubiquitin E3 ligases, usually composed of multiple protein subunits, have been classified into four groups on the basis of their subunit composition, indicating that the different combination of subunits can lead to a huge number of E3s. There are also 23 genes for the 20S catalytic core and 31 genes for the 19S regulatory particle of the 26S proteasome in *Arabidopsis*. It has been estimated that the ubiquitin/proteasome-mediated proteolysis regulates more than 10% of the total proteins in *Arabidopsis* (Vierstra 2003).

The COP/DET/FUS Proteins

The *cop/det/fus* mutants show similar constitutive photomorphogenic phenotypes, which are characterized by short hypocotyls and open and expanded cotyledons when grown in darkness (von Arnim and Deng 1996). Therefore, the COP/DET/FUS proteins were initially defined as negative regulators of photomorphogenesis (Figure 1). Later, it was found that six of the COP/DET/FUS proteins are incorporated into an approximately 450–550 kDa 8-subunit complex, named the COP9 signalosome (CSN) (Serino and Deng 2003, Wei and Deng 2003). Interestingly, the CSN is conserved from human to *Arabidopsis*, thus its function is not limited to repressing photomorphogenesis in darkness. Furthermore, null *csn* mutants show seedling lethality, suggesting that the CSN is essen-

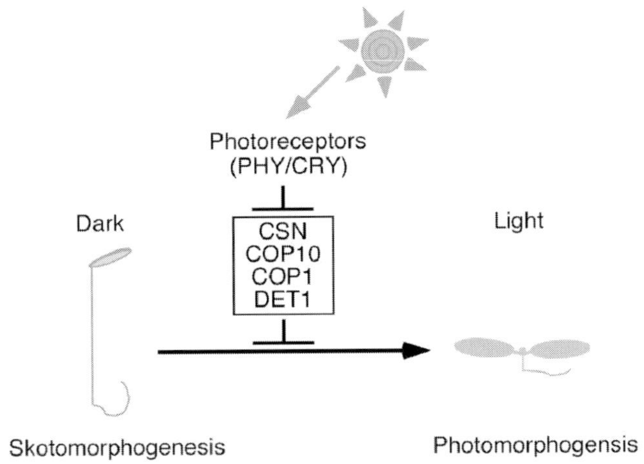

FIG. 1. The COP/DET/FUS proteins, including CSN, COP10, COP1, and DET1, are necessary for the light control of seedling photomorphogenesis in *Arabidopsis*. *PHY*, phytochrome; *CRY*, cryptochrome

tial for plant development in addition to photomorphogenesis. Indeed, the CSN has been reported to regulate flower development, hormone responses, and pathogen resistance in plants (Serino and Deng 2003).

Three other COP/DET/FUS proteins, COP1, COP10, and DET1, are not the components of the CSN (Serino and Deng 2003, Wei and Deng 2003). However, they presumably work together with the CSN in regulating photomorphogenesis, because their mutants display similar constitutive photomorphogenic phenotypes in darkness. The Ring finger protein COP1 was the first molecularly cloned and shown to play regulatory role in photomorphogenesis mediated by the COP/DET/FUS group of proteins (McNellis et al 1994). It has been shown that repartitioning of COP1 from the cytoplasm to the nucleus in darkness results in skotomorphogenesis, because COP1, together with other COP/DET/FUS proteins, promotes the degradation of photomorphogenesis-promoting transcription factors in the nucleus (von Arnim and Deng 1994, Osterlund et al 2000). COP1 acts as a large protein complex in vivo (Saijo et al 2003). COP10 is part of a nuclear protein complex of ~300 kDa with DET1 and UV-damaged DNA binding protein 1 (DDB1), designated CDD complex (Yanagawa et al 2004). This finding is consistent with an early report that DET1 and DDB1 are part of a protein complex (Schroeder et al 2002). COP10 has been shown to be able to interact with three CSN subunits and COP1 (Suzuki et al 2002). DET1 was also shown to interact with histone H2B (Benvenuto et al 2002), indicating that the chromatin remodeling might be involved in the regulation of photomorphogenesis.

Ubiquitin/Proteasome-Mediated Proteolysis in Photomorphogenesis

In plants, the ubiquitin/proteasome pathway has been reported to play important roles in many processes such as photomorphogenesis, flower development, hormone responses, plant pathogen resistance, and cell cycle control (Serino and Deng 2003, Vierstra 2003). Most of these pathways involve the COP9 signalosome.

Two conserved motifs are found in CSN subunits: the PCI (**P**roteasome, **C**OP9 signalosome and **I**nitiation factor 3) and MPN (**M**pr1, **P**ad1, **N**-terminal) domains. All of the CSN subunits have either PCI or MPN motif, analogous to the lid subcomplex of the 26S proteasome (Hofmann and Bucher 1998). The close resemblance of the CSN and the lid subcomplex, both of which are composed of eight subunits with PCI or MPN motif, implies that they have a common evolutionary ancestor, and that they might share similar functions. Moreover, it has been shown that the CSN interacts with the regulatory particle of the 26S proteasome in vitro, and with the 26S proteasome in vivo (Kwok et al 1999, Peng et al 2003). Thus, the CSN might regulate the function of the 26S proteasome. In addition, the CSN can interact with the SCF (SKP1/Cullin1/F-box protein) class of ubiquitin E3 ligase complexes, and catalyze the release of RUB/NEDD8, a ubiquitin-like protein, from the Cullin1 subunit of the SCF complexes (Schwechheimer et al 2002). Recently, the CSN has been shown to assemble into complexes containing DDB2 or CSA, with associated de-ubiquitination activity in mammals (Groisman et al 2003). Yet, it remains to be confirmed whether plant CSN has similar functions. CSN5 (also known as Jab1), a paralogue of Rpn11 in the lid subcomplex of the 26S proteasome, has a conserved JAMM (Jab1/Pad1/MPN) motif, which is believed to constitute the catalytic core of the deneddylation activity of the CSN (Cope et al 2002, Berndt et al 2002). Consistently, the JAMM motif is also essential for the deubiquitination activity of Rpn11 in the lid subcomplex. Interestingly, the JAMM-dependent deubiquitination activity of the lid subcomplex is active only when it is assembled in the 26S proteasome (Berndt et al 2002).

It is thought that the degradation of transcription factors such as HY5 and HYH are necessary for the repression of photomorphogenesis in darkness (Osterlund et al 2000, Holm et al 2002, Saijo et al 2003), because these proteins were prevented from degradation in *cop/det/fus* loss-of-function mutants. Recently, another transcription factor, LAF1, was found to be targeted by the COP1 and ubiquitin/proteasome-mediated proteolysis as well (Seo et al 2003). Interestingly, Seo et al (2004) also reported that COP1 interacts with phytochrome A and acts as a ubiquitin E3 ligase to target phytochrome A for degradation. Thus, these results indicate that COP1 has ubiquitin E3 ligase activity toward distinct target proteins involved in photomorphogenesis.

COP10 was shown to encode an E2 variant. Interestingly, COP10 has a higher homology to normal E2s than any known E2 variants such as MMS2 and UEV1, even though it lacks the conserved cysteine residue at the catalytic site (Suzuki et al 2002). Recently, recombinant COP10 and its native CDD complex were revealed to have the ability to enhance E2 activity (Yanagawa et al 2004). Since COP10 can interact with three CSN subunits and COP1, and the CDD complex is unstable in the *csn* mutants, thus the coordination of three factors, CSN, COP1 complex, and CDD complex, is necessary for the control of photomorphogenesis. Recently, human DET1 was suggested to be associated with COP1, DDB1, ROC1, and Cullin 4A, which forms a putative ubiquitin E3 ligase for the transcription factor c-Jun (Wertz et al 2004).

Concluding Remarks

In the last five years progress in this field has been remarkable. Significant breakthroughs have been made regarding the role of ubiquitination and proteasomal degradation of the photomorphogenesis-promoting transcription factors. Moreover, all of the known COP/DET/FUS proteins are related to ubiquitination, indicating that the ubiquitin/proteasome pathway plays a major role in photomorphogenesis. In short, it is the COP/DET/FUS protein-mediated degradation of the photomorphogenesis-promoting transcriptional factors, including HY5, HYH, and LAF1, which constitutes the regulatory switch for photomorphogenesis. Among them, the CSN and COP1 are defined as an E3 ligase modulator and an E3 ligase, respectively. Another factor, the CDD complex, consisting of COP10, DET1, and DDB1, is an E2 enhancer (Figure 2). The detailed

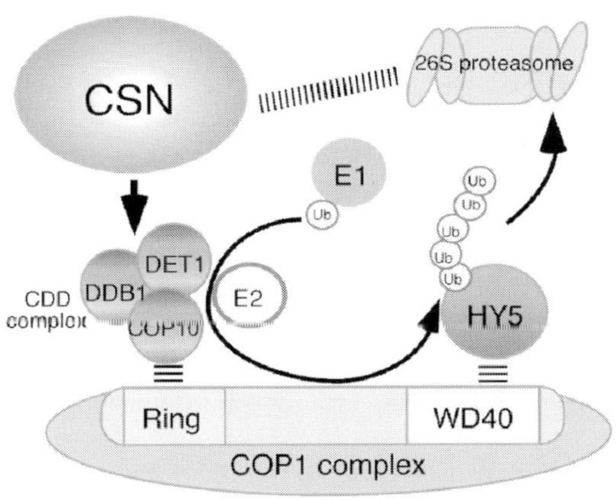

FIG. 2. A working model depicting the functional relationships among CSN, COP1, and CDD complexes, in the ubiquitin (Ub)/proteasome-mediated proteolysis. In darkness, CSN, COP1, and CDD complexes work together to promote the ubiquitination of photomorphogenesis-promoting transcription factors such as HY5. The CSN directly interacts with and stabilizes the CDD complex. The CDD complex has the activity to enhance E2 activity. COP10 interacts with the Ring finger domain of COP1. HY5 interacts with the WD-40 repeat domain of COP1 and is ubiquitinated by the ubiquitin E3 ligase activity of COP1. Polyubiquitinated HY5 is recognized and degraded by the 26S proteasome. The CSN might regulate the function of the 26S proteasome, CDD complex, and COP1 E3 ligase activity

mechanism of how the CSN, COP1, and CDD complexes work together to regulate the degradation of the transcription factors remains to be elucidated in the future.

References

Aguilar RC, Wendland B (2003) Ubiquitin: not just for proteasomes anymore. Curr Opin Cell Biol 15: 184–190

Benvenuto G, Formiggini F, Laflamme P, Malakhov M, Bowler C (2002) The photomorphogenesis regulator DET1 binds the amino-terminal tail of histone H2B in a nucleosome context. Curr Biol 12: 1529–1534

Berndt C, Bech-Otschir D, Dubiel W, Seeger M (2002) Ubiquitin system: JAMMing in the name of the lid. Curr Biol 12: R815–R817

Cope GA, Suh GS, Aravind L, Schwarz SE, Zipursky SL, Koonin EV, Deshaies RJ (2002) Role of predicted met alloprotease motif of Jab1/CSN5 in cleavage of Nedd8 from Cul1. Science 298: 606–611

Callis J, Vierstra RD (2000) Protein degradation in signaling. Curr Opin Plant Biol 3: 381–386

Glickman MH, Rubio DM, Coux O, Wefes I, Pfeifer G, Cjeka Z, Baumeister W, Fried VA, Finley D (1998) A subcomplex of the proteasome regulatory particle required for ubiquitin-conjugate degradation and related the COP9-signalosome and eIF3. Cell 94: 615–623

Groisman R, Polanowska J, Kuraoka I, Sawada J, Saijo M, Drapkin R, Kisselev AF, Tanaka K, Nakatani Y (2003) The ubiquitin ligase activity in the DDB2 and CSA complexes is differentially regulated by the COP9 signalosome in response to DNA damage. Cell 113: 357–367

Hofmann K, Bucher P (1998) The PCI domain: a common theme in three multiprotein complexes. Trends Biochem 23: 204–205

Holm M, Ma L, Qu L, Deng XW (2002) Two interacting bZIP proteins are direct target of COP1-mediated control of light-dependent gene expression in *Arabidopsis*. Genes Dev 16: 1247–1259

Kwok SF, Staub JM, Deng XW (1999) Characterization of two subunits of *Arabidopsis* 19S proteasome regulatory complex and its possible interaction with the COP9 complex. J Mol Biol 285: 85–95

McNellis TW, von Arnim AG, Deng XW (1994) Evidence for Arabidopsis COP1 as an autonomous light-inactivatable repressor of photomorphogenic development. Plant Cell 6: 1391–1400

Osterlund MT, Hardtke CS, Wei N, Deng XW (2000) Targeted destabilization of HY5 during light-regulated development of *Arabidopsis*. Nature 405: 462–466

Peng Z, Shen Y, Feng S, Wang X, Chitteti BN, Vierstra RD, Deng XW (2003) Evidence for a physical association of the COP9 signalosome, the proteasome, and specific SCF E3 ligases in vivo. Curr Biol 13: R504–R505

Saijo Y, Sullivan JA, Wang H, Yang J, Shen Y, Rubio V, Ligeng M, Hoecker U, Deng XW (2003) The COP1-SPA1 interaction defines a critical step in phytochrome A-mediated regulation of HY5 activity. Genes Dev 17: 2642–2647

Schroeder DF, Gahrtz M, Maxwell BB, Cook RK, Kan JM, Alonso JM, Ecker JR, Chory J (2002) De-etiolated 1 and damaged DNA binding protein 1 interact to regulate Arabidopsis photomorphogenesis. Curr Biol 12: 1462–1472

Schwechheimer C, Serino G, Callis J, Crosby WL, Lyapina S, Deshaies RJ, Gray WM, Estelle M, Deng XW (2002) Interactions of the COP9 signalosome with the E3 ubiquitin ligase SCF^{TIR1} in mediating auxin response. Science 292: 1379–1382
Seo HS, Yang JY, Ishikawa M, Bolle C, Ballesteros ML, Chua NH (2003) LAF1 ubiquitination by COP1 controls photomorphogenesis and is stimulated by SPA1. Nature 424: 995–999
Seo HS, Watanabe E, Tokutomi S, Nagatani A, Chua NH (2004) Photoreceptor ubiquitination by COP1 E3 ligase desensitizes phytochrome A signaling. Genes Dev 18: 1–6
Serino G, Deng XW (2003) The COP9 SIGNALOSOME: Regulating plant development through the control of proteolysis. Annu Rev Plant Biol 54: 165–182
Suzuki G, Yanagawa Y, Kwok SF, Matsui M, Deng XW (2002) Arabidopsis COP10 is a ubiquitin-conjugating enzyme variant that acts together with COP1 and the COP9 signalosome in repressing photomorphogenesis. Genes Dev 16: 554–559
Sullivan JA, Shirasu K, Deng XW (2003) The diverse roles of ubiquitin and the 26S proteasome in the life of plants. Nat Rev Genet 12: 948–958
Tanaka K, Chiba T (1998) The proteasome: a protein-destroying machine. Genes Cells 3: 499–510
Vierstra RD (2003) The ubiquitin/26S proteasome pathway, the complex last chapter in the life of many plant proteins. Trends Plant Sci 8: 135–142
von Arnim AG, Deng XW (1994) Light inactivation of *Arabidopsis* photomorphogenic repressor COP1 involves a cell-specific regulation of its nucleocytoplasmic partitioning. Cell 76: 1035–1045
von Arnim AG, Deng XW (1996) Light control of seedling development. Annu Rev Plant Physiol Plant Mol Biol 47: 215–243
Wei N, Deng XW (2003) The COP9 signalosome. Annu Rev Cell Dev Biol 19: 261–286
Wertz IE, O'Rourke KM, Zhang Z, Dornan D, Arnott D, Deshaies RJ, Dixit VM (2004) Human de-etiolated-1 regulates c-Jun by assembling a CUL4A ubiquitin ligase. Science 303: 1371–1374
Yanagawa Y, Sullivan JA, Komatsu S, Gusmaroli G, Suzuki G, Yin J, Ishibashi T, Saijo Y, Rubio V, Kimura S, Wang J, Deng XW (2004) Arabidopsis COP10 forms a complex with DDB1 and DET1 in vivo and enhances the activity of ubiquitin conjugating enzymes. Genes Dev 18: 2172–2181

CHAPTER 30

Phytochrome and COP1 Regulates Abundance of Phytochrome Interacting Factor 3

FERENC NAGY[1] and EBERHARD SCHÄFER[2]

Introduction

To sense the environmental factor light, plants have evolved different sensory photoreceptors (Kendrick and Kronenberg 1994). In *Arabidopsis* five members of a small gene family (PHYA to PHYE) encode the photoreceptor phytochromes (Clack et al 1994). Phytochromes are R/FR photoreversible chromoproteins, which form dimers with a molecular mass of ca. 120 kDa per monomer and in which an open-chain tetrapyrrole chromophore is autocatalytically attached to the apoprotein (Lagarias and Lagarias 1989, Eichenberg et al 2000). R induced formation of the FR absorbing active form of phytochrome (Pfr) initiates a signalling cascade which controls plant photomorphogenesis. Of these phytochromes, phyA has a very specific mode of action by controlling very low fluence responses (VLFR) and far-red high irradiance responses (HIR) (Furuya and Schäfer 1996). VLFR is initiated even by a few seconds of starlight and is saturated at about 1 µmol/m^2, whereas HIR requires prolonged irradiation with continuous far-red light (cFR). In contrast to phyA, phyB–E mediate responses to continuous red light (cR) and show the R/FR reversible induction responses. Between light absorption by photoreceptors and physiological and developmental responses lies a web of interacting factors and interacting pathways, either directly involved in or otherwise impinging upon light signal transduction.

Light Induces Nuclear Translocation of Phytochromes

In dark-grown seedlings phyA is exclusively, whereas phyB–E are predominantly localised in the cytosol. After irradiation, phyA–E are translocated into the nucleus in a light quality- and quantity-dependent manner: phyA requires either

[1] Biological Research Centre, Temesvari krt. 62, H-6726 Szeged, Hungary
[2] Institut für Botanik, University of Freiburg, Schanzlestrasse 1, D-7091 Freiburg, Germany

VLFR or HIR treatments (Kircher et al 1999, Kim et al 2000), whereas phyB–E need irradiation with R (Gil et al 2000, Kircher et al 2002). Induction of the nuclear import of phyB is mediated by R and reverted by subsequent pulses of FR (Kircher et al 1999). It has also been established that the majority of phy:GFP fusion proteins in the nucleus are not distributed randomly but converge at subnuclear foci (Yamaguchi et al 1999, Kircher et al 2002).

Phytochromes Interact with PIF3 in a Conformation-Dependent Fashion

Over the past few years several genes potentially involved in phyA- and phyB-controlled signal transduction have been identified. Most of these genes also encode proteins, which are localised in the nucleus. One of the best characterised of these proteins is PIF3, a transcription factor interacting with phyA and phyB. PIF3 is a basic helix-loop-helix (bHLH) protein (Ni et al 1998); it interacts in vitro with phyA and phyB in a conformation-specific manner (Ni et al 1999). PIF3 binds specifically to a cis-acting regulatory element (G-box) in the promoters of a variety of phytochrome-responsive genes. Simultaneous binding of PIF3 to promoters of light-responsive genes and to the Pfr form of phyB described by Martinez-Garcia et al (2000) indicates that PIF3 recruits phyB to the promoters of actively transcribed genes. Manipulation of PIF3 expression levels in transgenic plants resulted in altered photomorphogenesis. By characterising transgenic plants over-expressing the N-terminal truncated form of PIF3 or antisense PIF3, the physiological role of PIF3 has been classified as positive regulator of PHYB mediated signal transduction (Ni et al 1998). Features of the *poc1* mutant, which displays short hypocotyl phenotype and level of *PIF3* mRNA higher than those in wild-type seedlings in cR, were then interpreted as a phenotype associated with over-expression of PIF3 (Halliday et al 1999). These observations together with the microarray analyis of phytochrome-modulated gene transcription in *Arabidopsis* (Tepperman et al 2001) led to a hypothesis which postulated that (i) phytochromes, notably phyA, through PIF3 and other yet unidentified factors, regulate transcription of a master set of regulators like CCA1 (Wang and Tobin 1998), LHY1 (Schaffer et al 1998) TOC1-L, RT2, DOF, and CO (Teppermann et al 2001, Harmer et al 2000) and (ii) these regulators then control the transcription of genes encoding functions necessary for the terminal steps of the signalling cascade.

PIF3 Negatively Regulates phyB Signaling

The postulated positive regulatory role of PIF3 in phyB-mediated light signal transduction, however, has recently been challenged. Kim et al (2003) reported that T-DNA insertion mutant lines lacking a detectable amount of *PIF3* mRNA displayed hypersensitivity whereas transgenic lines over-expressing *PIF3* mRNA exhibited hyposensitivity to R regarding inhibition of hypocotyl growth. By analysing in detail additional photomorphogenic responses of these lines, Kim et al (2003) concluded that PIF3 acts mainly as a negative regulator of phyB-induced signalling. Matsushita et al (2003) demonstrated that a fusion protein consisting of the N-terminal part of phyB fused to GUS followed by the SV 40 NLS signal is capable of complementing the *phyB-5* mutant that lacks detectable amounts of phyB. This fusion protein does not contain the C-terminal domain of phyB, which has been shown to be important for interaction with PIF3 (Ni et al 1998, 1999) and other regulatory proteins (for a recent review see Gyula et al 2003). To analyse the mode of PIF3 action in planta, at molecular level, we produced transgenic plants that either expressed PIF3 fused to the red-shifted green fluorescent protein (rsGFP) or co-expressed the PIF3:CFP (CFP—cyan fluorescent protein) with various phy species fused to the yellow fluorescent protein (YFP). We then monitored the level and nucleo/cytoplasmic partitioning of PIF3 in these transgenic lines and other mutants grown under various light conditions.

We found that accumulation of both *PIF3* mRNA and protein is regulated developmentally. The level of the *PIF3* mRNA gradually increases up to 2–3 days after germination then declines. The PIF3 protein is not detectable in imbibed seeds; however, its accumulation increases after imbibition up to 4 days. The molecular mechanism responsible for the delayed accumulation of PIF3 protein as compared to that of *PIF3* mRNA remains to be elucidated. We showed that independent of its level PIF3, when it is detectable, is localised constitutively in the nuclei of etiolated seedlings and both the endogenous PIF3 as well as the PIF3:rsGFP and PIF3:CFP fusion proteins degrade after exposure to short pulses or continuous illumination. PIF3 degradation induced by R or FR is rapid; the half-life of PIF3 is about 10 min in R and controlled by the concerted action of phyA, phyB, and phyD. The PIF3 protein is not detectable in the nuclei of cells exposed to light longer than 1 h. Degradation of PIF3 takes place equally fast in dark and light after the inductive light treatment and PIF3 readily re-accumulates again to high levels in the dark. Taken together, these data indicate that the expression of PIF3 is negatively regulated by light at the level of protein degradation. Thus we propose that PIF3 is mainly required for phytochrome signalling during the developmental transition from etiolated growth to photomorphogenesis or during the transition from dark to light.

In addition we found that the *poc1* mutant, reported to over-express the *PIF3* mRNA (Halliday et al 1999), in fact does not contain detectable amounts of PIF3 protein either in dark or even after such extended cR irradiation that is sufficient for the manifestation of the characteristic *poc1* phenotype. Moreover, we demonstrated that transgenic plants over-expressing the PIF3 or PIF3:rsGFP protein are moderately hyposensitive to cR. *Poc1* seedlings display hypersensitivity to cR but not to cFR. Halliday et al (1999) showed by the analysis of *poc1/phyB* double mutant that *poc1* is epistatic to phyB. Kim et al (2003) reported that mutants lacking detectable amount of *PIF3* mRNA display a phenotype, very similar to that of *poc1*. Taking together these data, we conclude that (i) *poc1* is a PIF3 null mutant, (ii) PIF3 is a negative regulator of hypocotyl growth inhibition in cR, and (iii) light-induced degradation of PIF3 represents a key regulatory step in phyB-controlled signalling.

COP1 Promotes Accumulation of PIF3 in the Nucleus in the Dark

There is little evidence that regulated proteolysis plays a role in phyB-initiated signalling. In contrast, isolation of the *EID1* gene encoding an F-box protein (Büche et al 2000, Dieterle et al 2001) as well as the observation that SPA1 acts as a co-factor in COP1-mediated degradation of the transcription factor LAF1 (Seo et al 2003) provided evidence that phyA signalling is mediated, at least partly, by proteasome-related pathways. Light-induced rapid degradation of PIF3 and its re-accumulation in the dark suggests that the function of PIF3, similar to those of HY5 and LAF1, is regulated by proteolysis. There is, however, a significant difference between the modes of action of these transcriptional regulators. HY5 is targeted by COP1 to the COP9 signalosome and is degraded in the dark (Osterlund et al 2000), whereas FR induces transcription of the *HY5* gene (Oyama et al 1997) and accumulation of the HY5 protein in the nucleus. LAF1, similar to HY5, also acts as a positive regulator of phy-controlled signalling, and signalling by LAF1 in light is attenuated by the concerted action of SPA1 and COP1.

In sharp contrast, we showed that PIF3 accumulates in the dark only in the presence of COP1, and both FR and R treatments promote its degradation in a COP1-independent fashion. We found that the PIF3 protein accumulates to significantly lower levels in the dark in *cop1–4* and *eid6* (Dieterle et al 2003) mutants as compared to WT and that *poc1* seedlings similar to the *pif3* null mutant seedlings (Kim et al 2003) do not exhibit a dark phenotype. These data suggest that (i) PIF3 does not play a role in establishing the characteristic *cop1* phenotype and (ii) it is not required for the elevated level of transcription of light-responsive genes in the dark. Thus we suggest that COP1 promotes, directly, the degradation of positive regulators such as HY5 and probably indirectly the build-up of negative regulators such as PIF3 during skotomorphogenesis. PIF3 accumulation in the dark can be effected by COP1-mediated degradation of a PIF3 repressor, which may be a repressor of *PIF3* gene expression or a factor controlling PIF3 degradation. The latter scenario, that COP1 targets a factor controlling PIF3 degradation, is attractive because it is consistent with the observation that light induces exclusion of COP1 from the nucleus. Under this scenario, the absence of COP1 in the nucleus in the light would lead to COP1-independent degradation of PIF3, whereas COP1 nuclear localisation in the dark would lead to COP1-dependent accumulation of PIF3. The answer lies with identifying the COP1 target responsible for permitting the dark accumulation of PIF3.

However, independent of the mechanism by which COP1 protects PIF3 from degradation in darkness, we postulate that the action of COP1 could be important in determining the ratio of positive/negative acting regulatory proteins required during the early stage of photomorphogenesis.

Molecular Mechanism of phyB-Mediated Signaling Remains Elusive

While our work was in progress Matsushita et al (2003) reported that a chimeric gene containing the N-terminal domain of PHYB fused to GUS and SV 40 NLS is capable of complementing the *phyB-5* mutant lacking a functional phyB photoreceptor. These authors postulated that (i) the N-terminal domain of phyB positively regulates signalling, (ii) the C-terminal domain regulates translocation of phyB into the nucleus, and (iii) probably mediates the interaction of the photoreceptor with negative regulatory factors. These results together with the data presented here radically changed our view about phyB-mediated signalling (for a recent review see Nagy and Schäfer 2002). We show here that PIF3 is a negative regulator of phyB signalling, whereas all other interacting proteins except ARR4 (Sweere et al 2001) bind to the C-terminal domain of the photoreceptor. Thus we should conclude that, despite recent advances, the molecular nature of phyB-initiated signalling regulating photomorphogenesis still remains elusive.

PIF3 was shown to interact in vitro with the full-length phyA and phyB photoreceptors in a conformation-dependent fashion (Ni et al 1999). We observed that R but not FR pulses induced transient co-localisation of PIF3 with phyB in nuclear speckles. Extended R treatment led to the disintegration of early and the appearance of late phyB speckles, which differed in size and number and did not contain PIF3. Similar data were obtained by analysing the formation of phyA and phyD containing speckles after FR and R treatments, respectively. Experiments to determine co-localisation of PIF3 and phyC and phyE are in progress.

In contrast, Matsushita et al (2003) did not observe the formation of any phyB-containing speckles in transgenic plants exhibiting active phyB phototransduction, and concluded that speckle formation may not be required for phyB signalling. We found that detection of speckles is affected by a variety of factors including light conditions, the level of tagged proteins, the size of speckles, etc. The fact that formation of at least two types phyB speckles was induced by R treatment in WT background and the early ones were absent in seedlings lacking PIF3 indicates that the presence of PIF3 is essential for the detection of early phyB speckles. These data, together with observations showing that mutant versions of phyA and phyB also fail to form speckles or display aberrant speckles (Kircher et al 2002, Yanovsky et al 2002), lend credible support to the hypothesis that these subnuclear structures are required for or are characteristic of phytochrome signalling. To sort out conclusively the functional relevance of the possibly many types of phyA–E related speckles and to determine the factors influencing their appearance, however, will remain a challenging task.

The Functional Role of PIF3 in the Arabidopsis Circadian System

Recent findings indicated that PIF3 plays a role in regulating expression of CCA1 and LHY1 genes shown to be essential for a functional circadian clock in Arabidopsis (Martinez-Garcia et al 2000). Based on their observations, these authors postulated that PIF3 is likely to play a role in the entrainment of the circadian system by mediating phototransduction to the central clockwork. Although the hypothesis was attractive, there were no experimental data available to verify it. Our results presented in this article, together with observations reported by Kim et al (2003), significantly modified our view about the biological function of PIF3 in light-induced signaling. During our experiments we demonstrated that poc1 is a PIF3 null mutant, lacking a detectable amount of PIF3 protein, and we generated a series of transgenic lines over-expressing the PIF3 protein as compared to the wild type. These mutants provided a suitable material to test the potential function of PIF3 in the Arabidopsis circadian system. To this end we introduced the circadian clock-regulated Cab, CCR2, and CCA1 genes fused to the luciferase reporter and determined their oscillatory patterns in stably transformed transgenic lines. In addition, we produced transgenic plants expressing the PIF3 promoter fused to the luciferase reporter and determined the transcription profile of the transgene in plants grown under various light conditions. Our data obtained unambiguously show that (i) transcription of the PIF3 gene is not regulated by light or the circadian clock and (ii) neither the over-expression nor the lack of the PIF3 protein affects the period length and phase of circadian clock-regulated genes tested in these experiments. Taken together, these data strongly suggest that PIF3 does not play a significant role in the maintenance or light entrainment of the plant circadian clock. In addition, these experiments indicate that PIF3 is not required for those circadian clock-controlled output pathways that mediate the oscillating expression of Cab and CCR2 genes. Experiments to define the possible function of PIF3 in the circadian clock-controlled leaf movement, hypocotyl elongation, and flowering time are in progress in our laboratory.

Acknowledgements. Work in Freiburg, Germany was supported by the Wolfgang Paul Award to F.N. and SFB 592, and Fonds Chemical Industry grants to E.S. Work in Szeged, Hungary was supported by OTKA Grant T032565 and HHMI Grant #55000325 to F.N.

References

Büche C, Poppe C, Schäfer E, Kretsch T (2000) *eid1*: A new *Arabidopsis* mutant hypersensitive in phytochrome A-dependent high-irradiance responses. Plant Cell 12: 547–558

Clack T, Matthews S, Sharrock RA (1994) The phytochrome apoprotein family in *Arabidopsis* is encoded by five genes: the sequence and expression of *PHYD* and *PHYE*. Plant Mol Biol 25: 413–417

Dieterle M, Zhou YC, Schäfer E, Funk M, Kretsch T (2001) EID1, an F-box protein involved in phytochrome A-specific light signalling. Genes Dev 15: 939–944

Dieterle M, Buche C, Schäfer E, Kretsch T (2003) Characterization of a novel non-constitutive photomorphogenic *cop1* allele. Plant Physiol 133: 1557–1564

Eichenberg K, Baurle I, Paulo N, Sharrock RA, Rüdiger W, Schäfer E (2000) *Arabidopsis* phytochromes C and E have different spectral characteristics from those of phytochromes A and B. FEBS Lett 470: 107–112

Furuya M, Schäfer E (1996) Photoperception and signalling of induction reactions by different phytochromes. Trends Plant Sci 1: 301–307

Gil P, Kircher S, Adam E, Bury E, Kozma-Bognar L, Schäfer E, Nagy F (2000) Photocontrol of subcellular partitioning of phytochrome-B:GFP fusion protein in tobacco seedlings. Plant J 22: 135–145

Gyula P, Schäfer E, Nagy F (2003) Light perception and signalling in higher plants. Curr Opin Plant Biol 6: 446–452

Halliday KJ, Hudson M, Ni M, Qin M, Quail PH (1999) *poc1*: an *Arabidopsis* mutant perturbed in phytochrome signalling because of a T-DNA insertion in the promoter of PIF3, a gene encoding a phytochrome-interacting bHLH protein. Proc Natl Acad Sci USA 96: 5832–5837

Harmer SL, Hoegenesch JB, Straume M, Chang HS, Han B, Zhu T, Wang X, Kreps JA, Kay SA (2000) Orchestrated transcription of key pathways in *Arabidopsis* by the circadian clock. Science 290: 2110–2113

Kendrick RE, Kronenberg GMH (eds) (1994) Photomorphogenesis in plants, 2nd edn. Kluwer, Dordrecht

Kim L, Kircher S, Toth R, Adam E, Schäfer E, Nagy F (2000) Light-induced nuclear import of phytochrome-A:GFP fusion proteins is differentially regulated in transgenic tobacco and *Arabidopsis*. Plant J 22, 125–134

Kim J, Yi H, Choi G, Shin B, Song PS, Choi G (2003) Functional characterisation of phytochrome interacting factor 3 in phytochrome-mediated light signal transduction. Plant Cell 15: 2399–2407

Kircher S, Kozma-Bognar L, Kim L, Adam E, Harter K, Schäfer E, Nagy F (1999) Light quality-dependent nuclear import of the plant photoreceptors phytochrome A and B. Plant Cell 11: 1445–1456

Kircher S, Gil P, Kozma-Bognar L, Fejes E, Speth V, Husselstein-Muller T, Bauer D, Adam E, Schäfer E, Nagy F (2002) Nucleo-cytoplasmic partitioning of the plant photoreceptors phytochrome A, B, C, D and E is differentially regulated by light and exhibits a diurnal rhythm. Plant Cell 25: 1222–1232

Lagarias JC, Lagarias DM (1989) Selfassembly of synthetic phytochrome holoprotein *in vivo*. Proc Natl Acad Sci USA 86: 5778–5780

Martinez-Garcia JF, Huq E, Quail PH (2000) Direct targeting of light signals to a promoter element-bound transcription factor. Science 288: 859–863

Matsushita T, Mochizuki N, Nagatani A (2003) Dimers of the N-terminal domain of phytochrome B are functional in the nucleus. Nature 424: 571–574

Nagy F, Schäfer E (2002) Phytochromes control photomorphogenesis by differentially regulated, interacting signalling pathways in higher plants. Annu Rev Plant Biol 53: 329–355

Ni M, Tepperman JM, Quail PH (1998) PIF3, a phytochrome interacting factor necessary for normal photoinduced signal transduction, is a novel basic helix-loop-helix protein. Cell 9: 657–667

Ni M, Tepperman JM, Quail PH (1999) Binding of phytochrome B to its nuclear signalling partner PIF3 is reversibly induced by light. Nature 400: 781–784

Osterlund MT, Hardtke CS, Wie N, Deng XW (2000) Targeted destabilization of HY5 during light-regulated development of *Arabidopsis*. Nature 405: 462–466

Oyama T, Shimura Y, Okada K (1997) The *Arabidopsis HY5* gene encodes a bZIP protein that regulates stimulus-induced development of root and hypocotyl. Genes Dev 11 2983–2995

Schaffer R, Ramsay N, Samach A, Corden S, Putterill J, Carre IA, Coupland G (1998) The late elongated hypocotyl mutation of *Arabidopsis thaliana* disrupts circadian rhythms and the photoperiodic control of flowering. Cell 93: 1219–1229

Seo HS, Yang JY, Ishikawa M, Bolle C. Ballasteros ML, Chua NH (2003) LAF1 ubiquination by COP1 controls photomorphogenesis and is stimulated by SPA1. Nature 424: 995–999

Sweere U, Eichenberg K, Lohrmann J, Mira-Rodado V, Baurle I, Kudla J, Nagy F, Schäfer E, Harter K (2001) Interaction of the response regulator ARR4 with phytochrome B in modulating red light signalling. Science 942: 1108–1111

Tepperman JM, Zhu T, Chang HS, Wang X, Quail PH (2001) Multiple transcription-factor genes are early targets of phytochrome A signalling. Proc Natl Acad Sci USA 98: 9437–9442

Wang ZY, Tobin EM (1998) Constitutive expression of the circadian clock associated 1 (CCA1) gene disrupts circadian rhythms and suppresses its own expression. Cell 93: 1207–1217

Yamaguchi R, Nakamura M, Mochizuki N, Kay SA, Nagatani A (1999) Light-dependent translocation of a phytochrome B:GFP fusion protein to the nucleus in transgenic *Arabidopsis*. J Cell Biol 145 437–445

Yanovsky MJ, Luppi JP, Kirchenbauer D, Ogorodnikova OB, Sineshchekov VA, Adam E, Kircher S, Staneloni RJ, Schäfer E, Nagy F, Casal JJ (2002) Missense mutation in the PAS2 domain of phytochrome A impairs subnuclear localisation and a subset of responses. Plant Cell 14: 1591–1603

Chapter 31

Roles Played by Auxin in Phototropism and Photomorphogenesis

Moritoshi Iino and Ken Haga

Introduction

Plants respond to light to undergo adaptive changes in their growth patterns. The idea that these responses are mediated by plant hormones has long been investigated. In fact, the first-identified plant hormone auxin was discovered and the original plant hormone concept was formulated through the studies of coleoptile phototropism, a light-induced growth movement. The role for auxin has since been a central subject of phototropism research. The elongation growth of seedling organs such as mesocotyls and hypocotyls is subject to marked light-induced inhibition. Evidence has been provided that auxin and other plant hormones participate in these typical photomorphogenetic responses. This chapter reviews and discusses the mechanisms of phototropism and photomorphogenesis, focusing on the role played by the native auxin indole-3-acetic acid (IAA). Our understanding of the molecular mechanisms by which auxin mediates these physiological processes is far from complete, but recent molecular genetic studies have began to yield useful information.

Auxin and Phototropism

The hypothesis known as "the Cholodny–Went theory of tropisms" (Went and Thimann 1937) has provided the most specific and unified view concerning the relationship between auxin and phototropism. This hypothesis, which states that tropisms are induced by the lateral asymmetry of auxin generated by its lateral translocation, was supported by earlier auxin measurements with the *Avena* curvature test and by subsequent tracer experiments using radioisotope-labeled IAA. The applicability of the Cholodny–Went hypothesis for phototropism, however, has been a subject of repeated controversy. The most recent argument

Botanical Gardens, Graduate School of Science, Osaka City University, Kisaichi, Katano Osaka 576-0004, Japan

against this hypothesis was that no asymmetry in physicochemically determined IAA could be found in all the materials investigated, which included the coleoptiles of oats and maize (Togo and Hasegawa 1991 and references cited therein). The only exception is the result reported by Iino (1991), who was able to find an IAA asymmetry in phototropically stimulated maize coleoptiles. For a detailed treatment of the subject, the reader is referred to a review by Iino (2001).

While the results from auxin measurements remain controversial, recent molecular genetic studies with *Arabidopsis thaliana* have provided results that agree with the Cholodny–Went hypothesis. Blakeslee et al (2004) undertook immunochemical investigation of the cellular localization of PIN1, a putative auxin efflux carrier, in the hypocotyl. Following phototropic stimulation, the basal localization of PIN1 in cortical cells was disrupted more on the shaded side of the hypocotyl. This response was absent in the phot1-deficient mutant. Although the exact function of PIN1 remains to be elucidated, the results have supported the view that phototropic signal transduction involves auxin transport. Harper et al (2000) found that the *NPH4* gene, mutation of which results in severe defects in both phototropism and gravitropism, encodes the auxin-regulated transcriptional activator ARF7. The *nph4* mutant does not show any obvious phenotype in its growth behaviors and in its responsiveness to applied auxin. However, the curvature response to asymmetric auxin application, observable in WT hypocotyls, is impaired in this mutant. Subsequently, Tatematsu et al (2004) found that the *msg2* mutant, which shows similar phenotypes, is a mutant of the auxin-regulated gene *AUX/IAA19*, expression of which is likely to be regulated by ARF7. These results indicate that phototropism involves auxin-mediated growth asymmetry to which ARF7 and AUX/IAA19 make specific contributions.

We isolated a mutant of rice, named *cpt1* (*coleoptile phototropism 1*), that does not show coleoptile phototropism at all effective fluence rates of blue light, and cloned the *CPT1* gene (Haga et al 2005). The deduced amino acid sequence indicated that CPT1 is orthologous to *Arabidopsis* NPH3, a signaling component of hypocotyl phototropism that can physically interact with the photoreceptor phototropin (Motchoulski and Liscum 1999). We conducted tracer experiments with ^{3}H-IAA and found that phototropic stimulation causes an asymmetric distribution of IAA in WT coleoptiles but not in *cpt1* coleoptiles (Haga et al 2005). These results have demonstrated that, in agreement with the Cholodny–Went hypothesis, lateral translocation of auxin occurs downstream of the phototropism-limiting CPT1.

Figure 1 summarizes the phototropic signaling pathway that is likely to function in the two distinct materials, *Arabidopsis* hypocotyls and rice coleoptiles. It remains to be elucidated whether phot1 and phot2 (see Chapter 15) also play photoreceptor roles in rice. We searched the *japonica* rice genome database to find that rice has two copies of *PHOT1*, which we name *PHOT1a* (located in chromosome 11) and *PHOT1b* (located in chromosome 12), and *PHOT2*. The *PHOT1* rice homolog cloned by Kanegae et al (2000) corresponds to *PHOT1a*. The deduced amino acid sequences indicate that PHOT1a and PHOT1b differ only by four amino acids. All these genes are expressed in rice coleoptiles

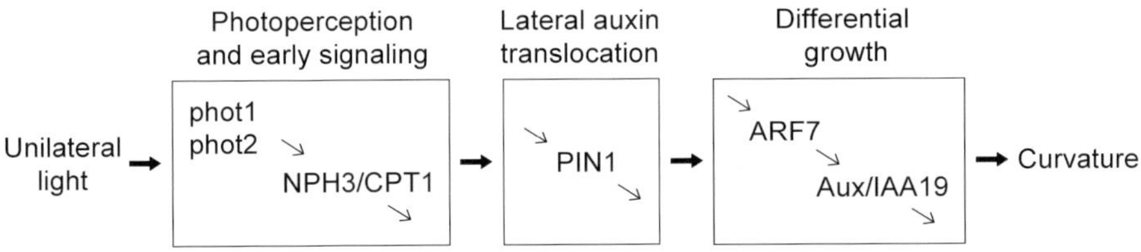

Fig. 1. The process of phototropism in hypocotyls and coleoptiles

(Yoshihara and Iino unpublished). A rice mutant that cannot express a *PHOT1* gene was independently isolated by the groups of An and Makoto Takano from T-DNA insertion and retrotransposon insertion lines, respectively (Gynheung). In collaboration with these groups, we analyzed the two mutants and found that both are mutants of *PHOT1a* and show no clear defect in phototropism (Yoshihara and Iino unpublished). Isolation of PHOT1b- and PHOT2-deficient mutants is now awaited. *Arabidopsis* RPT2, a NPH3 homolog, is another signaling component shown to participate in hypocotyl phototropism (see Chapter 20 by Sakai). Our search of the rice genome database indicated that, as for *PHOT1*, rice has two genes that are orthologous to *RPT2*.

Roots of many plants show negative phototropism (Iino 2001). Genetic evidence indicates that the phototropism of *Arabidopsis* primary roots is mediated by phototropins and involves the signaling components NPH3 and RPT2 (Motchoulski and Liscum 1999, Sakai et al 2000). Similarly, our mutant analysis indicated that CPT1 participates in the phototropism of rice primary roots (Haga et al 2005). As reported for root gravitropism, PIN3, a putative auxin efflux carrier of roots, might participate in root phototropism (Friml et al 2002). It is likely that root phototropism is mediated by a similar early signaling pathway. The phototropism of the *Phaseolus* leaf pulvinus is based on turgor-dependent changes in the volume of motor cells and differs from other growth-dependent phototropisms. During the phototropic curvature of the pulvinus, the cells on the irradiated side shrink and those on the shaded side swell (Koller and Ritter 1994). This feature and the finding that the pulvinar protoplasts swell in response to applied IAA (Iino et al 2001) have led to a hypothesis that the turgor-driven phototropic curvature might also involve lateral auxin redistribution (see discussion in Iino 2001).

Finally, for further generalization of the Cholodny–Went hypothesis, "the sunflower paradox" must be resolved. Although the hypocotyl and stem of sunflowers are highly phototropic, two independent groups could find no IAA asymmetry in this plant species. Iino (2001) hypothesized that the phototropic signal transduction generally involves a lateral asymmetry in apoplastic H^+, which precedes the induction of the auxin asymmetry and can itself cause growth differential, and that the former asymmetry plays a major role in the phototropism of sunflowers.

Auxin and Photomorphogenesis

Dark-grown seedlings of grasses such as oats and maize are marked by long elongated mesocotyls, and the most striking effect of light on these seedlings is the inhibition of mesocotyl growth. By the late 1930s, it was already clear that red light is the most effective in this response. After the discovery of phytochrome, the light inhibition of mesocotyl growth has been investigated as a typical phytochrome-mediated response. van Overbeek (1936) found that the amount of diffusible auxin obtained from the oat coleoptile tip is reduced by orange light and explained the light inhibition of mesocotyl growth in terms of a reduced supply of auxin from the coleoptile. Subsequent workers established with coleoptiles of maize (Briggs 1963) and oats (e.g. Huisinga 1976) that red light indeed reduces the amount of diffusible auxin obtained from the coleoptile tip. Furuya et al (1969) used rice seedlings to find a similar effect of red light on diffusible auxin. The result could be related to the red light-induced inhibition of coleoptile growth; in rice, the coleoptile is the major organ that undergoes dark elongation and light inhibition. The earliest explanation of the response found for diffusible auxin was that red light inhibits basipetal transport of auxin. This idea appeared to be attractive because it could also explain why the growth of oat coleoptiles is stimulated by red light (Huisinga 1976).

Iino and Carr (1982a,b) and Iino (1982a,b) conducted a series of experiments with maize seedlings to investigate the relationship between IAA and growth. It was first concluded that most of the free IAA present in the coleoptile is synthesized de novo in its tip and that the mesocotyl, which itself cannot produce IAA, receives IAA from the coleoptile and, to a lesser extent, from the primary leaves. After confirming that the amount of diffusible IAA obtained from the coleoptile tip is reduced in response to a pulse of red light (also of far-red light), Iino (1982a) found that the level of free IAA declines throughout the coleoptile and mesocotyl. No contribution of conjugated IAA could be indicated. These results led to the conclusion that what is inhibited by red light is the biosynthesis of IAA in the coleoptile tip. In fact, red light was found to inhibit the biosynthesis of IAA from applied ^{3}H-labeled L-tryptophan in the coleoptile tip. Comparisons with growth indicated that the reduction in the supply of IAA from the coleoptile accounts for a major part of the red light-induced inhibition of mesocotyl growth. The biosynthetic production of IAA from tryptophan and the inhibitory effect of red light on this biosynthesis were confirmed by Koshiba et al (1995).

The biosynthetic pathway for IAA in plants has yet to be elucidated. In the above-mentioned study (Iino 1982a), it was shown that the biosynthesis of IAA from applied ^{3}H-tryptophan is inhibited in the maize coleoptile tip with a time course that is very similar to the one obtained for the production of endogenous IAA in the same tip. This result and those reported by Koshiba et al (1995) indicate that most, if not all, of the IAA produced in the maize coleoptile tip is biosynthesized from tryptophan. This conclusion does not agree with the earlier conclusion of Cohen and his coworkers that the free IAA in the maize coleoptile derives from the IAA conjugates stored in the caryopsis, or with their later conclusion that IAA is synthesized de novo in maize coleoptiles but mainly through a tryptophan-independent pathway (Slovin et al 1999). Further investigation of the photocontrol of IAA biosynthesis in the coleoptile tip is expected to resolve not only the mechanism by which phytochrome controls IAA biosynthesis but also the biosynthetic pathway that operates in this classical system.

What elongate in dark-grown dicotyledonous seedlings are hypocotyls or epicotyls, and the elongation growth of these organs is subject to phytochrome-mediated inhibition. Behringer and Davies (1992) could detect no significant difference in the level of extractable IAA between dark-grown and red light-treated pea epicotyls (but see below for a difference found). Therefore, there was no sign that IAA biosynthesis is inhibited by red light in peas. Phytochrome-deficient mutants of *Nicotiana plumbaginifolia* had higher levels of IAA in the leaves of light-grown plants, suggesting that the level of IAA is down-regulated by phytochrome in this plant species (Kraepiel et al 1995).

It has long been assumed that the biosynthesis of IAA from tryptophan occurs through indole-3-acetaldehyde, but recent molecular genetic studies with *Arabidopsis* are revealing a new tryptophan pathway that involves indole-3-acetaldoxime as an intermediate. Two cytochrome P450s, CYP79B2 and CYP79B3, have been identified as the enzymes that catalyze conversion of tryptophan to indole-3-acetaldoxime, which can be metabolized to IAA (Hull et al 2000). Indole-3-acetaldoxime also serves as a precursor of indole glucosinolate biosynthesis and another cytochrome P450, CYP83B1, has been identified as the enzyme that metabolizes indole-3-acetaldoxime in the pathway for indole glucosinolates; evidence has been provided that the activity of CYP83 is negatively correlated with the production of IAA (Bak et al 2001). Interestingly, Hoecker et al (2004) found that hypocotyls of CYP83B1-deficient mutants elongate longer than wild-type hypocotyls under continuous red light and that the level of *CYP83B1* transcripts is enhanced by red light. These results indicated that red light enhances CYP83B1 expression, leading to a reduced level of IAA and thus to an inhibition of growth. The mutation in *CYP83B1* caused a defect in the red light-induced growth inhibition at high fluence rates, without affecting the response at low fluence rates (<1 $\mu mol\, m^{-2} s^{-1}$). It was anticipated that the red light-induced enhancement of *CYP83B1* transcripts is mediated by phyB, but the result obtained with a phyB-deficient mutant showed no significant contribution of this phytochrome. Further investigation is required to clarify whether any phytochrome species functions as a photoreceptor in this response. It would also be an attractive research direction to ask whether the indole-3-acetaldoxime pathway contributes to the phytochrome-mediated inhibition of IAA biosynthesis in grass coleoptiles (see above).

Clearly, biosynthesis is not the only possible target for photocontrol. Jones et al (1991) presented evidence that IAA becomes less distributed to the epidermis after red light treatment in maize coleoptiles and mesocotyls. Also, in the above-mentioned work with pea epicotyls, Behringer and Davies (1992) found that red light specifically reduced the level of IAA in epidermal peels. These results have suggested that lateral distribution or transport of auxin to the epidermis is controlled by phytochrome. At least in pea epicotyls, the reduced level of IAA in the epidermis may account for a large part of the phytochrome-mediated growth inhibition. There is no clear and consistent evidence for the idea that phytochrome controls growth by regulating the metabolism or basipetally polar transport of auxin. On the other hand, it has been repeatedly shown that blue light inhibits the basipetal transport of IAA in grass coleoptiles (for references and discussion, see Iino 2001). This response, probably mediated by either cryptochrome or phototropin, may participate in the inhibition of coleoptile growth known to be induced by blue light. It has been suggested that light may control the sensitivity or responsiveness to auxin. Any change in a cellular property that affects growth is likely to cause a change in auxin responsiveness. There is no evidence that light controls auxin sensitivity or responsiveness by specifically regulating auxin signaling processes. Finally, studies on *Arabidopsis* homologs of soybean *GH3*, initially identified as an auxin-inducible gene, are providing interesting results. The transcript level of a *GH3* homolog was shown to be regulated by phytochrome (Tanaka et al 2002). The products of this and several other homologs were found to have an enzymatic activity to adenylate IAA (Staswick et al 2002). Finding the function for IAA adenylation might uncover an important role played by the GH3 protein family in auxin-mediated photomorphogenesis.

Concluding Remarks

Physiological and genetic evidence now strongly indicates that phototropism involves asymmetric distribution of auxin. It is an important next task to resolve the cellular and molecular mechanisms by which auxin asymmetry is induced. How auxin asymmetry results in growth asymmetry is also an important question. We may find that phototropic growth asymmetry is not just a simple consequence of auxin asymmetry and local growth response to auxin (Iino 2001). There also seems to be no doubt that auxin plays important roles in photomorphogenesis. Physiological evidence indicates that inhibition of auxin biosynthesis is involved in phytochrome-mediated photomorphogenesis. Molecular genetic study on this mechanism is now awaited.

Clearly, the control of auxin biosynthesis is not the only mechanism by which auxin mediates photomorphogenesis (Iino 1982b, see above). Furthermore, genetic evidence indicates that brassinosteroids participate in photomorphogenesis of *Arabidopsis* seedlings (Clouse 2001). Recent results also indicate that jasmonic acid participates in photomorphogenesis of rice seedlings (Riemann et al 2003, Haga and Iino 2004). The rapid growth response to blue light is unlikely to be mediated by plant hormones (Wang and Iino 1997). It appears that photomorphogenesis involves multiple photoreceptors and multiple and interacting signaling pathways. How each signaling pathway contributes to photomorphogenesis might depend on the developmental stages of plants and the environmental conditions. There are many subjects to be studied before we understand the overall mechanisms of photomorphogenesis.

References

Bak S, Tax FE, Feldmann KA, Galbraith DW, Feyereisen R (2001) CYP83B1, a cytochrome P450 at the metabolic branch point in auxin and indole glucosinolate biosynthesis in Arabidopsis. Plant Cell 13: 101–111

Behringer FJ, Davies PJ (1992) Indole-3-acetic acid levels after phytochrome-mediated changes in the stem elongation rate of dark- and light-grown *Pisum* seedlings. Planta 188: 85–92

Blakeslee JJ, Bandyopadhyay A, Peer WA, Makam SN, Murphy AS (2004) Relocalization of the PIN1 auxin efflux facilitator plays a role in phototropic responses. Plant Physiol 134: 28–31

Briggs WR (1963) Red light, auxin relationship, and the phototropic responses of corn and oat coleoptiles. Am J Bot 50: 196–207

Clouse SD (2001) Integration of light and brassinosteroid signals in etiolated seedling growth. Trends Plant Sci 6: 443–445

Friml J, Wiśniewska J, Benková E, Mendgen K, Palme K (2002) Lateral relocation of auxin efflux regulator PIN3 mediates tropism in *Arabidopsis*. Nature 415: 806–809

Furuya M, Pjon C-J, Fujii T, Ito M (1969) Phytochrome action in *Oryza sativa* L. III. The separation of photoreceptive site and growth zone in coleoptiles, and auxin transport as effector system. Dev Growth Differ 11: 62–76

Haga K, Iino M (2004) Phytochrome-mediated transcriptional up-regulation of *ALLENE OXIDE SYNTHASE* in rice seedlings. Plant Cell Physiol 45: 119–128

Haga K, Takano M, Neumann R, Iino M (2005) The rice *COLEOPTILE PHOTOTROPISM1* gene encoding an ortholog of Arabidopsis NPH3 is required for phototropism of coleoptiles and lateral translocation of auxin. Plant Cell 17: 103–115

Harper RM, Stowe-Evans EL, Luesse DR, Muto H, Tatematsu K, Watahiki MK, Yamamoto K, Liscum E (2000) The NPH4 locus encodes the auxin response factor ARF7, a conditional regulator of differential growth in aerial Arabidopsis tissue. Plant Cell 12: 757–770

Hoecker U, Toledo-Oritz G, Beuder J, Quail PH (2004) The photomorphogenesis-related mutant *red1* is defective in *CYP83B1*, a red light-induced gene encoding a cytochrome P450 required for normal auxin homeostasis. Planta 219: 195–200

Huisinga B (1976) The export of auxin from tips and from sections of *Avena* coleoptiles as influenced by red light. Acta Bot Neerl 25: 313–320

Hull AK, Vij R, Celenza JL (2000) *Arabidopsis* cytochrome P450s that catalyze the first step of tryptophan-dependent indole-3-acetic acid biosynthesis. Proc Natl Acad Sci USA 97: 2379–2384

Iino M (1982a) Action of red light on indole-3-acetic-acid status and growth in coleoptiles of etiolated maize seedlings. Planta 156: 21–32

Iino M (1982b) Inhibitory action of red light on the growth of the maize mesocotyl: evaluation of the auxin hypothesis. Planta 156: 388–395

Iino M (1991) Mediation of tropisms by lateral translocation of endogenous indole-3-acetic acid in maize coleoptiles. Plant Cell Environ 14: 279–286

Iino M (2001) Phototropism in higher plants. In: Häder D-P, Lebert M (eds) Photomovement. Elsevier Science, Amsterdam, pp 659–811

Iino M, Carr DJ (1982a) Estimation of free, conjugated, and diffusible indole-3-acetic acid in etiolated maize shoots by the indolo-α-pyrone fluorescence method. Plant Physiol 69: 950–956

Iino M, Carr DJ (1982b) Sources of free IAA in the mesocotyl of etiolated maize seedlings. Plant Physiol 69: 1109–1112

Iino M, Long C, Wang X (2001) Auxin- and abscisic acid-dependent osmoregulation in protoplasts of *Phaseolus vulgaris* pulvini. Plant Cell Physiol 42: 1219–1227

Jones AM, Cochran DS, Lamerson PM, Evans ML, Cohen JD (1991) Red light-regulated growth. I. Changes in the abundance of indoleacetic acid and a 22-kilodalton auxin-binding protein in the maize mesocotyl. Plant Physiol 97: 352–358

Kanegae H, Tahir M, Savazzini F, Yamamoto K, Yano M, Sasaki T, Kanegae T, Wada M, Takano M (2000) Rice *NPH1* homologues, *OsNPH1a* and *OsNPH1b*, are differently photoregulated. Plant Cell Physiol 41: 415–423

Koller D, Ritter S (1994) Phototropic responses of the pulvinules and associated laminar reorientation in the trifoliate leaf of bean *Phaseolus vulgaris*. J Plant Physiol 143: 52–63

Koshiba T, Kamiya Y, Iino M (1995) Biosynthesis of indole-3-acetic acid from L-tryptophan in coleoptile tips of maize (*Zea mays* L.). Plant Cell Physiol 36: 1503–1510

Kraepiel Y, Marrec K, Sotta B, Caboche M, Miginiac E (1995) In vitro morphogenic characteristics of phytochrome mutants in *Nicotiana plumbaginifolia* are modified and correlated to high indole-3-acetic acid levels. Planta 197: 142–146

Motchoulski A, Liscum E (1999) *Arabidopsis* NPH3: a NPH1 photoreceptor-interacting protein for phototropism. Science 286: 961–964

Riemann M, Müller A, Korte A, Furuya M, Weiler EW, Nick P (2003) Impaired induction of the jasmonate pathway in the rice mutant *hebiba*. Plant Physiol 133: 1–11

Sakai T, Wada T, Ishiguro S, Okada K (2000) RPT2: a signal transducer of the phototropic response in Arabidopsis. Plant Cell 12: 225–236

Slovin JP, Bandurski RS, Cohen JD (1999) Auxin. In: Hooykaas PJJ, Hall MA, Libbenga KR (eds) Biochemistry and molecular biology of plant hormones. Elsevier Science, Amsterdam, pp 115–140

Staswick PE, Tiryaki I, Rowe ML (2002) Jasmonate response locus JAR1 and several related Arabidopsis genes encode enzymes of the firefly luciferase superfamily that show activity on jasmonic, salicylic, and indole-3-acetic acids in an assay for adenylation. Plant Cell 14: 1405–1415

Tanaka S, Mochizuki N, Nagatani A (2002) Expression of the *AtGH3a*, an Arabidopsis homologue of the soybean *GH3* gene, is regulated by phytochrome B. Plant Cell Physiol 43: 281–289

Tatematsu K, Kumagai S, Muto H, Sato A, Watahiki MK, Harper RM, Liscum E, Yamamoto KT (2004) *MASSUGU2* encodes Aux/IAA19, an auxin-regulated protein that functions together with the transcriptional activator NPH4/ARF7 to regulate differential growth responses of hypocotyl and formation of lateral roots in *Arabidopsis thaliana*. Plant Cell 16: 379–393

Togo S, Hasegawa K (1991) Phototropic stimulation does not induce unequal distribution of indole-3-acetic acid in maize coleoptiles. Physiol Plant 81: 555–557

van Overbeek J (1936) Growth hormone and mesocotyl growth. Rec Trav Bot Neerl 33: 333–340

Went FW, Thimann KV (1937) Phytohormones. Macmillan, New York London

Wang X, Iino M (1997) Blue light-induced shrinking of protoplasts from maize coleoptiles and its relationship to coleoptile growth. Plant Physiol 114: 1009–1020

CHAPTER 32

The Effect of Light and Gravity on Hypocotyl Growth Orientation

PATRICIA LARIGUET and CHRISTIAN FANKHAUSER

Introduction

Light energy capture by leaves, water, and mineral absorption by roots are crucial for plant survival. To guide their growth plant organs sense a variety of environmental cues, among which the direction of gravity and the direction of light are the most important. Other environmental factors dictating tropic responses like moisture and touch play minor roles in land plants and are discussed in a recent review (Blancaflor and Masson 2003).

Gravitropism and phototropism refer to the directional curvature of an organ in response to lateral differences in gravity or light, respectively. Plants respond to changes in gravity and light direction by modulating the rate of cellular elongation on opposite flanks of the stimulated organ. Such asymmetric growth leads to a curvature and subsequent realignment with the right orientation. The machinery that sustains tropism in roots, hypocotyls, and shoots of higher plants are at least partially distinct (Blancaflor and Masson 2003). This review focuses on the mechanisms underlying light interaction with gravitropism and phototropism, particularly in hypocotyls of higher plants. We will not discuss events occurring in root gravitropism or other phototropin-mediated responses. Recent reviews cover these fields more specifically (Boonsirichai et al 2002, Briggs and Christie 2002, Liscum 2003, Kiss et al 2003).

Gravitropism Signaling Pathway in Seedling Hypocotyls

In hypocotyls the primary site for gravity sensing covers the entire elongation zone (Blancaflor and Masson 2003). The current model proposes that gravity perception is initiated by sedimentation of amyloplasts in cells of the endodermis

Department of Molecular Biology, University of Geneva, Sciences III, 30 quai E. Ansermet, CH-1211 Geneva 4, Switzerland

(Boonsirichai et al 2002). The amyloplasts are surrounded by the vacuolar membrane inside of the vacuole. Vesicle transport to vacuoles mediated by a specific SNARE (Soluble NSF Attachment protein Receptors) complex appears to participate in the early events of gravity sensing. This specific SNARE complex includes SHOOT GRAVITROPIC 2 (SGR2), SGR3, and SGR4/ZIG in Arabidopsis (Blancaflor and Masson 2003).

An unknown process converts the physical movement of amyloplasts into a biochemical signal. The cytoskeleton could be involved in shoot gravitropic signal transduction because in Arabidopsis, mutations in *ALTERATED RESPONSE TO GRAVITY 1 (ARG1)* and *ARG1 LIKE 2 (ARL2)* specifically affect hypocotyl and root gravitropism. *ARG1* and *ARL2* encode dnaJ-like proteins that may interact with the cytoskeleton, suggesting that amyloplast interaction with the actin network may trigger downstream signaling events (Blancaflor and Masson 2003). The initial signaling stages required to generate the differential growth include changes in inositol 1,3,5-triphosphate and Ca^{2+} concentration. At a later stage the *GRAVITY PERSISTENCE SIGNAL (GPS)* loci contribute to the persistence of gravity perception or signal transduction (Blancaflor and Masson 2003).

The establishment of a lateral auxin gradient across the hypocotyl follows gravity stimulation. The auxin gradient is achieved by specific redistribution of the auxin carrier machinery. Auxin efflux is believed to involve two families of putative auxin efflux carriers: members of the PIN-FORMED (PIN) family and two related ABC transporters, GLYCOPROTEIN P1 (PGP1) and PGP19/MDR1. Indeed PIN3 displays a polar localization in the hypocotyl endodermis and it relocalizes in roots following gravity stimulation. Disruption of *PIN3* and of both *PGP1* and *PGP19* alter hypocotyl tropism (Friml 2003, Noh et al 2003). The dynamic nature of PIN proteins localization may allow rapid targeting to the appropriate location upon tropic stimulations (Friml 2003). Auxin transporter redistribution drives polar auxin transport across the elongation zones to cortical and epidermal cells to allow asymmetric growth (Blancaflor and Masson 2003).

Gravity stimulation modifies the level of expression of about 1.7% of the genes in Arabidopsis. Among them are *MASSUGU2/IAA19* (Tatematsu et al 2004), the auxin response factor *NON PHOTOTROPIC HYPOCOTYL 4* (*NPH4*) (Harper et al 2000), the secretory low molecular weight phospholipase A2 (Lee et al 2003), and ethylene-response element binding factors (Moseyko et al 2002). Gravistimulation also results in a differential expression of the gene encoding the auxin-induced K^+ channel ZMK1 in the two flanks of gravistimulated maize coleoptiles (Fuchs et al 2003). The activity of those gene products may contribute to differential cell elongation.

Light Modulation of Hypocotyl Gravitropism

In darkness, seedlings orient their growth according to the gravity vector. However, seedlings are rarely in the total absence of light. Therefore most often the combination of light and gravity defines the growth orientation of plant organs. Light has been shown to modulate graviresponses independently of phototropism in a number of plants. Red light reduces gravitropism in Arabidopsis (Hangarter 1997) and in the moss *Ceratodon purpureus* (Lamparter et al 1998). Far-red and blue light can limit hypocotyl gravitropism in Arabidopsis (Hangarter 1997, Lariguet and Fankhauser 2004). Therefore all light qualities negatively regulate the response of the hypocotyl to gravity.

In higher plants light is perceived by several families of photoreceptors, the UV-A/blue light sensors phototropins and cryptochromes, the phytochromes maximally absorbing red and far-red light, and by unknown UV-B photoreceptors (Quail 2002). Light modulation of gravitropism is due to the action of the phytochromes in Arabidopsis, tomato, and mosses (Lamparter et al 1996, Hangarter 1997, Behringer and Lomax 1999). In Arabidopsis the effect of red light on gravitropism is controlled by phyA and phyB and exclusively by phyA in far-red light (Hangarter 1997). Under low intensities of continuous blue light phyA is the predominant sensor that triggers gravitropism inhibition, with the cryptochromes playing a secondary role (Lariguet and Fankhauser 2004). The inhibition of gravitropism by far-red light increases with increasing fluence rates; however, under strong light the seedlings become gravitropic again (Fairchild et al 2000). The blue light-mediated limitation of hypocotyl gravitropism is also fluence rate dependent. The light intensity-dependent reduction of gravitropism could modulate the contribution of gravitropism and phototropism depending on the growth conditions.

We know little about the mechanisms involved downstream of phytochrome activation that result in inhibition of hypocotyl gravitropism. *shy2-1* is a gain-of-function mutant of IAA3 that undergoes photomorphogenesis in the dark. *shy2-1* exhibits reduced gravitropism and genetically interacts with phytochrome-deficient mutants. Interestingly, *SHY2* expression is light regulated and SHY2 can physically interact with phyA (Halliday and Fankhauser 2003). Other mutants that de-etiolate in the dark such as *cop1* also show an inhibition of gravitropism in the absence of light (Kim et al 2002). Far-red light-induced disruption of hypocotyl gravitropism is altered in *phyA* mutants and in mutants of phyA signaling components such as *hfr1* (*long hypocotyl in far-red*), *fin2* (*far-red insensitive*), *fhy1*, and *fhy3* (Fairchild et al 2000, Honsberger and Fankhauser unpublished). The fact that SHY2, FHY3, and HFR1 are nucleus localized suggests that nuclear events of phytochrome signaling are involved in the light-regulated inhibition of gravitropism.

Besides auxin, other hormones play a role in light modulation of gravitropism because cytokinin acting via ethylene can restore gravitropism in red light-induced agravitropic seedlings. The ethylene-insensitive mutant *ein2-1* is defective in gravitropism even in the dark (Hangarter 1997). This suggests that activation of the phytochromes could decrease the ethylene level or the responsiveness to ethylene, thereby inhibiting gravitropism.

Shoot Phototropism Signaling Pathway

Etiolated hypocotyls of higher plants are able to perceive the direction of UV-A/blue/green light through the phototropin photoreceptors (phot1 and phot2 in Arabidopsis). Recent reviews, including Winslow Briggs' in this volume (Chapter 15) cover phototropism in detail (Briggs and Christie 2002, Liscum et al 2003). Phototropins use an FMN (flavin mononucleoside) chromophore and are plasma membrane-associated serine/threonine protein kinases (Liscum et al 2003). They undergo autophosphorylation upon blue light irradiation (Briggs and Christie 2002, Liscum et al 2003). NPH3 (NON PHOTOTROPIC HYPOCOTYL3) and RPT2 (ROOT PHOTOTROPISM MUTANT 2) are the only early signal transducers acting in hypocotyl phototropism identified so far. *NPH3* and *RPT2* encode homologous proteins of unknown biochemical function. They are associated with the plasma membrane and can interact together and with phot1 (Inada et al 2004). The review by Tatsuya Sakai in this volume covers phototropin-signaling pathways in detail (Chapter 20).

Lateral blue light perception is followed by the establishment of differential auxin gradients within the hypocotyl of Arabidopsis or within the coleoptile of maize, with accumulation of auxin on the shaded side (Friml 2003, Fuchs et al 2003). phot1 and auxin transporters colocalize at the plasma membrane suggesting the possibility of direct interaction (Liscum 2003). Among the auxin carriers PIN3 appears to be of particular importance to establish lateral auxin gradients in response to changes in light direction (Friml 2003). Normal localization of PIN1 is also required for phototropism (Noh et al 2003, Blakeslee et al 2004). The altered phototropic phenotype of mutants affecting NPH4/ARF7 and MSG2/IAA19 supports the model that auxin-dependent changes in gene expression are required for phototropism (Tatematsu et al 2004). *ZMK1* is the best-characterized blue light-induced gene involved in hypocotyl tropism (Fuchs et al 2003). However, microarray analysis suggests that phototropins play minor roles in blue light-induced transcriptional regulation (Ohgishi et al 2004).

Other Photoreceptors Modulating Phototropism

Photosensory systems other than phototropins modulate phototropism in mono- and dicotyledonous plants (Liscum 2003). In maize and Arabidopsis phyA and phyB affect the magnitude of phototropic curvature (Hangarter 1997). Indeed, development of phototropic curvature in *phyA* and *phyB* single mutants is slower than in the wild type, and the amplitude of the curvature is even more severely attenuated in *phyA phyB* during the first hours of unilateral low blue light irradiation (Hangarter 1997). However, with long irradiations (24 h) of blue light, *phy* mutants differ little from the wild type. The cryptochromes also affect the magnitude of phototropic curvature by modulating signaling events downstream of the phototropins. At high intensity of blue light phototropism is attenuated. This decreased response is due to the co-action of phototropins and cryptochromes (Whippo and Hangarter 2003).

UV-A, blue, red and far-red light, applied before and/or during the lateral blue light treatment, can enhance the phototropic response in Arabidopsis or tomato (Hangarter 1997, Srinivas et al 2004). Red light and UV-A phototropic enhancement is due to the action of phytochromes (Hangarter 1997, Stowe-Evans et al 2001). The molecular mechanism of phytochrome-mediated phototropic enhancement is a matter of debate. Phytochromes appear not to be primary sensors of directional light in etiolated Arabidopsis seedlings (Liscum 2003). Our recent results indicate that phyA acts as a positive regulator of phototropism by inhibiting gravitropism, independently of the phototropins (Lariguet and Fankhauser 2004).

Interactions Between Phototropism and Gravitropism

When the light comes laterally an organism is subjected both to phototropic and gravitropic stimuli; it must be able to evaluate the two conflicting forces and find a compromise. Most data concerning the respective effects of gravitropism and phototropism come from studies in the fungus *Phycomyces blakesleemus* and in the moss *Ceratodon*. In *P. blakesleemus* the growth orientation of the sporangiophores appear to be governed by a feedback loop of mutual influence between gravitropism and phototropism (Grolig et al 2000). In *Ceratodon* red light can both inhibit negative gravitropism and induce positive phototropism. The phytochrome-mediated red light inhibition of gravitropism occurs independently of phototropism (Lamparter et al 1996). At low intensities of red light (<140 μmol $m^{-2} s^{-1}$) phototropism and gravitropism compete but above this threshold gravitropism is completely repressed (Kern and Sack 1999). These findings suggest the existence of an irradiance-dependent regulation of gravitropism in this moss species.

In higher plants, however, little is known concerning the interactions of gravitropism and phototropism. In Arabidopsis roots phototropic bending is exaggerated in graviresponse-impaired mutants (Kiss et al 2003). Maize coleoptiles with neutralized gravitropism exhibit greater phototropic bending than the controls (Fuchs et al 2003). It has been proposed that the decreased negative gravitropism could lead to the increased phototropic curvature (Hangarter 1997). The Arabidopsis *arl2-1* mutant exhibits reduced hypocotyl gravitropism and increased phototropism. The authors proposed that this "slight enhancement of phototropism may be the consequence of the gravitropism defect" (Guan et al 2003). The conclusion that arises from these experiments is the existence of a competition between gravitropism and phototropism in developing seedlings. However, the Arabidopsis *sgr1/scr*, *sgr2*, *sgr4*, *sgr7/shr*, and *arg1* mutants show reduced gravitropism but normal phototropism (Firn et al 2000). We assayed the reciprocal experiment consisting in eliminating phototropism and evaluating the consequence on gravitropism in Arabidopsis. Phototropism-deficient seedlings are randomly oriented under blue light because blue light simultaneously suppresses gravitropism through a phototropin-independent photosensory system. Under low blue light this is primarily a phyA response (see above; Lariguet and Fankhauser 2004).

Besides phytochrome signaling (and to a lesser extent cryptochrome signaling in blue light) that appear to directly limit hypocotyl gravitropism, other actors can play a role in integrating the two tropic signaling cascades. Indeed, although the perception systems for phototropism and gravitropism are distinct, the downstream signal transduction systems share a number of features. Auxin appears to be a central integrator of light and gravity signaling. Both tropisms indeed merge into auxin redistribution in the hypocotyl elongation zone and involve the action of the gene products of *NPH4/ARF7*, *PIN3*, *PGP19*, and *MSG2/IAA19* (Friml 2003, Noh et al 2003, Tatematsu et al 2004). Maize coleoptile tips laterally irradiated with blue light display a more pronounced lateral auxin gradient when gravitropism is neutralized (Fuchs et al 2003). Thus the asymmetric expression of the K^+ uptake channel ZMK1 is induced by blue light and gravity, presumably as a consequence of the auxin gradient generated by these two stimuli (Fuchs et al 2003). Ethylene may also play a role in the integration of signals from gravitropism and phototropism in tomato and Arabidopsis by modifying the auxin responses (Madlung et al 1999, Harper et al 2000). These common features between the phototropic and the gravitropic responses constitute potential points of interactions between both tropisms.

Concluding Remarks

Light-controlled inhibition of gravitropism in the hypocotyl is an important aspect of the de-etiolation response. This function is primarily achieved by the phytochromes under all light qualities. One of the future challenges of researchers in this field is to solve how phytochrome signaling interferes with gravity signaling. Does light signaling modulate early events of gravity perception/signaling such as amyloplast sedimentation or directly affect known components of this pathway such as ARG1? Does phytochrome signaling interact with auxin transport or metabolism processes? To date, SHY2/IAA3 represents the most direct link between phytochrome and auxin signaling (Halliday and Fankhauser 2003). Given the extent of cross-talk between the different plant hormones it may be wise not to limit the investigations to auxin. The answers to these questions are very important because ultimately it is the integration of signals coming from both gravity and light that will determine the growth orientation of plant organs.

Acknowledgments. We thank Paula Duek, Michel Goldsmith-Clermont, and Isabelle Schepens for reading the manuscript. We are grateful to the Swiss NSF (Marie Heim-Vögtlin fellowship to P.L., NCCR "plant survival" and grant PP00A-103005 to C.F.) for supporting our work.

References

Behringer FJ, Lomax TL (1999) Genetic analysis of the roles of phytochromes A and B1 in the reversed gravitropic response of the lz-2 tomato mutant. Plant Cell Environ 22: 551–558

Blakeslee JJ, Bandyopadhyay A, Peer WA, Makam SN, Murphy AS (2004) Relocalization of the PIN1 auxin efflux facilitator plays a role in phototropic responses. Plant Physiol 134: 28–31

Blancaflor EB, Masson PH (2003) Plant gravitropism. Unraveling the ups and downs of a complex process. Plant Physiol 133: 1677–1690

Boonsirichai K, Guan C, Chen R, Masson PH (2002) Root gravitropism: an experimental tool to investigate basic cellular and molecular processes underlying mechanosensing and signal transmission in plants. Annu Rev Plant Biol 53: 421–447

Briggs WR, Christie JM (2002) Phototropins 1 and 2: versatile plant blue-light receptors. Trends Plant Sci 7: 204–210

Fairchild CD, Schumaker MA, Quail PH (2000) HFR1 encodes an atypical bHLH protein that acts in phytochrome A signal transduction. Genes Dev 14: 2377–2391

Firn RD, Wagstaff C, Digby J (2000) The use of mutants to probe models of gravitropism. J Exp Bot 51: 1323–1340

Friml J (2003) Auxin transport—shaping the plant. Curr Opin Plant Biol 6: 7–12

Fuchs I, Philippar K, Ljung K, Sandberg G, Hedrich R (2003) Blue light regulates an auxin-induced K^+-channel gene in the maize coleoptile. Proc Natl Acad Sci USA 100: 11795–11800

Grolig F, Eibel P, Schimek C, Schapat T, Dennison DS, Galland PA (2000) Interaction between gravitropism and phototropism in sporangiophores of *Phycomyces blakesleeanus*. Plant Physiol 123: 765–776

Guan C, Rosen ES, Boonsirichai K, Poff KL, Masson PH (2003) The ARG1-LIKE2 gene of Arabidopsis functions in a gravity signal transduction pathway that is genetically distinct from the PGM pathway. Plant Physiol 133: 100–112

Halliday KJ, Fankhauser C (2003) Phytochrome-hormonal signalling networks. New Phytol 157: 449–463

Hangarter RP (1997) Gravity, light and plant form. Plant Cell Environ 20: 796–800

Harper RM, Stowe-Evans EL, Luesse DR, Muto H, Tatematsu K, Watahiki MK, Yamamoto K, Liscum E (2000) The NPH4 locus encodes the auxin response factor ARF7, a conditional regulator of differential growth in aerial Arabidopsis tissue. Plant Cell 12: 757–770

Inada S, Ohgishi M, Mayama T, Okada K, Sakai T (2004) RPT2 is a signal transducer involved in phototropic response and stomatal opening by association with phototropin 1 in *Arabidopsis thaliana*. Plant Cell 16: 887–896

Kern VD, Sack FD (1999) Irradiance-dependent regulation of gravitropism by red light in protonemata of the moss *Ceratodon purpureus*. Planta 209: 299–307

Kim YM, Woo JC, Song PS, Soh MS (2002) HFR1, a phytochrome A-signalling component, acts in a separate pathway from HY5, downstream of COP1 in *Arabidopsis thaliana*. Plant J 30: 711–719

Kiss JZ, Correll MJ, Mullen JL, Hangarter RP, Edelmann RE (2003) Root phototropism: how light and gravity interact in shaping plant form. Gravit Space Biol Bull 16: 55–60

Lamparter T, Esch H, Cove D, Hughes J, Hartmann E (1996) Aphototropic mutants of the moss *Ceratodon purpureus* with spectrally normal and with spectrally dysfunctional phytochrome. Plant Cell Environ 19: 560–568

Lamparter T, Hughes J, Hartmann E (1998) Blue light- and genetically-reversed gravitropic response in protonemata of the moss *Ceratodon purpureus*. Planta 206: 95–102

Lariguet P, Fankhauser C (2004) Hypocotyl growth orientation in blue light is determined by phytochrome A inhibition of gravitropism and phototropin promotion of phototropism. Plant J 40: 826–834

Lee HY, Bahn SC, Kang YM, Lee KH, Kim HJ, Noh EK, Palta JP, Shin JS, Ryu SB (2003) Secretory low molecular weight phospholipase A2 plays important roles in cell elongation and shoot gravitropism in Arabidopsis. Plant Cell 15: 1990–2002

Liscum E, Hodgson DW, Campbell TJ (2003) Blue light signaling through the cryptochromes and phototropins. So that's what the blues is all about. Plant Physiol 133: 1429–1436

Madlung A, Behringer FJ, Lomax TL (1999) Ethylene plays multiple nonprimary roles in modulating the gravitropic response in tomato. Plant Physiol 120: 897–906

Moseyko N, Zhu T, Chang HS, Wang X, Feldman LJ (2002) Transcription profiling of the early gravitropic response in Arabidopsis using high-density oligonucleotide probe microarrays. Plant Physiol 130: 720–728

Noh B, Bandyopadhyay A, Peer WA, Spalding EP, Murphy AS (2003) Enhanced gravi- and phototropism in plant mdr mutants mislocalizing the auxin efflux protein PIN1. Nature 424: 999–1002

Ohgishi M, Saji K, Okada K, Sakai T (2004) Functional analysis of each blue light receptor, cry1, cry2, phot1, and phot2, by using combinatorial multiple mutants in Arabidopsis. Proc Natl Acad Sci USA 101: 2223–2228

Quail PH (2002) Photosensory perception and signalling in plant cells: new paradigms? Curr Opin Cell Biol 14: 180–188

Srinivas A, Behera RK, Kagawa T, Wada M, Sharma R (2004) High pigment1 mutation negatively regulates phototropic signal transduction in tomato seedlings. Plant Physiol 134: 790–800

Stowe-Evans EL, Luesse DR, Liscum E (2001) The enhancement of phototropin-induced phototropic curvature in Arabidopsis occurs via a photoreversible phytochrome A-dependent modulation of auxin responsiveness. Plant Physiol 126: 826–834

Tatematsu K, Kumagai S, Muto H, Sato A, Watahiki MK, Harper RM, Liscum E, Yamamoto KT (2004) MASSUGU2 encodes Aux/IAA19, an auxin-regulated protein that functions together with the transcriptional activator NPH4/ARF7 to regulate differential growth responses of hypocotyl and formation of lateral roots in *Arabidopsis thaliana*. Plant Cell 16: 379–393

Whippo CW, Hangarter RP (2003) Second positive phototropism results from coordinated co-action of the phototropins and cryptochromes. Plant Physiol 132: 1499–1507

Chapter 33

Convergence of Phytochrome and Cryptochrome Signalling

Jorge José Casal

Introduction

Light signals perceived by phytochromes (phy) and cryptochromes (cry) control diverse growth and developmental decisions throughout the life cycle of plants. The vast majority of these processes are under the simultaneous influence of multiple photoreceptors, which share the control of expression of target genes (Ma et al 2001). Here we describe recent advances in our understanding of the mode of photoreceptor signalling convergence and the consequences of the interactive signalling network operating downstream the photoreceptors.

The Relationship Between Light Input and Physiological Output Depends on the Combined Action of Photoreceptors

If two photoreceptors together yield a response equivalent to the sum of their individual responses they are said to act additively, i.e., there is no evidence for the occurrence of interaction between them. When simultaneous or sequential activation of specific photoreceptors yields responses that cannot be accounted for by the response of each one of the photoreceptors in isolation, photoreceptor signalling is interactive. This section describes some patterns of interaction (see Casal 2000 for further detail).

There are conditions where two photoreceptors cause a response that is larger than the response expected based on the action of each one of the photoreceptors in isolation. One example is provided by phyB (a red light photoreceptor) and cry1 (a blue light photoreceptor). In *Arabidopsis thaliana* seedlings exposed to short periods of blue light added to a background of red light, the contribution of phyB to the inhibition of hypocotyl growth is larger when cry1 is present

Faculty of Agronomy, IFEVA, University of Buenos Aires, Av. San Martín 4453, 1417 Buenos Aires, Argentina

(i.e., the difference between the wild type and the *phyB* mutant) than when cry1 is absent (i.e., the difference between the *cry1* mutant and the *phyB cry1* double mutant) (Casal and Mazzella 1998). Complementarily, the contribution of cry1 is larger if phyB is present (i.e., the difference between the wild type and the *cry1* mutant) than if phyB is absent (i.e., the difference between the *phyB* mutant and the *phyB cry1* double mutant). This pattern illustrates a case of synergism. The synergistic interaction between phyB and cry1 operates under sub-optimal light inputs (short blue-light treatments) but not when the seedlings are exposed to prolonged blue plus red light (Casal and Mazzella 1998).

There are many examples where a photoreceptor mutation has little or no effect on its own and makes a more obvious contribution to the phenotype when combined with other photoreceptor mutations. These are cases of genetic redundancy. In tomato, for instance, single, double or triple mutants combining *phyA*, *phyB1*, *phyB2* and/or *cry1* show morphological phenotypes but only the quadruple *phyA phyB1 phyB2 cry1* mutant fails to achieve reproductive development (Weller et al 2001). In *A. thaliana* plants grown under white light, the rate of leaf production is reduced by the *phyB* mutation and unaffected by the *phyA*, *cry1* or *cry2* single mutations. However, the quadruple *phyA phyB cry1 cry2* mutant shows a much more dramatic reduction in leaf production than the *phyB* single mutant (Mazzella et al 2001). This indicates that phyA, cry1 and cry2 have the potential to control this trait but they fail to do so in the presence of phyB.

Redundancy can be the result of negative regulation of the action mediated by one photoreceptor by the action of another. For instance, under red light the *phyB* mutant de-etiolates poorly and the *phyA phyB* double mutant exhibits virtually no morphological response to red light, indicating that the residual response to red light observed in the *phyB* mutant is mediated by phyA. However, under red light the *phyA* mutant is somewhat more de-etiolated than the wild type itself, indicating that phyA also represses phyB-mediated photomorphogenesis (Cerdán et al 1999).

As described above for phyB and cry1, the interaction between phyA and phyB is also light conditional. The negative regulation of phyB signalling by phyA occurs under red light but if phyA is activated by prolonged far-red light followed by a pulse of red light (to activate phyB), the interaction among these photoreceptors is synergistic rather than antagonistic (e.g. Cerdán et al 1999).

Noteworthy, there are several examples where photoreceptor interaction does not require seedling exposure to light signals normally activating each one of the photoreceptors. The Cvi allele of *A. thaliana* cry2 interacts synergistically with phyA in the control of cotyledon unfolding under pulses of far-red light (Botto et al 2003). COP1, a nuclear repressor of photomorphogenesis, migrates to the cytosol when dark-grown seedlings are transferred to light. The effect of far-red light is impaired not only by the *phyA* but also by the *cry1* mutation (Osterlund and Deng 1998). Normal signalling of phyA to the clock under red light also requires cry1 (Devlin and Kay 2000).

Ecological Significance

Under natural radiation, the relationship between light input and physiological output depends on the combined action of multiple photoreceptors. One example of such regulation is provided by the analysis of seedlings of the *phyA*, *phyB* and *cry1* mutants in all possible combinations, grown in darkness or under different levels of sunlight irradiance. During de-etiolation, the interactions among photoreceptors enhance the impact of low irradiances compared to darkness and simultaneously reduce the differences between high and low irradiances (Mazzella and Casal 2001). Clearly, de-etiolation must be initiated when a seedling emerges from the soil even if it faces reduced irradiance levels. The interactions among photoreceptors help to achieve this goal by enhancing the sensitivity to light compared to darkness and reducing the impact of different irradiance levels.

A second example is provided by the response of vegetative growth to the presence of neighbour plants. Before mutual shading among plants is established in sparse canopies, the red to far-red ratio is subtly reduced due to the selective light reflection by the green foliage. The response to neighbour seedlings is initiated by the small reductions in the levels of active phyB caused by this early warning signal of impending competition. phyA does not play a direct role in the promotion of stem growth caused by low red to far-red ratios. However, seedlings of the *phyA* mutant fully ignore the small reductions in red to far-red ratio typical of sparse canopies (Casal 1996). These observations have been interpreted as a down-regulation of phyB-mediated inhibition of stem growth by phyA. This regulation would render the system sensitive to the small reductions in the levels of active phyB caused by slightly decreased red to far-red ratios (Casal 1996).

A third example is provided by the transition between vegetative and reproductive development. The *cry2* mutant of *A. thaliana* is late flowering under long days but the *phyB cry2* double mutant flowers as early as the *phyB* mutant or very nearly (Mockler et al 1999, Mazzella et al 2001). The contribution of cry2 is larger in the presence than in the absence of phyB. Thanks to this interaction between phyB and cry2, the transition towards the reproductive program can be accelerated by any one of two different light signals. Long days perceived mainly by cry2 promote flowering even if the red to far-red ratio is high (Guo et al 1998). Complementarily, even if the days are short, flowering can also be accelerated by low red to far-red ratios that reduce the levels of active phyB (Cerdán and Chory 2003). Long days and low red to far-red ratios provide cues about the most favourable season and about the presence of neighbour plants that can compete for resources, respectively.

Early Convergence: Physical Interaction Between Photoreceptors

A first step towards understanding how the concerted action of multiple photorectors modulates input–output relationships is to uncover the points of signalling convergence. Physical interaction occurs between phyA and cry1 (Ahmad et al 1998) as well as between phyB and cry2 (Más et al 2000). Whether this physical interaction is required for normal light responses is unknown. Since the effect of cry2 on phyA-mediated cotyledon unfolding is allele-specific (Botto et al 2003) this physiological interaction could be due to direct interaction between phyA and cry2 but other interpretations cannot be ruled out.

COP1 as a Node of Convergence and Divergence

COP1 is another point of convergence of photoreceptor signalling. COP1 acts as an ubiquitin-protein ligase that promotes the proteasome-mediated degradation of transcription factors required for normal photomorphogenesis such as HY5 (Saijo et al 2003) and LAF1 (Seo et al 2003). The morphology, physiology and transcriptome patterns of *cop1* mutants grown in darkness resemble the patterns of wild-type seedlings grown in the light (Ma et al 2002). COP1 is nuclear localised in darkness. In the light, the action of multiple photoreceptors causes gradual mobilisation of COP1 to the cytosol (Osterlund and Deng 1998). Depletion of the COP1 nuclear pool by light allows the accumulation of HY5 and LAF1 and therefore photomorphogenesis proceeds (Figure 1). Light-induced dissociation of protein complexes containing COP1 could also be involved in a more rapid inactivation of COP1 activity (Saijo et al 2003).

Despite the convergent action of phyA, phyB and cry1 on COP1, some of the COP1 downstream targets are involved in de-etiolation under selective light conditions (Figure 1). LAF1 is required for normal phyA-mediated responses (Seo et al 2003) and HYH is required for normal de-etiolation under blue light (Holm et al 2002). PIF3 is a repressor of phyB-mediated responses and its abundance in darkness depends on the presence of COP1, which likely down-regulates a factor (X in Figure 1) that promotes de-stabilisation of PIF3 (Bauer et al 2004). Other factors, such as HY5 and CPI4 (Yamamoto et al 2001) are involved in photomorphogenesis under diverse light conditions. The occurrence of light-quality selective roles downstream of COP1 indicates that in addition to their common role in down-regulating COP1 activity, the different photoreceptors retain a more selective action with targets downstream of COP1. The levels of HYH mRNA are dramatically increased by red or blue light compared to darkness but not by far-red light (Holm et al 2002). This could only partially account for the selective action of HYH under blue light. The abundance of the PIF3 protein is down-regulated under red or far-red light via the action of phyA, phyB and phyD (Bauer et al 2004), but selectivity of PIF3 activity can be accounted for by its preferential binding to phyB compared to phyA (Zhu et al 2000).

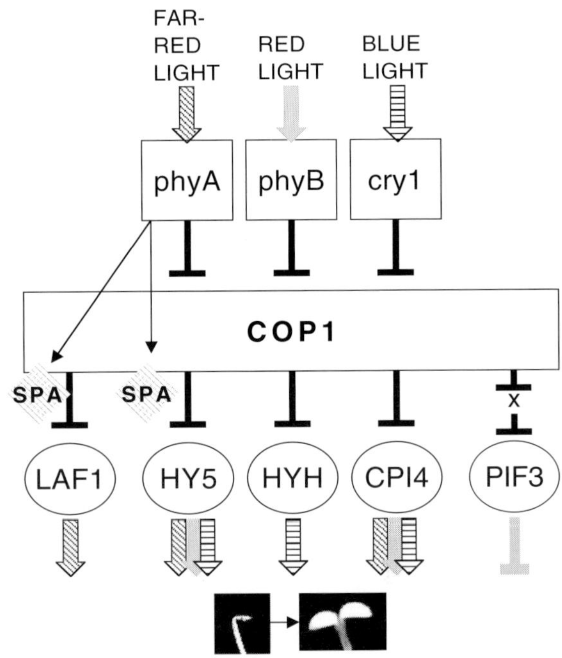

FIG. 1. COP1 is a point of signalling convergence and divergence. Phytochromes and cryptochromes converge to down-regulate COP1 at least partially by inducing its migration from the nucleus to the cytosol. Despite this convergence, some downstream targets of COP1 retain photoreceptor-selective roles. Different arrow fills identify far-red, red or blue light. *Arrow*, promotion; *"T" line*, inhibition

The role played by the residual nuclear pool of COP1 can also be selective because its ubiquitin-protein ligase activity is regulated by factors like SPA1, which controls phyA-mediated responses (Saijo et al 2003, Seo et al 2003). SUB1 is a calcium-binding protein enriched in the nuclear periphery that negatively regulates phyA-mediated responses under far-red and blue light, at least partially by reducing HY5 protein levels (Guo et al 2001).

Partial Convergence of phyA- and cry1 Signalling on HFR1

HFR1 is a putative bHLH transcription factor involved in phyA- and cry1-mediated responses. The analysis of the *hfrs1* mutant under blue light revealed a phenotype still present in the *phyA*-mutant background and very weak in the *cry1*-mutant background (Duek and Fankhauser 2003). The *HFR1* gene is expressed in darkness and its mRNA levels are slightly increased by far-red or blue light and strongly decreased by red light (Duek and Fankhauser 2003). Thus, there is coincidence between the wavebands where *HFR1* is expressed and the *hfr1* phenotype is observed. As noted by Duek and Fankhauser (2003), although HFR1 represents a point of convergence of phyA and cry1 signalling, its role in photoreceptor interactions remains to be established.

Antagonistic Convergence of Photoreceptor Signalling on the Regulation of CONSTANS Abundance

In *A. thaliana*, the perception of long days depends on the coincidence between high levels of expression of the *CO* gene, which is under the control of a circadian clock (Suarez-López et al 2001) and light perceived mainly by cry2 and phyA (Yanovsky and Kay 2002). The occurrence of this coincidence promotes the expression of the flowering gene *FT*. One point of action of plant photoreceptors on the control of flowering is the light input to the clock that regulates *CO* expression (Figure 2). phyA enhances *CO* expression in etiolated seedlings (Tepperman et al 2001). The waveform of *CO* expression is somewhat affected by long compared to short days (Suarez-López et al 2001) but the specific role of different photoreceptors has not been elucidated. A second and more dramatic target of photoreceptor action involves the regulation of CO protein stability (Valverde et al 2003). In darkness, CO is degraded by the proteasome. Light perceived by cry2, cry1 and phyA stabilises CO whereas light perceived by phyB causes the opposite effect (Figure 2). The circadian control of *CO* expression, the stabilising effect of cry2, phyA and cry1, and the de-stabilising effect of phyB create a balance where CO protein levels build up towards the final part of long days (Valverde et al 2003). Since the effects of the *cry1 cry2* double mutation are larger on the expression of *FT* than on CO protein stability, an additional role in

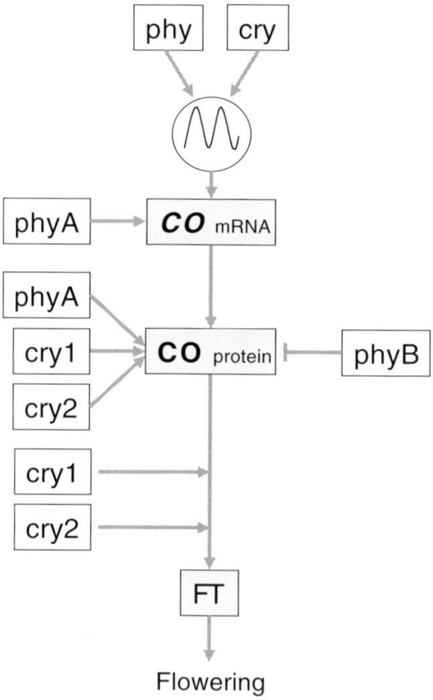

FIG. 2. Phytochromes and cryptochromes converge to control CO activity at different points and in antagonistic direction. *Arrows*, promotion; *"T" line*, inhibition

activating CO after its accumulation has been proposed for cry1 and/or cry2 (Valverde et al 2003) (Figure 2). The negative regulation of CO protein stability by phyB correlates with the enhanced *FT* expression observed in the *phyB* mutant (Cerdán and Chory 2003).

Summary and Perspectives

The occurrence of an interactive network downstream of the photoreceptors is crucial for the regulation of the relationship between light input and physiological output. Some points of signalling convergence downstream the photoreceptors have been uncovered. Convergence is not complete and other points of action of the photoreceptors help to maintain specificity. One of the future challenges is to establish causal links between points of convergence and the observed physiological interactions.

References

Ahmad M, Jarrillo JA, Smirnova O, Cashmore A (1998) The CRY1 blue light photoreceptor of *Arabidopsis* interacts with phytochrome A in vitro. Mol Cell 1: 939–948

Bauer D, Viczián A, Kircher S, Nobis T, Nitschke R, Kunkel T, Panigrahi KCS, Ádám E, Fejes E, Schäfer E, Nagy F (2004) Constitutive Photomorphogenesis 1 and multiple photoreceptors control degradation of phytochrome interacting factor 3, a transcription factor required for light signalling in Arabidopsis. Plant Cell 16: 1433–1445

Botto JF, Alonso Blanco C, Garzarón I, Sánchez RA, Casal JJ (2003) The Cvi allele of cryptochrome 2 enhances cotyledon unfolding in the absence of blue light in Arabidopsis. Plant Physiol 133: 1539–1546

Casal JJ (1996) Phytochrome A enhances the promotion of hypocotyl growth caused by reductions of phytochrome B Pfr levels in light-grown *Arabidopsis thaliana*. Plant Physiol 112: 965–973

Casal JJ (2000) Phytochromes, cryptochromes, phototropin: Photoreceptor interactions in plants. Photochem Photobiol 71. 1–11

Casal JJ, Mazzella MA (1998) Conditional synergism between cryptochrome 1 and phytochrome B is shown by the analysis of *phyA*, *phyB* and *hy4* simple, double and triple mutants in Arabidopsis. Plant Physiol 118: 19–25

Cerdán PD, Chory J (2003) Regulation of flowering time by light quality. Nature 423: 881–885

Cerdán PD, Yanovsky MJ, Reymundo FC, Nagatani A, Staneloni RJ, Whitelam GC, Casal JJ (1999) Regulation of phytochrome B signaling by phytochrome A and FHY1 in *Arabidopsis thaliana*. Plant J 18: 499–507

Devlin PF, Kay SA (2000) Cryptochromes are required for phytocarome signaling to the clock but not for rhythinicity. Plant Cell 12: 2499–2510

Duek PD, Fankhauser C (2003) HFR1, a putative bHLH transcription factor, mediates both phytochrome A and cryptochrome signalling. Plant J 34: 827–836

Guo H, Yang H, Mockler TC, Lin C (1998) Regulation of flowering time by *Arabidopsis* photoreceptors. Science 279: 1360–1363

Guo H, Mockler T, Duong H, Lin C (2001) SUB1, an Arabidopsis Ca^{2+}-binding protein involved in cryptochrome and phytochrome coaction. Science 19: 487–490
Holm M, Li-Geng M, Li-Jia Q, Deng XW (2002) Two interacting bZIP proteins are direct targets of COP1-mediated control of light-dependent gene expression in *Arabidopsis*. Genes Dev 16: 1247–1259
Ma L, Li J, Qu L, Hager J, Chen Z, Zhao H, Deng XW (2001) Light control of Arabidopsis development entails coordinated regulation of genome expression and cellular pathways. Plant Cell 13: 2589–2607
Ma L, Gao Y, Qu L, Chen Z, Li J, Zhao H, Deng XW (2002) Genomic evidence for COP1 as a repressor of light-regulated gene expression and development in Arabidopsis. Plant Cell 14: 2383–2398
Más P, Devlin PF, Panda S, Kay SA (2000) Functional interaction of phytochrome B and cryptochrome 2. Nature 408: 207–211
Mazzella MA, Casal JJ (2001) Interactive signalling by phytochromes and cryptochromes generates de-etiolation homeostasis in *Arabidopsis thaliana*. Plant Cell Environ 24: 155–162
Mazzella MA, Cerdán PD, Staneloni R., Casal JJ (2001) Hierarchical coupling of phytochromes and cryptochromes reconciles stability and light modulation of *Arabidopsis* development. Development 128: 2291–2299
Mockler T, Guo H, Yang H, Duong H, Lin C (1999) Antagonistic actions of Arabidopsis cryptochromes and phytochrome B in the regulation of floral induction. Development 126: 2073–2082
Osterlund MK, Deng XW (1998) Multiple photoreceptors mediate the light induced reduction of GUS-COP1 from Arabidopsis hypocotyl nuclei. Plant J 16: 201–208
Saijo Y, Sullivan JA, Wang H, Yang J, Shen Y, Rubio V, Ma L, Hoecker U, Deng XW (2003) The COP1-SPA1 interaction defines a critical step in phytochrome A-mediated regulation of HY5 activity. Genes Dev 17: 2642–2647
Seo HS, Yang JY, Ishikawa M, Bolle C, Ballesteros ML, Chua NH (2003) LAF1 ubiquitination by COP1 controls photomorphogenesis and is stimulated by SPA1. Nature 2003: 995–999
Suárez-López P, Wheatley K, Robson F, Onouchi H, Valverde F, Coupland G (2001). CONSTANS mediates between the circadian clock and control of flowering in Arabidopsis. Nature 410: 1116–1120
Tepperman JM, Zhu T, Chang HS, Wang X, Quail PH (2001) Multiple transcription-factor genes are early targets of phytochrome A signaling. Proc Natl Acad Sci USA 98: 9437–9442
Valverde F, Mouradov A, Soppe W, Ravenscroft D, Samach A, Coupland G (2003) Photoreceptor regulation of CONSTANS protein and the mechanism of photoperiodic flowering. Science 303: 1003–1006
Weller JL, Perrotta G, Schreuder MEL, van Tuinen A, Koornneef M, Giuliano G, Kendrick RE (2001) Genetic dissection of blue-light sensing in tomato using mutants deficient in cryptochrome 1 and phytochromes A, B1 and B2. Plant J 25: 427–440
Yamamoto YY, Deng XW, Matsui M (2001) CIP4, a new COP1 target, is a nucleus-localized positive regulator of Arabidopsis photomorphogenesis. Plant Cell 13: 399–411
Yanovsky MJ, Kay SA (2002) Molecular basis of seasonal time measurement in Arabidopsis. Nature 419: 308–312
Zhu Y, Tepperman JM, Fairchild CD, Quail PH (2000) Phytochrome B binds with a greater apparent affinity than phytochrome A to the basic helix-loop-helix factor PIF3 in a reaction requiring the PAS domain of PIF3. Proc Natl Acad Sci USA 97: 13419–13424

Chapter 34

Downstream Integrators of Red, Far-Red, and Blue Light Signaling for Photomorphogenesis

Min Ni

Introduction

Plants have evolved several classes of photoreceptors to monitor their light environment. The photoreceptors include red and far-red light-absorbing phytochromes and UV-A/blue light-absorbing cryptochromes and phototropins (Cashmore 2003, Quail 2002, Wada et al 2003). Optimal performance of a plant therefore depends on coordination among the light signaling pathways. Among the photoreceptors, phytochromes and cryptochromes regulate seedling de-etiolation responses, photoperiodic flowering, and circadian rhythm (Guo et al 1999, Mas et al 2003), whereas phototropins regulate phototropic response, stomatal opening, and chloroplast relocation movement to improve the efficiency of photosynthesis (Wada et al 2003).

Genetic approaches have been used in the last decade to understand the signaling processes downstream of phytochromes and cryptochromes. The genetic screens are largely based on seedling de-etiolation responses and have identified many far-red light signaling mutants (Quail 2002) and several red light signaling mutants (Huq and Quail 2002, Liu et al 2001, Staiger et al 2003). Recently, a blue light signaling component, PP7, has been identified through a reverse genetic approach (Møller et al 2003). Those studies together suggest the existence of major distinct red, far-red, or blue light signaling pathways. However, genetic screens have also identified mutants with defective seedling de-etiolation responses under more than a single wavelength. For example, *sub1* has a hypersensitive response to both blue and far-red light (Guo et al 2001), indicating an integration of blue and far-red light signaling pathways at some branch points. This review thus intends to describe a few such components identified in recent years. In addition, photoreceptor signaling also appears to be integrated with the regulation of photoperiodic flowering and circadian rhythm, and a few such examples will be briefly presented in this review.

Department of Plant Biology, University of Minnesota, St. Paul, MN 55108, USA

Integration of Photoreceptor Signaling with Other Photophysiological Responses

Many mutants, although isolated for their defects in photoperiodic flowering and circadian rhythm, also exhibit certain defective seedling de-etiolation responses or vice versa (Cerdan and Chory 2003, Mas et al 2003, Staiger et al 2003). Some components such as GI, ELF3, and ELF4 appear to integrate light signals to control photoperiodic flowering response. Mutation in *GI* leads not only to a late flowering phenotype under inductive long-day conditions, but also to a long hypocotyl phenotype under red light (Huq et al 2000). In contrast, mutations in *ELF3* and *ELF4* result in an early flowering phenotype, but a longer hypocotyl under red light (Khanna et al 2003, Liu et al 2001). In addition, both *ELF3* and *ELF4* genes are also required for normal circadian regulation. For example, *elf4* has elevated expression of *CO*, a gene that promotes floral induction, and also shows attenuated expression of *CCA1*, a gene that may function as a central oscillator (Doyle et al 2002). A newly isolated mutant, *pft1*, shows a mild hypocotyl growth phenotype, but displays a strong late-flowering phenotype under long-day conditions (Cerdan and Chory 2003). Mutation in *PFT1* also completely suppresses the early-flowering phenotype of *phyB*, suggesting that PFT1 mainly functions to regulate flowering downstream of phyB in a photoperiod-independent pathway. Interestingly, almost all of the flowering genes have a circadian clock-regulated expression.

Other components appear to integrate light signals to control both morphogenic and circadian responses. For example, a newly identified mutant, *srr1*, is altered not only in phyB-controlled hypocotyl elongation but also in multiple outputs of the circadian clock (Staiger et al 2003). On the other hand, many of the clock function-associated genes such as *CCA1*, *LHY*, *TOC1*, *FKY1*, and *ZTL* have been implicated in integration of light signals to control photomorphogenesis although they were initially isolated from a number of recessive mutations that alter the free-running period of the *Arabidopsis* circadian clock (Mas et al 2003).

Integration of Red and Far-Red Light Signaling

Genetic screens have identified several red/far-red light mutants such as *pef1, psi2, cog1-D, prr7,* and *rfi2* (Quail 2002, Park et al 2003, Kaczorowski and Quail 2003, Chen and Ni, unpublished). A dominant mutant, *cog1-D*, showed defects in both phyA- and phyB-mediated light responses (Park et al 2003). The mutation is caused by activation of a Dof domain-containing transcription factor, and transgenic lines expressing antisense *COG1* results in a hypersensitive response to red and far-red light. In contrast, *prr7* has reduced response to both red and far-red light (Kaczorowski and Quail 2003). The absence of PRR7 or PSEUDO-RESPONSE REGULATOR 7 also causes a coordinated 3- to 6-h shift in the phasing of the oscillatory expression of *CCA1*, *LHY*, and *TOC1*, the central components of the circadian clock. *PRR7* belongs to a small gene family called *TOC1*/*APRR1* that includes *TOC1* or *TIMING-OF-CAB1*. The proteins in this family lack the conserved phospho-accepting Asp of the bacterial response regulators.

Recently, TOC1 has also been reported for its involvement in red and far-red light control of hypocotyl elongation and red light-mediated regulation of *CCA1* and *LHY* expression during early seedling development (Mas et al 2003). Apparently, both TOC1 and PRR7 are able to integrate red and far-red light signals to control circadian and morphogenic responses. We have recently isolated a mutant with a long hypocotyl phenotype under red and far-red light, and we named it **r**ed and **f**ar-red light **i**nsensitive **2** or *rfi2* (Chen and Ni unpublished). The mutant resembles many defective light responses of *phyA* and *phyB*, including reductions in end-of-day far-red light response, far-red light block of greening, cotyledon expansion, and expression of a few light-regulated genes. Mutation in *RFI2* not only causes a defective photomorphogenic development, but also leads to a very early flowering under both long-day and short-day conditions. In *rfi2*, the expression of *CO* seems to be up-regulated over an entire circadian period.

Integration of Blue and Far-Red Light Signaling

It has been realized that a minimal level of active phytochrome seems to be necessary for full activity of cryptochromes or phototropins, and one classic example is the enhancement by red light of phototropic bending toward unilateral blue light (Ahmad et al 1998, Casal 2000). The dependence of the blue responses on active phytochromes may result from a direct interaction of both photoreceptors as indicated in early studies that cry1 and cry2 can be phosphorylated by a phytochrome A-associated kinase activity (Ahmad et al 1998). The enzymatic interaction of phyA with cry1 was further confirmed in targeted yeast two-hybrid assays. The studies suggest that photoactivated phytochrome can phosphorylate a number of substrates including crys, and subsequent exposure of crys to blue light would enhance the signaling activities of crys.

Equally possible, the dependence of blue responses on active phytochromes may occur at a common intermediate step of their signaling pathways. SUB1, a cytoplasmic calcium-binding protein, is such a candidate (Guo et al 2001). Mutation in *SUB1* has a stronger hypersensitive hypocotyl growth response and much enhanced *CHS* and *CHI* expression under relatively low fluence rates of blue and far-red light. Genetic analysis indicates that SUB1 functions downstream of crys and modulates phyA-mediated far-red light responses. SUB1 localizes in the nuclear periphery region surrounding the nucleus, and may regulate light responses by suppressing light-dependent accumulation of HY5 protein. Another candidate is HFR1, a bHLH transcription factor. HFR1 was initially isolated based on a defect in a subset of phyA-mediated far-red light responses in *hfr1* mutant (Duek and Fankhauser 2003). Recently, *hfr1* has been shown to have reduced de-etiolation responses, including hypocotyl elongation, cotyledon expansion, and anthocyanin accumulation, under high fluence rates of blue light (Duek and Fankhauser 2003). Genetic analysis indicates that HFR1 function in cry1 signaling pathway since cry1 is the major photoreceptor responsible for de-etiolation under high fluence rates of blue light. Although SUB1 and HFR1 function in both blue and far-red light signaling pathways, they may use quite different mechanisms to integrate phyA and cry signaling.

Integration of Red and Blue Light Signaling

Recently, genetic interaction has been implicated between phyB and cry2 in control of hypocotyl elongation, flowering time, and circadian rhythm (Mas et al 2000). In vivo coimmunoprecipitation, colocalization, and FRET analysis showed a direct interaction of phyB with cry2 in nuclear speckles, suggesting a possible integration of blue and red light signaling at photoreceptor level. Evidence also suggests an integration of blue and red light signals at other steps of the signaling cascades. For example, we have recently isolated an *Arabidopsis* mutant that displays a short hypocotyl phenotype under both red and blue light and we name it *hrb1* for **h**ypersensitive to **r**ed and **b**lue **1** (Kang et al unpublished). Mutation in *HRB1* also enhances end-of-day far-red light response, inhibits leaf expansion and petiole elongation, and represses expression of *CAB3* and *CHS* under red and blue light. Double-mutant analysis indicates that expression of *hrb1* phenotype requires a functional phyB molecule, and HRB1 may regulate hypocotyl elongation response to blue light downstream of crys. *HRB1* expression is enhanced by red, far-red, and blue light, and HRB1 belongs to a protein family of Drought induced 19 or Di19. HRB1 and the other family members contain a ZZ-type zinc finger domain, which in other organisms is implicated in protein–protein interactions between dystrophin and calmodulin, and between transcriptional adaptors and activators. Thus, HRB1 may interact with other light signaling components through its ZZ-type zinc finger domain.

We also detected a reduced expression of *PIF4*, a gene encoding a phyB interacting protein, under red and blue light in *hrb1*, suggesting that HRB1 may be directly involved in the control of *PIF4* expression in the nucleus. Loss-of-function *pif4* has a similar hypersensitive hypocotyl growth response as *hrb1* to red light (Huq and Quail 2002), and also to blue light (Kang et al, unpublished). The hypersensitive response of *hrb1* to red and blue light may be partially attributed to a reduced *PIF4* expression since *pif4* is a semi-dominant mutation and a half reduction in *PIF4* message can result in a noticeable change in hypocotyl growth response (Huq and Quail 2002). Thus, the roles of HRB1 and PIF4 together in regulating both red and blue light responses may represent points where red light signaling and blue light signaling cross-talk.

We have isolated another *Arabidopsis* mutant, **l**ight **i**nsensitive **r**esponse **1**-**1** or *lir1-1*, with a long hypocotyl phenotype under red, far-red, and blue light (Kang and Ni, unpublished). The long hypocotyl phenotype is caused by an overaccumulation of LIR1 in *lir1-1*, and is recapitulated by overexpression of *LIR1* in transgenic *Arabidopsis*. However, *lir1-2*, a knockout allele of *LIR1*, exhibits a short hypocotyl phenotype under blue light. LIR1 thus functions in blue light signaling, but overexpression of LIR1 expands its signaling activity to red and far-red light. Studies on both *lir1-1* and *lir1-2* indicate that LIR1 negatively regulates inhibition of hypocotyl elongation, cotyledon opening, and leaf expansion. LIR1 also positively regulates cotyledon expansion, inhibition of petiole elongation, pigment accumulation, and the expression of *CAB3* and *CHS*. The regulation of LIR1 on the responses may involve HFR1, a basic helix-loop-helix protein, since *lir1-2* and *hfr1* have overlapping phenotypes and LIR1 is required for the proper expression of *HFR1* under blue light. LIR1 localizes to cytoplasmic speckles, and contains an N-terminal SPX and a C-terminal EXS domain found in members of the SYG1 protein family from fungi, worm, fly, mammals, and *Arabidopsis*.

Perspectives

Mechanisms for integration of red, far-red, and blue light signaling in control of photomorphogenesis remain largely unknown. This review has described a few such components involved in the integration processes. However, the nature and biochemical consequences of the integration at either photoreceptor level or subsequent signal transduction steps still remain unclear. Considering the various subcellular locations of the integrators, we anticipate the existence of multi-mechanisms that plants may use to efficiently integrate red, far-red, and blue light signaling. A possible mechanism may operate in a way that an integrating component functions independently in two different light signaling pathways, and the signaling activity of the component is coordinately regulated. On the other hand, an integration component may function in one light signaling pathway, but modulate the activity of another light signaling pathway. Such mechanisms may involve SUB1 and HRB1. In these cases, the modulation can be achieved through either a direct post-translational control of protein stability or a direct transcriptional control on gene expression. Other ways of modulation include post-transcriptional control on mRNA stability, and post-translational controls such as activation or sequestration through protein–protein interactions and protein modification through phosphorylation.

References

Ahmad M, Jarrilo JA, Smirnova O, Cashmore AR (1998) The CRY1 blue light photoreceptor of *Arabidopsis* interacts with phytochrome A in vitro. Mol Cell 1: 939–948

Casal JJ (2000) Phytochromes, cryptochromes, phototropin: photoreceptor interactions in plants. Photochem Photobiol 71: 1–11

Cashmore AR (2003) Cryptochromes: enable plants and animals to determine circadian time. Cell 114: 537–543

Cerdan PD, Chory J (2003) Regulation of flowering time by light quality. Nature 423: 881–885

Doyle MR, Davis SJ, Bastow RM, McWatters HG, Kozma-Bognar L, Nagy F, Millar AJ, Amasino RM (2002) The ELF4 gene controls circadian rhythms and flowering time in *Arabidopsis thaliana*. Nature 419: 74–77

Duek PD, Fankhauser C (2003) HFR1, a putative bHLH transcription factor, mediates both phytochrome A and cryptochrome signaling. Plant J 34: 827–836

Guo H, Duong H, Ma N, Lin C (1999) The *Arabidopsis* blue light receptor cryptochrome 2 is a nuclear protein regulated by a blue light-dependent post-translational mechanism. Plant J 19: 279–287

Guo H, Mockler T, Duong H, Lin C (2001) SUB1, an *Arabidopsis* Ca^{2+}-binding protein involved in cryptochrome and phytochrome coaction. Science 291: 487–490

Huq E, Quail PH (2002) PIF4, a phytochrome-interacting bHLH factor, functions as a negative regulator of phytochrome B signaling in *Arabidopsis*. EMBO J 21: 2441–2450

Huq E, Tepperman JM, Quail PH (2000) GIGANTEA is a nuclear protein involved in phytochrome signaling in *Arabidopsis*. Proc Natl Acad Sci USA 97: 9789–9794

Kaczorowski KA, Quail PH (2003) Arabidopsis *PSEUDO-RESPONSE REGULATOR7* is a signaling intermediate in phytochrome-regulated seedling deetiolation and phasing of the circadian clock. Plant Cell 15: 2654–2665

Khanna R, Kikis EA, Quail PH (2003) *EARLY FLOWERING 4* functions in phytochrome B-regulated seedling de-etiolation. Plant Physiol 133: 1530–1538

Liu XL, Covington MF, Fankhauser C, Chory J, Wagner DR (2001) *ELF3* encodes a circadian clock-regulated nuclear protein that functions in an *Arabidopsis* phyB signal transduction pathway. Plant Cell 13: 1293–1304

Mas P, Devlin PF, Panda S, Kay SA (2000) Functional interaction of phytochrome B and cryptochrome 2. Nature 408: 207–211

Mas P, Alabadi D, Yanovsky MJ, Oyama T, Kay SA (2003) Dual role of TOC1 in the control of circadian and photomorphogenic responses in *Arabidopsis*. Plant Cell 15: 223–236

Møller SG, Kim YS, Kunkel T, Chua NH (2003) PP7 is a positive regulator of blue light signaling in *Arabidopsis*. Plant Cell 15: 1111–1119

Park DH, Lim PO, Kim JS, Cho DS, Hong SH, Nam HG (2003) The *Arabidopsis COG1* gene encodes a Dof domain transcription factor and negatively regulates phytochrome signaling. Plant J 34: 161–171

Quail PH (2002) Phytochrome photosensory signaling network. Nat Rev Mol Cell Biol 3: 85–93

Staiger D, Allenbach L, Salathia N, Fiechter V, Davis SJ, Millar AJ, Chory J, Fankhauser C (2003) The *Arabidopsis SRR1* gene mediates phyB signaling and is required for normal circadian clock function. Genes Dev 17: 256–268

Wada M, Kagawa T, Sato Y (2003) Chloroplast movement. Annu Rev Plant Biol 54: 455–468

CHAPTER 35

Signal Transduction Pathways Regulating Chromatic Adaptation

RICHARD M. ALVEY[1], LINA LI[1], BARBARA E. BALABAS[1], LAURA SEIB[1], EMILY L. STOWE-EVANS[1,2], and DAVID M. KEHOE[1]

Introduction

Over 100 years ago, dramatic changes in the color phenotypes of certain filamentous cyanobacteria in response to changing ambient light conditions were first described. This acclimation process, known as complementary chromatic adaptation (CCA), leads to cell coloration that can range from brick red to blue-green. CCA is a photoreversible process with features that are similar to processes controlled by plant phytochrome photoreceptors. This review provides an overview of the physiology of CCA as well as a summary of recent findings concerning the nature and function of the signal transduction pathways used to regulate CCA in the filamentous cyanobacterium *Fremyella diplosiphon*, which has been used as a model system to study this process for over four decades. For historical perspectives on CCA, the reader is referred to other reviews (Bogorad 1975, Tandeau de Marsac 2003).

Complementary Chromatic Adaptation

The light harvesting structures used by cyanobacteria are called phycobilisomes (PBS). PBS are composed primarily of two protein classes, chromophorylated phycobiliproteins (which contain an α and β subunit) and non-pigmented structural proteins called linkers. In *F. diplosiphon*, CCA involves changes in the abundance of these proteins that is effectively triggered by red light (RL, 650–660 nm) and green light (GL, 540–550 nm). During growth in RL, a blue-colored phycobiliprotein called phycocyanin (PC) and its associated linker proteins are produced and incorporated into PBS. However, if the same cells are instead grown in GL, a red-colored phycobiliprotein called phycoerythrin (PE) and its corresponding linkers are synthesized and added to PBS (Figure 1). Thus the cells are

[1] Department of Biology, Indiana University, 1001 East 3rd Street, Bloomington, IN 47405, USA
[2] Current address: Biology Department, Bucknell University, Lewisburg, PA 17837, USA

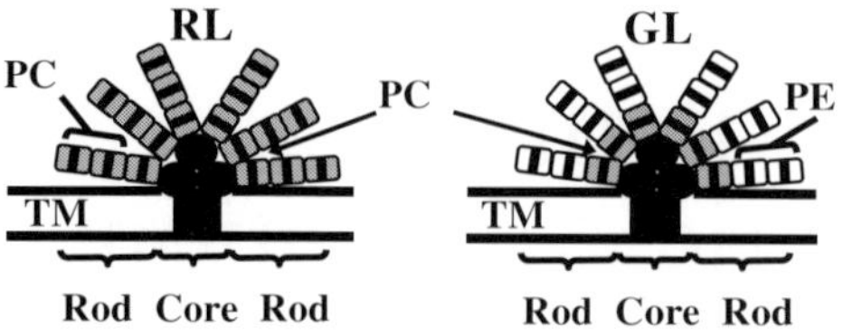

Fig. 1. *Fremyella diplosiphon* PBS contain phycocyanin (*gray*) in red light and phycoerythrin (*white*) in green light. *Black box*, photosynthetic reaction center; *TM*, thylakoid membrane

blue-green when grown in RL but are a brick red color if they are grown in GL. Since PC maximally absorbs RL (~620 nm) and PE maximally absorbs GL (~560 nm), CCA allows cyanobacteria to optimize photosynthetic efficiency by spectrally tuning the absorption characteristics of its PBS to the predominant wavelength of ambient light.

CCA only occurs in species that contain both PC and PE, but not all cyanobacteria that contain these pigments undergo CCA. Tandeau de Marsac (1977) noted that PC and PE abundance in some species (called Group 1) is unaffected by ambient light color and classified chromatically adapting species into two groups. Group 2 species produce PC at constant levels regardless of ambient light color, while PE levels increase when cells are grown in GL. Group 3 species, such as *F. diplosiphon*, control both PE and PC levels in response to light color as described above.

Several important cellular events that occur during CCA have been elucidated. A number of laboratories demonstrated that GL and RL strongly affect the expression of the genes encoding the α and β subunits of PE (*cpeBA*), the PE linkers (*cpeCDE*), as well as the α and β subunits of PC and its associated linkers (*cpcB2A2H2I2D2*, abbreviated as *cpc2*). In GL, the expression of the *cpeBA* and *cpeCDE* operon are up regulated and *cpc2* expression is shut down. In RL, the converse occurs: *cpc2* is more highly expressed and both *cpeCDE* and *cpeBA* are down regulated (see Grossman 2003).

CCA Signal Transduction

Subsequent studies have focused on the signal transduction pathway(s) controlling CCA. Analyses of the promoter regions of *cpeBA*, *cpeCDE*, and *cpc2* failed to uncover any clearly shared *cis*-acting elements that might participate in coordinating the light-regulated expression of these operons. However, deletion analysis demonstrated that a region from –76 to +25 of the *cpc2* promoter was sufficient for light responsiveness (Casey and Grossman 1994). Electrophoretic mobility shift assays detected binding activity to this region exclusively in protein extracts from RL-grown cells, which was eliminated when a DNA fragment from –37 to +25 was added at 5–10 fold molar excess. This region of the *cpc2* promoter contains one direct repeat. Several groups have also detected major DNA binding activity in protein extracts from RL grown cells further upstream in this promoter (Casey and Grossman 1994, Sobczyk et al 1994). DNase I footprinting

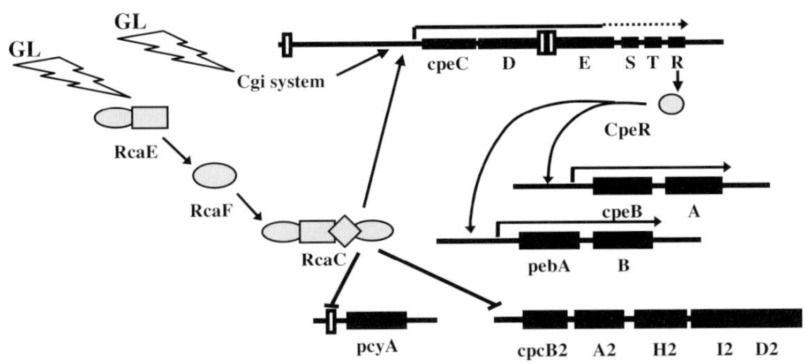

FIG. 2. Model of CCA regulation in *F. diplosiphon* during growth in green light. Regulatory proteins are noted. *Arrows* above operons denote active transcription. In red light, *pcyA* and *cpc2* are highly expressed while the *cpe* and *peb* operons are not. Each *small rectangle* is an R Box (see text). Only some light color-responsive genes are shown

studies demonstrated that a protein named RcaD bound to the regions from –265 to –245 and from –186 to –166. Phosphatase treatment led to a loss of this activity. Recently RcaD was biochemically purified, and *rcaD* was cloned and mutated via allelic exchange. The mutant had a wild-type phenotype during growth in RL and GL but differed from wild type just after light color shifts (Noubir et al 2002). RcaD was proposed to coordinate *cpcB2A2*, *cpeBA*, and other PBS-encoding genes whose expression is not CCA regulated during transitions between GL and RL conditions. Footprinting studies of the *cpeBA* promoter identified a region of unknown function from –66 to –45 that contained a direct repeat (different from that in the *cpc2* promoter) and bound a protein called RcaA/PepB (Sobzyk et al 1993, Schmidt-Goff and Federspiel 1993).

The generation and complementation of CCA mutants resulted in the isolation of three genes that encode CCA signal transduction proteins that were named RcaC, RcaE, and RcaF (Figure 2) (Chaing et al 1992, Kehoe and Grossman 1996, 1997). Sequence analyses suggested that these three proteins were elements of a complex type of a two component regulatory system called a "multi-step phosphorelay." RcaC and RcaF were both response regulator class proteins, while RcaE was a hybrid protein with its C-terminal domain related to output domains of sensor histidine kinases and its N-terminal domain similar to the chromophore-binding domain of plant phytochromes. Genetic analyses suggested that these three proteins functioned in the same pathway and that RcaE acted before RcaF, which acted prior to RcaC.

It has recently been confirmed that RcaE is a CCA photoreceptor that is present in cells at equal levels in RL and GL (Terauchi et al 2004). In addition, RcaE has been shown to have a covalently attached bilin chromophore in vivo and require a cysteine at position 198 for this attachment. Interestingly, attempts to correctly attach bilin chromophores to RcaE, either in vitro or in bilin-producing *Escherichia coli* cells, were not successful. Thus it is possible that for RcaE, correct chromophore attachment requires a separate lyase enzyme. Such a requirement would make RcaE different from any previously analyzed prokaryotic or eukaryotic phytochrome, all of which have been found to possess autolyase activity.

An in vivo functional study of RcaC has also been carried out. At 75 kDa, RcaC is an unusually large response regulator that contains at least four domains (Chaing et al 1992, Kehoe and Grossman 1997). There are two receiver domains, one at each end of the protein. Each contains an aspartate residue (D51 and D576) that is typically conserved in such domains and is the site of reversible phosphorylation during signal transduction in other systems. A DNA binding domain is adjacent to the N-terminal receiver domain and is also adjacent to a histidine phosphotransfer domain, which contains a histidine (H316) that is conserved in such domains and is also reversibly phosphorylated during signal transduction in other systems. Mutants with amino acid substitutions at D51, H316, and D576 were created and introduced into an *rcaC* null mutant to test the role of each of these residues in the control of CCA. The majority of CCA regulation was found to operate through H316 and D51, while D576 had a minor role. The phenotypes of the substitution mutants also suggested that RcaC is phosphorylated during growth in RL and dephosphorylated when grown in GL (Li and Kehoe 2004).

In addition, Western blot analyses using antibodies raised against RcaC demonstrated that this protein was approximately fivefold more abundant in RL-grown cells than in GL-grown cells (Li and Kehoe unpublished). Preliminary data on *rcaC* RNA levels in RL and GL (Stowe-Evans et al 2004) suggest that the RcaC abundance difference is regulated post-transcriptionally. Interestingly, RcaC is equally abundant in RL and GL both in an *rcaE* (photoreceptor) null mutant background as well as when it contains a mutation at either D51 or H316 (Li and Kehoe unpublished). These data suggest that either the interaction of RcaC with RcaE and/or the presence of D51 and H316 within RcaC are responsible for the light color-mediated changes in RcaC abundance measured in wild-type cells. Future studies will establish whether the changes in RcaC levels are due to differential translation or protein stability. In either case, this result provides a unique example of a response regulator whose abundance is regulated by the presence of the conserved aspartate (or histidine) residue, and suggests a new and perhaps widespread mechanism for controlling the activity of response regulators in both prokaryotes and eukaryotes.

The regulation of CCA involves components in addition to RcaE, RcaF, and RcaC. Detailed analysis of the RNA accumulation patterns in an *rcaE* null mutant led to the identification of a second light-responsive pathway that controls CCA (Figure 2). Interestingly, while the Rca system exerts its effect on both RL- and GL-induced genes, the second system appears to control only GL-induced genes (Seib and Kehoe 2002). Here we will call this the Cgi (**C**ontrol of **g**reen light **i**nduction) pathway. The Cgi and Rca pathways both contribute to the regulation of GL-induced genes. We have found that the 20- to 30-fold GL induction of the *cpe* operons that occurs in wild-type cells is reduced approximately

10 times, to 2- to 3-fold, in the absence of the Rca system (Seib and Kehoe 2002, Alvey et al 2003). The majority of this reduction is the result of a decrease in the shutdown of *cpe* operon expression in RL. No components of the Cgi system have been identified thus far.

The identification of the Cgi system and how it controls CCA may provide an explanation, at the molecular level, for the phenotypic differences between Group 2 and Group 3 chromatic adapting species noted by Tandeau de Marsac (1977). The Group 2 species, which light regulate only PE production, may use only the Cgi system, while in Group 3 species (such as *F. diplosiphon*), which light regulate both PE and PC, the Rca system may operate in conjunction with the Cgi system so that both pathways control GL-induced genes, but only the Rca system controls genes induced by RL.

The generation and complementation of additional *F. diplosiphon* mutants led to the isolation of a novel activator in the CCA regulatory pathway called CpeR, which is a protein required for the expression of *cpeBA* but not *cpeCDE* (Seib and Kehoe 2002, Cobley et al 2002). This work provided the first evidence that these two operons were not regulated through the same mechanism. John Cobley's group proposed that *cpeR* was cotranscribed with *cpeCDE*, which provided a potential mechanism to coordinate the expression of *cpeCDE* and *cpeBA* in series. CpeR is a small protein with limited similarity to PP2C-class protein phosphatases, and its function is currently unknown. A number of new GL-induced genes have been isolated and analyzed in *F. diplosiphon* recently (see below), and it has been found that some of these require CpeR for their expression, while others do not (Alvey et al 2003, Stowe-Evans et al 2004, Stowe-Evans and Kehoe unpublished). Those that require CpeR appear to be involved in PBS biosynthesis in GL. This suggests that CpeR may both coordinate *cpeCDE* and *cpeBA* expression and be a global activator of genes required for PBS production in GL.

Another recently described *F. diplosiphon* mutant class only accumulated 30%–40% of wild-type amounts of both PE and PE linkers and their corresponding RNAs during growth in GL. These mutants had normal expression of *cpeBA* and *cpeCDE* in RL, and normal *cpc2* RNA and protein levels in both RL and GL (Balabas et al 2003). The deduced protein sequence of the complementing gene, called *cotB* (complementation of tan), is modestly related to the lyase-class protein NblB, which is required for proper bleaching during nutrient limitation in the cyanobacterium *Synechococcus* sp. PCC 7942. However, *cotB* mutants bleach normally during nutrient limitation. The role, if any, of CotB in the control of CCA is not yet clear.

Identification of New Light Color-Responsive Genes

For decades this organism's acclimation to changes in the ratio of RL to GL has been generally known to involve much more than just PBS restructuring (Bogorad 1975, Tandeau de Marsac et al 1988). Recently, microarray technology was developed for *F. diplosiphon*, which does not yet have a sequenced genome. This work led to the identification of 17 new genes that were differentially expressed in RL and GL in *F. diplosiphon* (Stowe-Evans et al 2004). These included genes whose translated proteins were similar to a tryptophan-rich sensory protein, the PBS degradation protein NblA, a ribosomal subunit protein, a subunit of the dark-operative form of protochlorophyllide reductase, a NADH dehydrogenase subunit, carbonic anhydrase, an ATPase subunit, and several two-component sensor kinases. The expression patterns of several of these suggested that they have novel roles, while others may have highly specialized, unique uses in this species. Some of these genes are regulated by RcaE and/or CpeR, while others are not (Stowe-Evans and Kehoe unpublished).

Traditional molecular approaches have also led to a better understanding of light color acclimation in *F. diplosiphon*. We hypothesized that the expression of the genes encoding the enzymes responsible for the synthesis of the PE and PC chromophores might also respond to changes in light color. Thus, we cloned and analyzed the light color-responsive expression of *pebAB* and *pcyA*, the genes responsible for the production of the chromophores that are covalently attached to PE and PC, respectively. The primary reason for this effort was to identify an element(s) that was also present in promoters of genes known to be controlled by CCA. We found that the expression of both of these operons is controlled by light color (Alvey et al 2003, Alvey et al unpublished).

The *pebAB* operon is GL induced and is under the control of CpeR (Alvey et al 2003). It contains a region of DNA in its promoter that is highly similar to the region of the *cpeBA* promoter that has been shown to be the binding site for RcaA/PepB; this sequence is also present upstream of the novel GL-induced genes identified in the microarray studies described above that are controlled by CpeR. These results also demonstrated that the coordination of the GL-induced expression of the *cpeBA* and *pebAB* operons is not through any feedback systems that sense the absence of PE or its chromophore, phycoerythrobilin, but rather appears to be via a common transcriptional control pathway that involves CpeR.

The cloning of *pcyA* and analysis of its expression demonstrated that it is RL induced (Alvey et al unpublished). No clear promoter elements are present upstream of both *pcyA* and *cpc2*; however, a 28 bp region (R Box) upstream of *pcyA* was also found in three copies near the *cpeCDESTR* operon. One was approximately 1 kbp upstream of *cpeC* and two were in the intergenic region between *cpeD* and *cpeE* (Figure 2). All of these are greater than 90% identical. The R Box upstream of *pcyA* is currently being deleted in order to test its role in the regulation of CCA. If the R Box does play a role in CCA regulation of *pcyA* expression, we will test the effect of deleting the R Boxes in the *cpeCDESTR* region on CCA regulation. The discovery of R Boxes is exciting because it may help us to understand the long sought-after mechanism through which the integration of GL- and RL-induced gene expression occurs during CCA.

Summary and Conclusion

In the past decade, significant progress has been made in unraveling the mechanisms that control CCA in cyanobacteria. While the identification and analysis of two CCA photosensory pathways has resolved long-standing questions on this subject, the nature of the Cgi system remains to be described. A number of new light-responsive genes have also been identified; some of these are clearly controlled by CpeR and RcaE, while others appear to be controlled through other, as yet unknown, mechanisms. How CpeR functions is another unresolved issue. The binding site (if any) of RcaC remains unknown. The discovery of such a site is likely to contribute significantly to our understanding of the currently mysterious mechanism through which RcaC acts to control the expression of *cpeCDE*, *cpeBA*, and *cpcB2A2*, and other recently identified light-responsive genes under the control of the Rca pathway. It is possible that the R Box may play a role in mediating this aspect of CCA regulation.

References

Alvey RM, Karty JA, Roos E, Reilly JP, Kehoe DM (2003) Lesions in phycoerythrin chromophore biosynthesis in *Fremyella diplosiphon* reveal coordinated light regulation of apoprotein and pigment biosynthetic enzyme gene expression. Plant Cell 15: 2448–2463

Balabas BE, Montgomery BL, Ong LE, Kehoe DM (2003) CotB is essential for complete activation of green-light induced genes during complementary chromatic adaptation in *Fremyella diplosiphon*. Mol Microbiol 50: 781–793

Bogorad L (1975) Phycobiliproteins and complementary chromatic adaptation. Annu Rev Plant Physiol 26: 369–401

Casey ES, Grossman AR (1994) *In vivo* and *in vitro* characterization of the light-regulated *cpcB2A2* promoter of *Fremyella diplosiphon*. J Bacteriol 176: 6362–6374

Chiang GG, Schaefer MR, Grossman AR (1992) Complementation of a red-light indifferent cyanobacterial mutant. Proc Natl Acad Sci USA 89: 9415–9419

Cobley JG, Clark AC, Weerasurya S, Queseda FA, Xiao JY, Bandrapali N, D'Silva I, Thounaojam M, Oda J, Sumiyoshi T, Chu M (2002) CpeR is an activator required for the expression of the phycoerythrin operon (*cpeBA*) in the cyanobacterium *Fremyella diplosiphon* and is encoded in the phycoerythrin linker-polypeptide operon (*cpeCDESTR*). Mol Microbiol 44: 1517–1531

Grossman AR (2003) A molecular understanding of complementary chromatic adaptation. Photosynth Res 76: 207–215

Kehoe DM, Grossman AR (1996) Similarity of a chromatic adaptation sensor to phytochrome and ethylene receptors. Science 273: 1409–1412

Kehoe DM, Grossman AR (1997) New classes of mutants in complementary chromatic adaptation provide evidence for a novel four-step phosphorelay system. J Bacteriol 179: 3914–3921

Li L, Kehoe DM (2004) *In vivo* analysis of the roles of conserved aspartate and histidine residues within a complex response regulator. Mol Microbiol in press

Noubir S, Luque I, Ochoa de Alda JAG, Perewoska I, Tandeau de Marsac N, Cobley JG, Houmard J (2002) Co-ordinated expression of phycobiliprotein operons in the chromatically adapting cyanobacterium *Calothrix* PCC 7601, a role for RcaD and RcaG. Mol Microbiol 43: 749–762

Schmidt-Goff CM, Federspiel NA (1993) *In vivo* and *in vitro* footprinting of a light-regulated promoter in the cyanobacterium *Fremyella diplosiphon*. J Bacteriol 175: 1806–1813

Seib LO, Kehoe DM (2002) A turquoise mutant genetically separates expression of genes encoding phycoerythrin and its associated linker polypeptides. J Bacteriol 184: 962–970

Sobczyk A, Schyns G, Tandeau de Marsac N, Houmard J (1993) Transduction of the light signal during complementary chromatic adaptation in the cyanobacterium *Calothrix* sp. PCC 7601: DNA-binding proteins and modulation by phosphorylation. EMBO J 12: 997–1004

Sobczyk A, Bely A, Tandeau de Marsac N, Houmard J, Links J (1994) A phosphorylated DNA-binding protein is specific for the red-light signal during complementary chromatic adaptation in cyanobacteria. Mol Microbiol 13: 875–885

Stowe-Evans EL, Ford J, Kehoe DM (2004) Genomic DNA microarray analysis: identification of new genes regulated by light color in the cyanobacterium *Fremyella diplosiphon*. J Bacteriol 186: 4338–4349

Tandeau de Marsac N (1977) Occurrence and nature of chromatic adaptation in cyanobacteria. J Bacteriol 130: 82–91

Tandeau de Marsac N (2003) Phycobiliproteins and phycobilisomes: the early observations. Photosynth Res 76: 197–205

Tandeau de Marsac N, Mazel D, Damerval T, Guglielmi G, Capuano V, Houmard J (1988) Photoregulation of gene expression in the filamentous cyanobacterium *Calothrix* sp. PCC 7601: light-harvesting complexes and cell differentiation. Photosynth Res 18: 99–132

Terauchi K, Montgomery BL, Grossman AR, Lagarias JC, Kehoe DM (2004) RcaE is a complementary chromatic adaptation photoreceptor required for green and red light responsiveness. Mol Microbiol 51: 567–577

CHAPTER 36

Components of Light-Induced Signal Transduction in Cyanobacteria

WOLFGANG GÄRTNER

Introduction

Sensing the quality of light with respect to intensity, spectral composition, duration, and direction and polarization is of outstanding importance for photosynthetic organisms, being them immobilized as the plants or motile as many bacteria or unicellular algae (Smith 2000). This capability might also be of value for non-photosynthetic parasitic bacteria searching and invading plants. Phytochromes, the ubiquitous plant photoreceptors, absorb light around 660nm, and are converted by light into the signaling state, absorbing around 730nm, called P_r (*red*-) and P_{fr}- (*far red* absorbing) forms (Braslavsky et al 1997). Plant phytochromes carry a covalently bound open-chain tetrapyrrole chromophore (phytochromobilin, PΦB, in only few cases phycocyanobilin (PCB) (Jorissen et al 2002a, Wu et al 1997) that undergoes a photoisomerization at one of its double bonds. The covalent attachment of the chromophore to the protein is accomplished in all plant phytochromes via a thioether formed between the thiol group of a cysteine residue and the 3′-position of the ethylidene substituent at ring A of the bilin (Figure 1).

Phytochromes in Cyanobacteria

The existence of photosensors with absorbances similar to the phytochromes has long been postulated for the prokaryote kingdom. However, their molecular structure has long escaped a detailed characterization. This was until the gene causing a phenotype with altered complementary chromatic adaptation (CCA) was identified in the cyanobacterium *Fremyella diplosiphon* (Kehoe and Grossman 1996) (*F. diplosiphon* is a derivative of *Calothrix* PCC7601). The sequence of the gene product of *rcaE* showed strong similarities to plant phytochromes, however, exhibited some significant deviations in the chromophore

Max-Planck-Institute for Bioinorganic Chemistry, Stiftstr. 34–36, D-45470 Mülheim, and Department of Organic Chemistry, University of Duisburg Essen, Duisburg, Germany

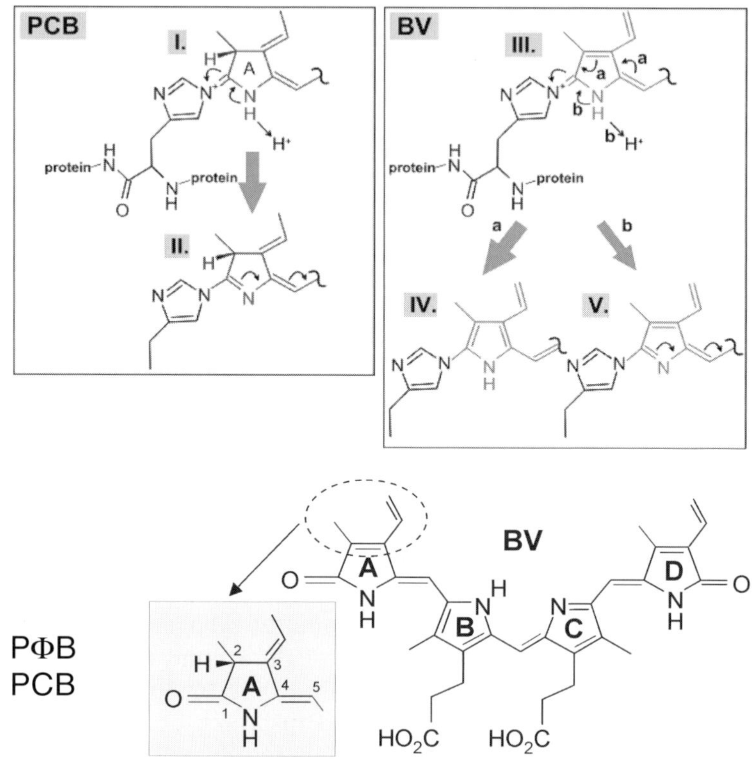

FIG. 1. Possible electronic rearrangements following chromophore–protein contact that cause increased stability of the interactions between the histidine side chain and biliverdin (*BV*), compared to those of phycocyanobilin (*PCB*). The same primary complex, a Schiff base-like interaction, is formed initially. A stable aromatic pyrrole structure can be formed only for the BV compound, but not for PCB

binding region. Incidentally with the identification of RcaE, the first completed genome of a cyanobacterium, *Synechocystis* PCC6803, was released (Kaneko et al 1996). Sequence alignments immediately revealed a strong sequence similarity between several open reading frames (ORFs) from *Synechocystis* and highly conserved domains of the plant phytochromes. The protein encoded by one of these ORFs, slr0473, was called Cph1 (**c**yanobacterial **ph**ytochrome **1**). It consists of 748 amino acids (molecular weight of ca. 84 kDa) and shows a remarkably high overall sequence similarity to plant phytochromes, indicating that Cph1 might represent the prototype of the long-searched bacterial long-wavelength photoreceptor (Hughes et al 1997, Yeh et al 1997). As in the plant phytochromes, a cysteine residue (cys259) as the putative chromophore binding site is located in a highly conserved stretch of amino acids, including a histidine following the cysteine. This histidine was demonstrated to be essential for chromophore attachment in the plant phytochromes (Remberg et al 1999). As a second salient

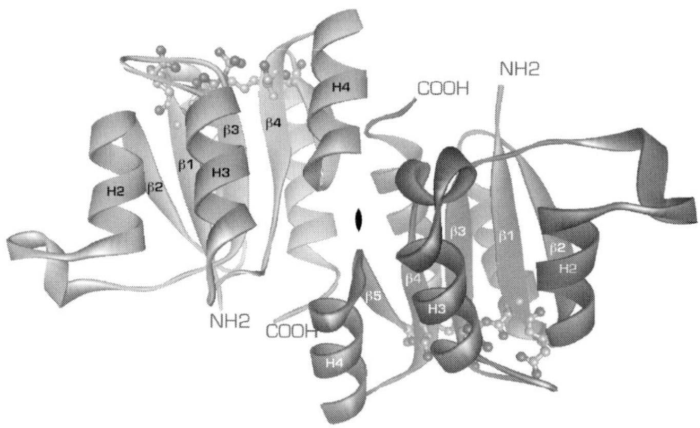

FIG. 2. Structure of homodimeric response regulators from *Calothrix* (shown for RcpA). Phosphorylation sites arrange at opposite ends of the homodimer

feature, Cph1 carries in its C-terminal part a histidine kinase domain. This finding drew further attention to the bacterial phytochromes (Bphs), since histidine kinases are key motifs in the sensor proteins of the so-called two-component signal transduction system that, in its central part, consists of a histidine kinase and an interacting response regulator protein (RR) (Chang 1996, Parkinson and Kofoid 1992). The interaction between the sensor and its cognate response regulator is highly conserved. Upon activation, the histidine-attached phosphate group is transferred during a protein-protein interaction to an aspartate of the response regulator (Figure 2). This aspartate residue, together with other two acidic amino acids (asp or glu) and one lysine forms the binding site for the phosphate, and holds the transferred phosphate group via a carboxylate–phosphate anhydride bond.

This finding of a stimulus- (=light-) sensing input domain, connected to a well-known structural motif of signal transduction in prokaryotes, makes the bacterial phytochromes a light-inducible subgroup of the widely spread two-component system. A second class of prokaryotic photoreceptors with blue light sensitivity has only recently been identified (Losi et al 2002). In fact, an ongoing investigation revealed that the gene product of *rcp1* (Rcp1), which follows *cph1*, exhibits significant sequence similarities to CheY-type response regulators. These RRs are small protein of ca. 18 kDa with a highly conserved $(\beta\alpha)_5$-folding motif.

The Family of Prokaryotic Phytochromes

Following the first two identified cyanobacterial genes encoding phytochrome-like proteins, RcpE and Cph1, a polymerase chain reaction (PCR)-based search for the highly conserved chromophore-binding domain of plant phytochromes revealed the presence of this protein motif in a wide number of cyanobacterial strains (Herdman et al 2000), and ongoing research yielded the presence of phytochromes in other phototrophic and even non-phototrophic bacteria. Arrangement of the found sequences into a phylogenetic tree showed a significant relationship to the plant phytochromes and gave strong evidence that the (cyano)bacterial phytochromes are the ancestors of plant phytochromes. Strongest similarity to Cph1 is found for a protein from *Calothrix sp.* PCC7601, CphA, and for AphA from *Anabaena sp*. PCC7120. An additional result from the phylogenetical arrangement was the identification of two families of bacterial phytochromes, exhibiting a strong sequence similarity to each other, of which only one, however, carries the essential cysteine residue being identified as the chromophore attachment site. The other group carried (in the case of the cyanobacteria) either a leucine or an isoleucine. In the following, these two groups will be called as A- and B-type Bphs (either carrying or lacking the phytochrome-specific cysteine). Also, these cyanobacterial ORFs from *Anabaena* and *Calothrix*, encoding Bphs are accompanied by RRs, for *Calothrix* called RcpA and RcpB.

In the genome of *Deinococcus radiodurans*, only a gene encoding a B-type Bph, called *Dr*BphP, was found (Davis et al 1999). Furthermore, solely a gene encoding a heme oxygenase was found, and no other enzyme activity for further modification of the generated biliverdin (BV) could be assigned (Bhoo et al 2001). Accordingly, in case *Dr*BphP were a functional photoreceptor, BV had to serve as chromophore, either being incorporated noncovalently or bound by an unknown mechanism. Both results, BV being apparently the sole bilin in *D. radiodurans* and the finding of bacterial phytochromes lacking the covalently binding cysteine residue, caused confusion on the type of chromophore–protein interaction in bacterial phytochromes. This finding was originally explained as an interaction of the imidazole side chain of the histidine residue with the carbonyl group of ring A in a Schiff base type manner (Davis et al 1999). This suggestion gained additional attention, as some of these A- and B-type phytochromes could be homologously expressed in their native host (Hübschmann et al 2001a, Gärtner and Quest unpublished). They also could be phosphorylated in a light-dependent manner and interacted very specifically with their cognate response regulators (Hübschmann et al 2001b). As mentioned, the position following the putative chromophore-binding cysteine is in all phytochromes a histidine. It was thus suggested that probably this amino acid interacts with the bilin (see Figure 1), since its essential involvement during the covalent attachment had already been proven (Deforce et al 1993, Remberg et al 1999).

Homologously Expressed Bphs

Homologous expression would be the preferred method to address the question after the genuine chromophore of bacterial phytochromes. However, the very low cellular concentration of bacterial phytochromes (between 10 and 200 copies were estimated per cell, T Lamparter personal communication), would render this approach difficult. Homologous expression of affinity tagged proteins was successfully performed with Cph1 in *Synechocystis* (Hübschmann et al 2001a) and with CphB in *Fremyella* (Quest and Gärtner unpublished). In both cases, the recombinant gene was inserted via by triparental homologous recombination. This approach yielded chromoproteins that showed spectral features reminiscent to the PCB–apo Cph1 adduct for the protein from *Synechocystis*. In case of CphB from *Calothrix*, a difference absorption spectrum identical to the BV–apo CphB adduct was recorded. The absorption spectrum of the homologously expressed CphB was strongly bathochromic to those of "normal" phytochromes (702, 754 nm for P_r/P_{fr}, respectively), giving evidence for a chromophore–protein interaction, clearly different from formerly characterized Cph1 or CphA. The homologously expressed CphB was identified by mass spectroscopy (MALDI-TOF). The high degree of sequence similarity of Cph1 and CphA and their nearly identical spectra upon assembly with PCB allows to propose that PCB is probably the genuine chromophore also of CphA. The existence of two phytochromes in *Calothrix*, both being expressed (Jorissen et al 2002b) and one carrying (CphA) and the other lacking the binding cysteine (CphB), revealed a possible rationale for the simultaneous employment of both types of Bphs. The incorporation of PCB into the CphA-type proteins, and of BV into the CphB-type proteins, yields remarkably different absorption properties and constitutes a simple color discrimination system.

Steady-State and Time-Resolved Spectroscopy of Bacterial Phytochromes

Assembly of heterologously expressed Cph1 with PCB and PΦB yielded chromoproteins with absorption maxima and spectral shape very similar to those of the plant phytochromes (λ_{max} = 654/658 and 702/706 nm for PCB, P_r and P_{fr}, respectively (Hughes et al 1997, Remberg et al 1997), and λ_{max} = 668 and 717 nm for PΦB, P_r and P_{fr}, respectively (Remberg et al 1997). The PCB-derived P_{fr}-absorption maximum of Cph1 (λ_{max} = 710–715 nm) is slightly hypsochromic to PCB-assembled plant phytochromes. However, in contrast to plant phyA or phyB that convert thermally from the light induced P_{fr} state back into the P_r form, the P_{fr} forms of Cph1, CphA and CphB remain stable in the dark for hours and even days (Jorissen et al 2002b). CphA from *Calothrix* showed very similar absorption maxima to Cph1, when assembled with PCB. However, the addition of PCB to apo-CphB caused strongly red shifted absorption maxima, compared to that of CphA–PCB (685, 734 nm for P_r, P_{fr}), which was even more pronounced when BV was employed (702, 754 nm for P_r, P_{fr}). Interestingly, the single mutation of L266C, introducing a cysteine residue, converted the red-shifted absorption maxima to those found for the CphA–PCB adduct (Quest and Gärtner 2004).

The preference of CphB for biliverdin has been extensively studied, also performing competition experiments (Quest and Gärtner 2004). It was found that pre-incubation of CphB with PCB gave a complete exchange of the chromophore within less than one hour, when BV was added. On the other hand, if the L266C mutant of CphB (capable to covalently bind PCB) was pre-incubated with BV and PCB was added, it took more than 40 h to expel only half of the BV from the binding site, despite of the binding of PCB. This extreme selectivity of CphB for BV can readily be explained if the histidine directly following the putative binding position is included in the chromophore attachment. Only in the case of BV, a set of mesomeric structure can be formulated that include the conversion of the A-ring of the bilin into a pyrrole structure (Figure 1). Such hetero aromatic structure cannot be formed with PCB (Quest and Gärtner 2004).

Nanosecond excitation of PCB-assembled CphA from *Calothrix* revealed conversion kinetics of 24 μs, 1.4, 13 and 66 ms, and slower processes of ca. 0.5 and 3 s (Jorissen et al 2002b). The major contribution for the formation of the P_{fr} form is found for the 1.4, 13, and 66 ms processes. The changed binding site of CphB caused by the lack of the covalently binding cysteine became also apparent in the flash photolysis of the PCB-assembled holoprotein. Laser-flash excitation revealed an entirely different kinetic behavior with only two lifetimes of 1.9 and 12.8 ms being detected (Jorissen et al 2002b).

The Reactions of Bacterial Phytochromes During the Early Steps of Signal Transduction

Being members of the bacterial two-component signal transduction system, the bacterial phytochromes should undergo conformational changes after sensing the incoming light stimulus that activate the histidine kinase activity and lead to "autophosphorylation." The amino acids involved in this phosphate transfer could be identified by site-directed mutagenesis (Yeh et al 1997). The phosphate transfer occurred remarkably specifically. Neither did phosphate transfer occur when phosphorylated CphA was incubated with RcpB (or vice versa), nor did the presence of RcpB influence the phosphate transfer from CphA to RcpA (and vice versa) (Hübschmann et al 2001b). Also, the above-described homologously expressed CphB carrying BV showed autophosphorylation, which was more pronounced in the P_{fr}-form. Addition of RcpB led to immediate transfer of the phosphate group to the response regulator protein.

Crystal Structures of RcpA and RcpB

No three-dimensional structure of any phytochrome has been reported up to now; however, the interacting response regulators should be more prone to crystallization as has been demonstrated for several RRs and for CheY from *Escherichia coli* (Stock et al 1989). Crystallization has been accomplished for Rcp1 (Im et al 2002) and also for RcpA and RcpB (Benda et al 2004). Inspection of their folding pattern revealed a strong similarity to the formerly reported CheY-type response regulators, consisting of an alternating $(\beta\alpha)_5$ motif that arranges three acidic and one basic amino acid residue as the phosphorylation site at the protein periphery.

For Rcp1 a phosphorylation-dependent monomer-dimer equilibrium has been suggested from a gel filtration experiment, however, only the apoform has been crystallized. For the RRs from *Calothrix*, as well the apo- as also the phosphorylated forms have been crystallized (Figure 2) (Benda et al 2004). This fortunate situation makes a comparison of the active (i.e. phosphorylated) and the inactive state of these cyanobacterial response regulators possible. RcpA and RcpB crystallize as homodimers, irrespective of their phosphorylation state. The high resolution for both RcpA and RcpB (1.8 Å) allow a detailed inspection of the phosphate-binding site and the dimer contact. Both RcpA and RcpB exhibit a large interdomain contact area of 1000 and 1175 $Å^2$ (quite in agreement to contact sites in other dimeric RRs), consisting of a hydrophobic core with very specific amino acid interactions, which is surrounded by a rim of hydrophilic amino acids and water molecules.

Outlook

Advanced methods like the polymerase chain reaction and the vast amount of genome sequencing information has revealed the existence of prokaryotic phytochromes. The architecture of these photosensory pigments, being composed of a chromophore-bearing domain and, in most cases, of a histidine kinase motif has decluded the mechanism, how light perception might induce a physiological response in microorganisms. However, a detailed analysis of the function and their physiological relevance of the prokaryotic phytochromes is, in many cases, still sparse. An identification of the members of the signal transduction chain, as well in vitro by recombinant proteins as also in vivo, employing knockout mutants in a proteomics approach, and the three-dimensional structure of the bacterial phytochromes still awaits its disclosure.

References

Benda C, Scheufler C, Tandeau de Marsac N, Gartner W (2004) A new dimerization motif in the structures of two light-inducible response regulators from the cyanobacterium *Calothrix* PCC7603. Biophys J 87: 476–487

Bhoo SH, Davis SJ, Walker JM, Karniol B, Vierstra RD (2001) Bacteriophytochromes are photochromic histidine kinases using a biliverdin chromophore. Nature 414: 776–779

Braslavsky SE, Gärtner W, Schaffner K (1997) Phytochrome photoconversion. Plant Cell Environ 20: 700–706

Chang C (1996) The ethylene signal transduction pathway in *Arabidopsis*: an emerging paradigm? Trends Biochem Sci 21: 129–133

Davis SJ, Vener AV, Vierstra RD (1999) Bacteriophytochromes: Phytochrome-like photoreceptors from nonphotosynthetic eubacteria. Science 286: 2517–2520

Deforce L, Furuya M, Song PS (1993) Mutational analysis of the pea phytochrome A chromophore pocket: chromophore assembly with apophytochrome A and photoreversibility. Biochemistry 32: 14165–14172

Herdman M, Coursin T, Rippka R, Houmard J, Tandeau de Marsac N (2000) A new appraisal of the prokaryotic origin of eukaryotic phytochromes. J Mol Evol 51: 205–213

Hübschmann T, Börner T, Hartmann E, Lamparter T (2001a) Characterization of the Cph1 holo-phytochrome from *Synechocystis* sp PCC 6803. Eur J Biochem 268: 2055–2063
Hübschmann T, Jorissen HJMM, Börner T, Gärtner W, Tandeau de Marsac N (2001b) Phosphorylation of proteins in the light-dependent signalling pathway of a filamentous cyanobacterium. Eur J Biochem 268: 3383–3389
Hughes J, Mittmann F, Wilde A, Gärtner W, Börner T, Hartmann E, Lamparter T (1997) A prokaryotic phytochrome. Nature 386: 663
Im YJ, Rho SH, Park CM, Yang SS, Kang, JG, Lee JY, Song PS, Eom SH (2002) Crystal structure of a cyanobacterial phytochrome response regulator. Protein Sci 11: 614–624
Jorissen HJMM, Braslavsky SE, Wagner G, Gärtner W (2002a) Heterologous expression and characterization of recombinant phytochrome from the green alga *Mougeotia scalaris*. Photochem Photobiol 76: 457–461
Jorissen HJMM, Quest B, Remberg A, Coursin T, Braslavsky SE, Schaffner K, Tandeau de Marsac N, Gärtner W (2002b) Two independent light-sensing two-component systems in a filamentous cyanobacterium. Eur J Biochem 269: 2671
Kaneko T, Sato S, Kotani H, Tanaka A, Asamizu E, Nakamura Y, Miyajima N, Hirosawa M, Sugiura M, Sasamoto S, Kimura T, Hosouchi T, Matsuno A, Muraki A, Nakazaki N, Naruo K, Okumura S, Shimpo S, Takeuchi C, Wada T, Watanabe A, Yamada M, Yasuda M, Tabata S (1996) Sequence analysis of the genome of the unicellular cyanobacterium *Synechocystis* sp. strain PCC6803. II. Sequence determination of the entire genome and assignment of potential protein-coding regions. DNA Res 3: 109–136
Kehoe DM, Grossmann AR (1996) Similarity of a chromatic adaptation sensor to phytochrome and ethylene receptors. Science 273: 1409–1412
Losi A, Polverini E, Quest B, Gärtner W (2002) First evidence for phototropin-related blue-light receptors in prokaryotes. Biophys J 82: 2627–2634
Parkinson JS, Kofoid EC (1992) Communication modules in bacterial signaling proteins. Annu Rev Genet 26: 71–112
Quest B, Gärtner W (2004) Chromophore selectivity in bacterial phytochromes: dissecting the process of chromophore attachment. Eur J Biochem 271: 1117–1126
Remberg A, Lindner I, Lamparter T, Hughes J, Kneip C, Hildebrand P, Braslavsky SE, Gärtner W, Schaffner K (1997) Raman spectroscopic and light-induced kinetic characterization of a recombinant phytochrome of the cyanobacterium *Synechocystis*. Biochemistry 36: 13389–13395
Remberg A, Schmidt P, Braslavsky SE, Gärtner W, Schaffner K (1999) Differential effects of mutations in the chromophore pocket of recombinant phytochrome on chromoprotein assembly and P_r-to-P_{fr} photoconversion. Eur J Biochem 266: 201–208
Smith H (2000) Phytochromes and light signal perception plants- an emerging synthesis. Nature 407: 585–591
Stock AM, Mottonen JM, Stock JB, Schutt CE (1989) Three-dimensional structure of CheY, the response regulator of bacterial chemotaxis. Nature 337: 745–749
Wu SH, Mcdowell MT, Lagarias JC (1997) Phycocyanobilin is the natural precursor of the phytochrome chromophore in the green alga *Mesotaenium caldariorum*. J Biol Chem 272: 25700–25705
Yeh KC, Wu SH, Murphy JT, Lagarias JC (1997) A cyanobacterial phytochrome two-component light sensory system. Science 277: 1505–1508

Chapter 37

Light Signal Transduction Coupled with Reactive Oxygen Species in *Neurospora crassa*

Kohji Hasunuma, Yusuke Yoshida, and Bumkyu Lee

Introduction

The development of the filamentous fungus *Neurospora crassa* is subject to light regulation at various stages of its life cycle. The light regulated processes include the following. (1) Induction of carotenoid synthesis in the mycelia. (2) Development of aerial hyphae and conidium from mycelia. (3) Phase shift of circadian rhythm of conidiation. (4) Protoperithecium formation under nitrogen limited condition. (5) Formation of perithecium beak under directional light. (6) Bending of beak toward directional light.

In recent research, at least three candidates for photoreceptors have been reported. The first is the well known WC-1/WC-2 complex (Talora et al 1999), the second is VVD (Heintzen et al 2001), and the third is NOP-1 (Bieszke et al 1999). Among the processes described above, the WC-1/WC-2 complex was shown to mediate (1), (2), (4), (5), (6), and partly (3) (Lakin-Thomas and Brody 2000). In the *vvd* mutant, the light-induced accumulation of carotenoids in the mycelia was enhanced and the light-induced phase shift of circadian rhythm became abnormal, suggesting that VVD controls the function of the WC-1/WC-2 complex repressively (Heintzen et al 2001). The *nop-1* knockout mutant showed no apparent phenotype and, therefore, the function of *nop-1* is not clear.

To investigate the molecular mechanism of light signal transduction in *N. crassa*, we have developed an in vitro system. In a mycelial membrane fraction, blue light illumination increased the phosphorylation of a 15 kDa protein (Oda and Hasunuma 1994). This protein was identified as nucleoside diphosphate kinase (NDK; EC 2.7.4.6) and was designated NDK-1. We confirmed the existence of single gene for a *ndk-1* in *N. crassa* from a DNA sequence data base (Galagan et al 2003).

Kihara Institute for Biological Research, Graduate School of Integrated Science, Yokohama City University, Maioka-cho, Totsuka-ku, Yokohama 244-0813, Japan

Early Stages of Light Signal Transduction

A membrane fraction from dark-grown mycelia of strain *band* (*bd*) was illuminated with blue light for 1 s and, 5 s after this illumination, the reaction was stopped by adding SDS sample buffer. After the proteins had been separated by SDS–PAGE, the gel was exposed to X-ray film (Oda and Hasunuma 1994). We detected an increase in the radioactivity of the 15 kDa protein, which was designated NDK-1. The genomic DNA and cDNA were cloned. The gene included two introns and one alternative splicing (Ogura et al 1999). NDKs are known to be multifunctional proteins, occurring in a wide range of organisms. The putative functions of NDK from various organisms are summarized in Table 1 (Hasunuma et al 1998). In humans, nm23-H1 and nm23-H2 were detected as factors controlling the metastasis of tumors. We have used genetic and biochemical methods to analyze the function of NDK-1. We isolated a point mutant of *ndk-1*^{Pro72His} (Oda and Hasunuma 1997, Ogura et al 2001), which showed exceedingly reduced autophosphorylation activity as well as phosphotransferring activity for MBP (myelin basic protein) (Ogura et al 2001). The mutant lacked the ability to show light-induced polarity of perithecia. The cDNAs of *ndk-1* and *ndk-1*^{Pro72His} were employed to produce NDK-1 fused with GST, and these proteins were purified. The purified preparations of GST-NDK-1 and GST-NDK-1$^{\mathrm{Pro72His}}$ were used to

TABLE 1. Putative functions of nucleoside diphosphate kinases (NDKs) (Hasunuma et al 1998)

1. NDK is known to catalyze the following reactions:
   $N_1TP + E \leftrightarrows N_1DP + E\text{-}P$
   $E\text{-}P + N_2DP \leftrightarrows N_2TP + E$
   $N_1TP + N_2DP \leftrightarrows N_1DP + N_2TP$
   In *Escherichia coli*, NDK has His 177, which is autophosphorylated and can function as a histidine kinase. NDK is also autophosphorylated at Ser 112 and at Ser 115. NDK phosphorylates histone H1, ovalbumin and myelin basic protein. NDK forms a tetramer in prokaryotic cells and a hexamer in eukaryotic cells
2. In membrane fractions prepared from dark-grown seedlings of *Pisum sativum*, *Arabidopsis thalinana*, *Oryza sativa*, and *Neurospora crassa*, light increases the phosphorylation of cytosolic NDKs
3. NDK activates G proteins Gs, Gi and Gt, by supplying GTP in the vicinity of G protein. NDK interacts with Gβ. K^+ channels controlled by G protein can be activated by NDK
4. NDK can interact with Ras protein. In *Drosophila melanogaster*, *pn* (*prune*) encodes a protein with high homology to GTPase activating protein interacting with Ras. *pn* causes *k-pn* (killer of prune) mutation in *Drosophila* NDK leading to lethality. *k-pn* is a mutation in NDK allelic with *awd* (abnormal wing disc). Small G protein, rho/rac, and NDK localize to the cytoskeleton. and interact with each other
5. NDK(nm23-H2) can function as a transcription factor for *c-myc*
6. NDK located in mitochondria is encoded by the nuclear genome. Arabidopsis NDK3 is localized in the inter space between the outer membrane and cristae of mitochondria
7. Two different NDKs localized in the chloroplast are identified in *Spinacia oleracea*
8. In human, rat, and mouse, NDK(nm23-H1) can suppress the metastasis of the tumor

check (i) NDK activity for catalyzing the conversion of ATP + GDP to ADP + GTP, (ii) autophosphorylation activity, and (iii) protein kinase activity for MBP phosphorylation. Although the mutant GST-NDK-1^{Pro72His} showed a wild-type level of nucleoside diphosphate kinase activity, the autophosphorylation activity of the mutant protein was 5% that of the wild-type protein and the Vmax of protein kinase activity for the mutant protein was 2% that of the wild type. Thus, the mutant protein is severely impaired in kinase activity, including the activity for autophosphorylation. Furthermore, the *ndk-1*^{Pro72His} mutant lacked light-induced perithecial polarity (Ogura et al 2001).

Recent results on early light signaling from various laboratories are summarized on the right side of Figure 1. Light energy perceived by the WC-1/WC-2 photoreceptor complex is transduced for the translocation of WC-1, which contains a FAD-binding LOV domain, and WC-2 into the nucleus to form the WC-1/WC-2 complex. The nucleus-localizing WC1/WC-2 complex functions as a transcription factor that binds the C-box located upstream of carotenoid synthetic genes, *al-1*, *al-2*, and *al-3*. The WC-1/WC-2 complex stimulates the transcription, resulting in the accumulation of carotenoid in mycelia. The WC-1/WC-2 complex can also bind the C-box located upstream of the *vvd* gene. The protein product, VVD, has the ability to suppress the function of WC-1/WC-2. In this

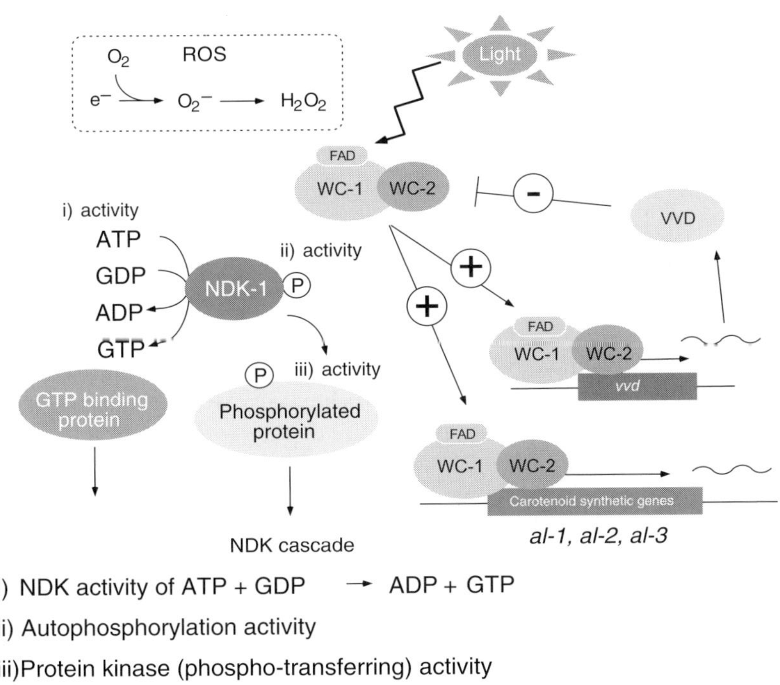

FIG. 1. Molecular model of the light signal transduction involving reactive oxygen species (*ROS*)

well-accepted model of light signal transduction, NDK-1 can be considered to function downstream of WC-1 and WC-2.

On the left side of Figure 1, our hypothetical model is illustrated (Hasunuma et al 2003). In *Raffia rhodozyma*, accumulation of carotenoid was enhanced in the presence of rose Bengal exposed to 550 nm light. However, in its absence, carotenoid accumulation could not be observed (Schroeder and Johnson 1995). Upon reception of light by a photoreceptor or by a pigment, the energy may be transduced to generate an electrogenic radical or to release an electron, which may be captured by solubilized dioxygen to form a superoxide anion radical, $O_2^{\cdot-}$. $O_2^{\cdot-}$ can be converted into hydrogen peroxide (H_2O_2) by the function of superoxide dismutase (SOD). H_2O_2 can then be converted into H_2O and O_2 by the function of catalase and peroxidase. NDK-1, forming a protein complex in vivo, may be activated in the presence of these reactive oxygen species (ROSs). At least in *Arabidopsis thaliana*, AtNDK-1 forms a protein complex with the catalases AtCat-1, AtCat-2, and AtCat-3 (Fukamatsu et al 2003). Even in *N. crassa*, the *ndk-1*^{Pro72His} mutant showed high sensitivity to paraquat (methyl viologen) producing ROS and this sensitivity was the same level as that of the *sod-1* mutant. Reactive oxygen species produced upon perception of light by the WC-1/WC-2 complex and also by other flavin-containing photoreceptors may affect the state of the NDK-1 molecular complex and may activate the latter complex. The enhanced NDK-1 activity will provide GTP in the vicinity of G protein, thereby activating G protein. Autophosphorylation of NDK-1, induced by light, may also lead to an induction or enhancement of its kinase activity, phosphorylating target proteins downstream of the signal transduction. On the basis of the result that the *ndk-1*^{Pro72His} mutant shows no light induction of the perithecial polarity, we have proposed a new signal transduction pathway, designated the NDK cascade (Hasunuma 2000).

Carotenoid Accumulation in Relation to the Concentration of ROS

In *N. crassa*, two SODs have been reported. One is Mn SOD in mitochondria and the other is Cu,Zn SOD in the cytosol. We have investigated the effect of ROS on the accumulation of carotenoid in mycelia using a *sod-1* mutant lacking Cu,Zn SOD. The rate of accumulation of carotenoid in the mycelia of *sod-1* after illumination with white light was twice that in the wild type. However, the rate of accumulation in *ndk-1*^{Pro72His} was reduced to one half of that in the wild type (K Hasunuma unpublished). In the case of the *vvd* mutant, the accumulation was enhanced close to the rate for *sod-1* and lasted 15 h. From these results it is evident that ROS regulates the accumulation of carotenoid in the mycelia. We investigated the effect of an antioxidant added to the mycelial culture medium and confirmed that, in the presence of the antioxidant, the accumulation rate of carotenoid in both wild-type and *sod-1* mycelia was effectively reduced (Yoshida and Hasunuma 2004).

The enhanced accumulation of carotenoid in the mycelia of the *sod-1* and *vvd* mutants was further analyzed based on expression levels of carotenoid synthetic genes, *al-1*, *al-2*, and *al-3*. The accumulation of *al-1* mRNA in the mycelia of the wild type was transient, showing a maximum at 30 min after white light illumination. However, the accumulation of *al-1* mRNA in *sod-1* and *vvd* was sustained (Yoshida and Hasunuma 2004). These results suggest that the change in the distribution of ROS in the mycelia affects functional change of the WC-1/WC-2 complex caused by the oxido-reductive state.

The Effect of Mutations Related to Light Signal Perception and Transduction on Perithecial Polarity

Light-induced perithecial polarity of the wild type and mutants including *vvd*, *ndk-1*^{Pro72His}, *sod-1*, *wc-1*, *wc-2* and their double mutants was analyzed. In the case of the wild type directional light induced perithecial polarity of 82% upward (18% random direction) and 40% upward (60% random direction) in darkness. In *vvd* under light, 85% of perithecia showed upward and 35% of it showed upward in darkness, showing enhanced stimulation of light-induced perithecial polarity. In the *ndk-1*^{Pro72His} mutant, only 20% of perithecia showed upward and in darkness 23% of them become upward with no stimulation by light. In *sod-1*, we detected 20% of them showing upward under directional light and 18% under darkness with no detectable stimulation by light. The *ndk-1*^{Pro72His}, *sod-1* double mutant showed 15% of them upward both in darkness and under the directional light. These results indicate that the signal transduction pathways controlled by these two mutants may differ and the result may be additive. Compared with the *sod-1* mutant, the *sod-1*, *vvd* double mutant showed a small enhancement of light-induced perithecial polarity. The photoreceptor mutants, *wc-1* and *wc-2*, became 10% of perithecia with upward polarity. The result was not affected by adding *sod-1* mutation forming *wc-1*, *sod-1* and *wc-2*, *sod-1*. From these results (Yoshida and Hasunuma 2004) we conclude that ROS is certainly included in the signal transduction pathway not only for the synthesis of carotenoid but also during the morphogenesis of perithecia inducing the polarity.

Conclusions

The hypothesis that the photoreceptor WC-1/WC-2 complex will produce ROS upon the perception of light was proposed, and also that ROS generated by flavin-containing pigments (photoreceptors) may affect the function of the WC-1/WC-2 complex. The *vvd* mutation caused an enhanced accumulation of carotenoid in the mycelia, suggesting that *vvd* functions to reduce the formation of the WC-1/WC-2 complex, although in the present state the mode of action to control WC-1/WC-2 complex by *vvd* or the product is not clear. The accumulation of the mRNAs of carotenoid synthetic genes, *al-1*, *al-2*, and *al-3*, occurred transiently in the wild-type mycelia. However, the accumulation was sustained in the *sod-1* mutant. The *sod-1* mutant showed a lack of light-induced perithecial porality, as observed in the *ndk-1*^{Pro72His} mutant. These results lead us to conclude that ROS is involved in the light signal transduction.

The mutant in Cu,Zn SOD, *sod-1*, showed an enhanced accumulation of carotenoid in the mycelia, while *ndk-1*^{Pro72His} showed a reduced accumulation. The accumulation of carotenoid in the mycelia of the wild type and the *sod-1* mutant was reduced in the presence of antioxidant reagents, suggesting that wild-type NDK-1 may function to reduce intracellular ROS. In *A. thaliana*, AtNDK-1 interacted with catalases, AtCat-1, AtCat-2, and AtCat-3, and over-expression of AtNDK-1 caused the plants to become resistant to oxidative stress. These results led us to conclude that NDK-1 controls the intracellular activity of catalases in *N. crassa*.

The localization of AtNDK-3 in the inter space of the outer membrane and cristae of mitochondria suggested to us that the inter space of mitochondria not only accumulates H^+ but also releases electrons during electron transport. The released electrons readily form superoxides, which may be converted into hydrogen peroxide by the function of Mn SOD. The hydrogen peroxide thus produced will be converted into H_2O and O_2 by the action of the putative AtNDK-3/AtCat complex. AtNDK localizing in chloroplasts may also function to reduce the hydrogen peroxide produced in the process of electron transport during photosynthesis.

Acknowledgments. We are grateful to Winslow R. Briggs and Charles Yanofsky for valuable discussions throughout this work. This work was supported by a Grant-in-Aid from the Ministry of Science and Culture, Yamada Science Foundation, and Mishima Kaiun Memorial Foundation.

References

Bieszke JA, Braun EL, Bean LE, Kang S, Natviq DO, Borkovich KA (1999) The *nop-1* gene of *Neurospora crassa* encodes a seven transmembrane helix retinal-binding protein homologous to archaeal rhodopsins. Proc Natl Acad Sci USA 96: 8034–8039

Fukamatsu Y, Yabe N, Hasunuma K (2003) *Arabidopsis* NDK-1 is a component of ROS signaling by interacting with three catalases. Plant Cell Physiol 44: 982–989

Galagan JE, Calvo SE, Borkovich KA, Selker EU, Read ND, Jaffe D, FitzHugh W, Ma LJ, Smirnov S, Purcell S, Rehman B, Elkins T, Engels R, Wang S, Nielsen CB, et al (2003) The genome sequence of the filamentous fungus *Neurospora crassa.* Nature 422: 859–868

Hasunuma K (2000) Signal transduction of light through NDP kinase inducing the morphogenesis of perithecia in *Neurospora crassa.* Plant Morphol 12: 39–51

Hasunuma K, Ogura Y, Yabe N (1998) Early events occurring during light signal transduction in plants and fungi. J Photosci 5: 73–81

Hasunuma K, Yabe N, Yoshida Y, Ogura Y, Hamada T (2003) Putative functions of nucleoside diphosphate kinase in plants and fungi. J Bioenerg Biomembr 35: 57–65

Heintzen C, Loros JJ, Danlap JC (2001) The PAS protein VIVID defines a clock-associated feedback loop that repress light input, modulates gating, and regulates clock resetting. Cell 104: 453–464

Lakin-Thomas P, Brody S (2000) Circadian rhythms in *Neurospora crassa*: lipid deficiencies restore robust rhythmicity to null frequency and white collar mutants. Proc Natl Acad Sci USA 97: 256–261

Oda K, Hasunuma K (1994) Light signal is transduced to the phosphorylation of 15 kDa protein in *Neurospora crassa*. FEBS Lett 345: 162–166

Oda K, Hasunuma K (1997) Genetic analysis of signal transduction through light induced protein phosphorylation in *Neurospora crassa* perithecia. Mol Gen Genet 256: 593–601

Ogura Y, Yoshida Y, Ichimura K, Aoyagi C, Yabe N, Hasunuma K (1999) Isolation and characterization of *Neurospora crassa* nucleoside-diphosphate kinase NDK-1. Eur J Biochem 266: 709–714

Ogura Y, Yoshida Y, Yabe N, Hasunuma K (2001) A point mutation in nucleoside diphosphate kinase results in a deficient light response for perithecial polarity in *Neurospora crassa*. J Biol Chem 276: 21228–21234

Schroeder WA, Johnson EA (1995) Singlet oxygen and peroxyl radicals regulate carotenoid biosynthesis in *Phaffia rhodozyma*. J Biol Chem 270: 18374–18379

Talora C, Franchi L, Linden H, Ballario P, Macino G (1999) Role of white collar-1-white collar-2 complex in blue-light signal transduction. EMBO J 18: 4961–4968

Yoshida Y, Hasunuma K (2004) Reactive oxygen species affect photomorphogenesis in *Neurospora crassa*. J Biol Chem 279: 6986–6993

Part VII
Photoperiodism and Circadian Rhythm

CHAPTER 38

Light Regulation of Flowering Time in Arabidopsis

XUHONG YU and CHENTAO LIN

Introduction

Plant development is dependent on not only endogenous conditions but also environmental factors. One of the best examples of environmental regulation of plant development is photoperiodic flowering, by which plant flower in response to changes of day length (Garner and Allard 1920). The predictability conferred by the seasonal changes in photoperiod enables plants to flower at the most favorable time of the year. The question of what photoreceptors mediate photoperiodic flowering has been one of the focuses in our efforts to understand the underlying mechanisms of photoperiodism. An action spectrum for the photoperiodic regulation of flowering time was reported in as early as 1945, which showed that red light was the most effective spectrum of light used in the night-break experiments to inhibit flowering of SD plants, suggesting a red light-absorbing pigment in the photoperiodic response (Parker et al 1945). It was later found that the red light effect could be reversed by far-red light which, together with a similar effect of light on germination, contributed to the discovery of phytochrome (Borthwick et al 1952). In addition to red light, blue/UV-A light has also been found to affect flowering time in some of the early works, but most of these light effects were attributed to phytochromes (Parker et al 1946, Meijer 1959, Brown and Klein 1971). We now know that, in addition to phytochromes, blue/UV-A light receptors also play important roles in the light regulation of flowering time (Guo et al 1998, Imaizumi et al 2003). In the last 5 years, significant progress has been made in the study of plant photoreceptors and the molecular mechanisms underlying light regulation of flowering time. Most of these studies were carried out in the model plant *Arabidopsis thaliana*. Although it has been clearly shown in the earlier physiological studies that photoperiodic flowering in different plant species responds to light in different ways, the studies in Arabidopsis nevertheless provides a good framework of how photoreceptors generally work, and it is likely that the observed variations among different plants

Department of Molecular, Cell, and Developmental Biology, University of California, Los Angeles, CA 90095, USA may represent modifications of the basic mechanisms revealed in Arabidopsis. In this short review, we focus on our current understanding of how photoreceptors regulate flowering time in Arabidopsis. Readers are suggested to also read recent review articles covering the related topics (Lin 2000, Mouradov et al 2002, Yanovsky and Kay 2003), and other chapters in this volume for additional discussions of phytochromes, cryptochromes, and other photoreceptors.

Photoreceptors

The Arabidopsis genome encodes at least ten different photosensory receptors, including five phytochromes (phyA to phyE), three cryptochromes (cry1 to cry3), and two phototropins (Cashmore 1997, Briggs and Huala 1999, Nagy and Schafer 2002, Quail 2002, Lin and Shalitin 2003). This list is likely to grow as more LOV-domain proteins other than phototropins may also act as photoreceptors (Imaizumi et al 2003). All of these photoreceptors, except phototropins, have been shown to play roles in the regulation of flowering time. Our current view with respect to how different photoreceptors regulate flowering time in Arabidopsis has been shaped largely by the physiological and genetics studies of Arabidopsis mutants. Several recent review articles have provided a detailed account of these studies (Koornneef et al 1998, Lin 2000, Mouradov et al 2002, Yanovsky and Kay 2003). Mutations in a photoreceptor gene may cause delayed or accelerated flowering. Among different Arabidopsis photoreceptor mutants, *phyB*, *phyC*, *phyE*, and *phyD* mutants showed accelerated flowering under various experimental conditions tested, and they are most likely negative regulators of floral initiation. On the other hand, *phyA*, *cry1*, and *cry2* mutants exhibit delayed flowering phenotype, so they are positive regulators of flowering. Using different combinations of these photoreceptor mutations to test flowering time in plants grown under different light conditions, it has been shown that different phytochromes and cryptochromes act antagonistically as well as redundantly to influence the developmental transition from vegetative growth to reproductive development (Mockler et al 2003) (Figure 1).

The complex interactions of different phytochromes and cryptochromes are interpreted in a model in Figure 1. One may expect that in young seedlings, the major function of photosensory receptors should be to promote vegetative growth and accumulation of photosynthetic products. In doing so, these photoreceptors may also act to suppress reproductive development until plants are mature enough. This view is certainly consistent with the finding that most phytochromes are negative regulators of floral initiation. For example, mutations of *PHYB*, *PHYC*, *PHYD*, and *PHYE* genes all cause the mutant plants to flower earlier than the wild type (Reed et al 1993, Devlin et al 1998, 1999, Franklin et al 2003). Like its function in the regulation of stem elongation, the phyB function in the regulation of flowering time is dependent on red light (Lin 2000, Quail 2002). It is possible that phyC, phyD, and phyE also mediate red light inhibition of floral initiation. The action of phyB in the suppression of floral initiation is

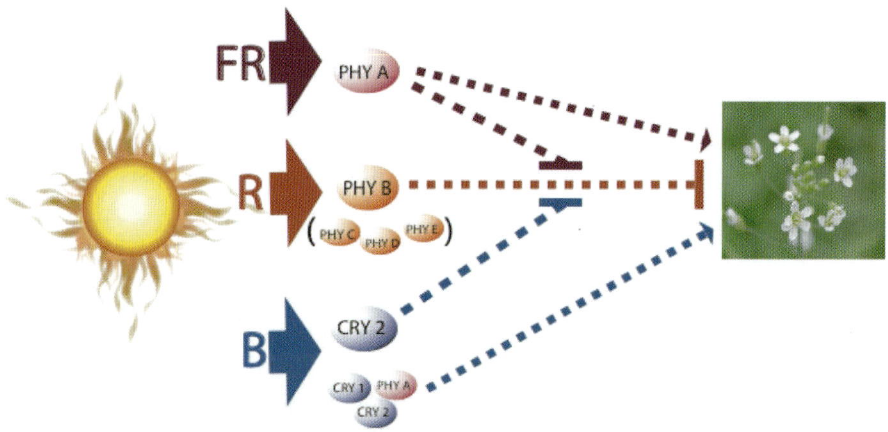

Fig. 1. A model depicting roles of different photoreceptors and their interactions. *Arrows* indicate positive effect on floral initiation, *bar-headed lines* depict negative effect on flowering. *Dashed lines* indicate incomplete understanding of the molecular mechanisms

antagonized by two other photoreceptors, phytochrome A and blue light receptor cry2. PhyA mediates FR light promotion of flowering that is antagonistic to the phyB function; cry2 mediates blue light suppression of phyB activity. In addition, cry1, cry2, and phyA also mediate, in a partially redundant manner, blue light promotion of flowering that is independent from their activity in the suppression of the phyB function (Mockler et al 1999, 2003) (Figure 1). The interactions among different photoreceptors responsive to different spectra of light would presumably allow plants to fine-tune the timing of their developmental transition in adaptation to different light environments. For example, phyB, phyD, and phyE can apparently act in response to a decreased R/FR ratio of light received by plants grown under the shade of canopies of neighboring plants (Devlin et al 1998, 1999, Franklin et al 2003). In the absence of shade, these three phytochromes promote vegetative growth and suppress flowering. In the presence of shade, the activity of these photoreceptors decreases, allowing floral initiation to take place. There seems an apparent advantage for a plant that grows under the shade from the canopies of surrounding plants to complete its life cycle before deprivation of light, water, and nutrients by its competing neighbors. Different photoreceptors sensing different spectral regions of light may also help discriminate photoperiods. For example, it is known that the relative light spectral composition changes throughout a day: blue and far-red spectra are relatively more abundant in twilight, whereas the red spectrum is relatively more abundant in daylight (Hart 1988). Therefore, different photoreceptors acting in response to different spectra of light may provide a more accurate measurement of the day length, although it is not immediately obvious what adaptive advantages plants may have by possessing different photoreceptors that act antagonistically. One outcome of the antagonistic actions between phyB, phyA, and cry2 has been discovered recently in the control of protein stability of a flowering-time regulator, CO (*CONSTANS*), as discussed later. Interactions among different photoreceptors acting in different spectral ranges of solar radiation may also help plants to adapt to certain light conditions yet to be recognized. It will be interesting to examine whether such antagonistic interactions between phytochromes and cryptochromes are also present in plant species other than Arabidopsis.

Mechanisms

Photoreceptors may exert a different effect on light regulation of reproductive development in plants through their roles in the regulation of photosynthetic gene expression, metabolite partitioning, nutrient uptake and distribution, and hormone biosynthesis. However, the question of how light regulates flowering time has been traditionally focused on its role in sensing the change of day length. How plants "recognize," "remember," and respond to day-length changes have challenged plant biologists for the last 80 years or so. Among various hypotheses based on early physiological studies, the external coincidence model has gained most of the experimental support in recent years (Thomas and Vince-Prue 1997). According to this hypothesis, photoperiodism is governed by two interacting mechanisms: one controlled by the circadian clock and the other regulated by the photoreceptors (Yanovsky and Kay 2003). The circadian clock is entrained according to environmental signals such as light and temperature. Phytochromes and cryptochromes are apparently the major photoreceptors mediating light entrainment of the circadian clock in plants (Somers et al 1998). The role of light in the photoperiodic flowering is more than the entrainment of the clock. It is the interactions between the circadian clock-dependent processes called photoperiodic response rhythm (PRR) and the photoreceptor-dependent reactions independent of the clock that allow plants to distinguish a long day from a short day and to trigger or suppress floral initiation. The molecular nature of the PRR and how PRR interact with the photoreceptor-regulated reactions have remained elusive until recently (also see Chapters 39 and 41 by Izawa and Somers, respectively). Several studies have demonstrated that the photoperiod-dependent circadian rhythm of mRNA expression of the flowering-time gene *CO* and the photoreceptor-dependent light regulation of CO protein level form a basis for the external coincidence mechanism underlying photoperiodism in Arabidopsis (Figure 2). It is now clear that the expression of certain flowering-time genes such as *FT* is, at least partially, controlled by light regulation of the amount of CO protein (Valverde et al 2004). The CO protein is ubiquitinated and degraded by the 26S proteosome in darkness, but CO protein is relatively stable in white light. Analysis of the CO protein in the photoreceptor mutants demonstrates that cry1/cry2 and phyA stabilize CO protein in response to blue light and far-red light, respectively, and that phyB promotes CO degradation in red light. As described previously, phyB mediates red light suppression of flowering, whereas cry2 and phyA mediate blue and far-red light promotion, respectively (Figure 1).

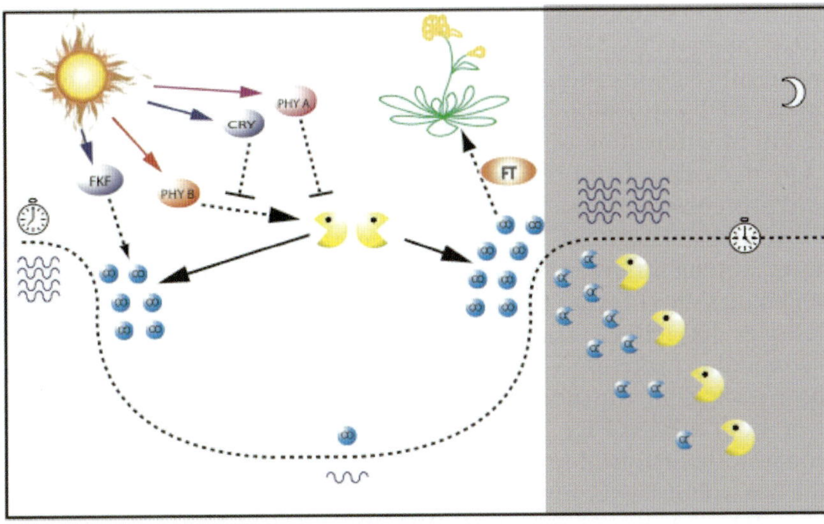

FIG. 2. Photoreceptors and the circadian clock exert functional interaction in the regulation of cellular level of CO protein. *CO* mRNA levels (depicted by the number of *wavy lines*) is regulated by the circadian clock, which is entrained via the action of phytochromes and cryptochromes. The peak of the circadian rhythm of *CO* mRNA expression (*dashed curve*) runs from the late afternoon to the early morning. CO protein levels (depicted by the number of *spheres*) are determined by not only its mRNA expression, but also protein degradation by the proteosome. CO degradation is promoted by phyB (*oval and arrow*), but inhibited by cryptochromes and phyA during the day. During the night, the CO mRNA level remains high, but little CO protein accumulates due to proteolysis. During long days depicted, phyA and cryptochromes help maintain higher CO protein level to promote flowering. *Arrows* indicate stimulatory actions, *lines with bar-heads* represent inhibitory actions, and *dashed lines* suggest the involvement of additional proteins

The discovery of different roles of the three photoreceptors in the control of CO protein stability revealed an important mechanism by which photoreceptors regulate flowering time. The photoreceptor regulation of CO degradation, coupled with the photoperiodic response rhythms of *CO* transcription, enables plants to decrease the amount of CO protein in short days, and to gradually increase the level of CO protein when the day length gets longer. Similar mechanisms are probably also used by rice, a short-day plant, wherein CO acts as a transcription suppressor of FT, to inhibit flowering in long days (Hayama et al 2003).

Perspective

How photoreceptors mediate light regulation of CO protein stability is apparently one of the questions that remains to be answered. Genes that are known to be involved in the light regulation of FT expression but do not affect the expression of *CO* mRNA expression would likely play a role in the light regulation of CO degradation. *PFT1* (*PHYTOCHROME AND FLOWERING TIME 1*) is apparently a good candidate (Cerdan and Chory 2003). PFT1 mediates phyB regulation of FT expression independent from regulation of CO mRNA expression. It will be interesting to see whether *pft1* mutation affects CO protein stability. On the other hand, E3 ubiquitin ligase must play critical role in the ubiquitin/proteosome-mediated degradation of CO in darkness. E3 ubiquitin ligase is responsible for the substrate recognition of the ubiquitin-proteosome apparatus. Among different types of E3 ubiquitin ligase, the RNIG E3 and SCF (SKP1, Cullin, F-box) E3 are the most versatile families. The Arabidopsis genome encodes over 400 RING proteins, or 21 SKP1-like, 10 Cullin-like, and over 700 F-box-containing proteins (Vierstra 2003). It is not known what type of E3 ligase may be involved in the ubiquitination and degradation of CO in darkness. However, both a RING E3 protein (COP1) and an F-box protein (ZTL) have been found to be involved in protein degradation in the absence of light as well as in the control of flowering time (Osterlund et al 2000, Mas et al 2003). *COP1* encodes a RING-finger protein with WD-40 repeats that was originally identified in the constitutive photomorphogenesis mutant *cop1* (Deng et al 1989). Mutations in the *COP1* gene caused accelerated flowering, in addition to its well-known constitutive photomorphogenesis phenotypes. COP1 has been found to act as the E3 ubiquitin ligase in the proteosome-mediated degradation of the transcription factor HY5 in darkness (Osterlund et al 2000). COP1 may also be involved in the light-dependent degradation of photoreceptors such as cry2 and phyA (Shalitin et al 2002, Seo et al 2004). It will be interesting to find out whether COP1, ZTL, or related proteins might be involved in the degradation of CO.

References

Borthwick HA, Hendricks SB, Parker MW (1952) The reaction controlling floral initiation. Proc Natl Acad Sci USA 38: 929–934

Briggs WR, Huala E (1999) Blue-light photoreceptors in higher plants. Annu Rev Cell Dev Biol 15: 33–62

Brown JAM, Klein WH (1971) Photomorphogenesis in *Arabidopsis thaliana* (L.) Heynh, threshold intensity and blue-far-red synergism in floral induction. Plant Physiol 47: 393–399

Cashmore AR (1997) A cryptochrome family of photoreceptors. Plant Cell Environ 20: 764–767

Cerdan PD, Chory J (2003) Regulation of flowering time by light quality. Nature 423: 881–885

Deng XW, Caspar T, Quail PH (1989) COP1: a regulatory locus involved in light-controlled development and gene expression in Arabidopsis. Genes Dev 5: 1172–1182

Devlin PF, Patel SR, Whitelam GC (1998) Phytochrome E influence internode elongation and flowering time in Arabidopsis. Plant Cell 10: 1479–1488

Devlin PF, Robson PR, Patel SR, Goosey L, Sharrock RA, Whitelam GC (1999) Phytochrome D acts in the shade-avoidance syndrome in Arabidopsis by controlling elongation growth and flowering time. Plant Physiol 119: 909–915

Franklin KA, Praekelt U, Stoddart WM, Billingham OE, Halliday KJ, Whitelam GC (2003) Phytochromes B, D, and E act redundantly to control multiple physiological responses in Arabidopsis. Plant Physiol 131: 1340–1346
Garner WW, Allard HA (1920) Effect of the relative length of day and night and other factors of the environment on growth and reproduction in plants. J Agric Res 18: 553–606
Guo H, Yang H, Mockler TC, Lin C (1998) Regulation of flowering time by Arabidopsis photoreceptors. Science 279: 1360–1363
Hart JW (1988) Light and plant growth. Unwin Hyman, London
Hayama R, Yokoi S, Tamaki S, Yano M, Shimamoto K (2003) Adaptation of photoperiodic control pathways produces short-day flowering in rice. Nature 422: 719–722
Imaizumi T, Tran HG, Swartz TE, Briggs WR, Kay SA (2003) FKF1 is essential for photoperiodic-specific light signalling in Arabidopsis. Nature 426: 302–306
Koornneef M, Alonso-Blanco C, Peeters AJM, Soppe W (1998) Genetic control of flowering time in Arabidopsis. Annu Rev Plant Physiol Plant Mol Biol 49: 345–370
Lin C (2000) Photoreceptors and regulation of flowering time. Plant Physiol 123: 39–50
Lin C, Shalitin D (2003) Cryptochrome structure and signal transduction. Annu Rev Plant Biol 54: 469–496
Mas P, Kim WY, Somers DE, Kay SA (2003) Targeted degradation of TOC1 by ZTL modulates circadian function in *Arabidopsis thaliana*. Nature 426: 567–570
Meijer G (1959) The spectral dependence of flowering and elongation. Acta Bot Neerl 8: 189
Mockler TC, Guo H, Yang H, Duong H, Lin C (1999) Antagonistic actions of Arabidopsis cryptochromes and phytochrome B in the regulation of floral induction. Development 126: 2073–2082
Mockler T, Yang H, Yu X, Parikh D, Cheng YC, Dolan S, Lin C (2003) Regulation of photoperiodic flowering by Arabidopsis photoreceptors. Proc Natl Acad Sci USA 100: 2140–2145
Mouradov A, Cremer F, Coupland G (2002) Control of flowering time: interacting pathways as a basis for diversity. Plant Cell 14 suppl:S111–S130
Nagy F, Schäfer E (2002) Phytochromes control photomorphogenesis by differentially regulated, interacting signaling pathways in higher plants. Annu Rev Plant Physiol Plant Mol Biol 53: 329–355
Osterlund MT, Hardtke CS, Wei N, Deng XW (2000) Targeted destabilization of HY5 during light regulated development of Arabidopsis. Nature 405: 462–466
Parker MW, Hendricks SB, Borthwick HA, Scully NJ (1945) Action spectrum for photoperiodic control of floral initiation in Biloxi soybean. Science 102: 152–155
Parker MW, Hendricks SB, Borthwick HA, Scully NJ (1946) Action spectrum for the photoperiodic control of floral initiation of short-day plants. Bot Gaz 108: 1–26
Quail PH (2002) Phytochrome photosensory signalling networks. Nat Rev Mol Cell Biol 3: 85–93
Reed JW, Nagpal P, Poole DS, Furuya M, Chory J (1993) Mutations in the gene for the red/far-red light receptor phytochrome B alter cell elongation and physiological responses throughout Arabidopsis development. Plant Cell 5: 147–157
Seo HS, Watanabe E, Tokutomi S, Nagatani A, Chua NH (2004) Photoreceptor ubiquitination by COP1 E3 ligase desensitizes phytochrome A signaling. Genes Dev 18: 617–622
Shalitin D, Yang H, Mockler TC, Maymon M, Guo H, Whitelam GC, Lin C (2002) Regulation of Arabidopsis cryptochrome 2 by blue-light-dependent phosphorylation. Nature 417: 763–767

Somers DE, Devlin PF, Kay SA (1998) Phytochromes and cryptochromes in the entrainment of the Arabidopsis circadian clock. Science 282: 1488–1490
Thomas B, Vince-Prue D (1997) Photoperiodism in plants. Academic, New York
Valverde F, Mouradov A, Soppe W, Ravenscroft D, Samach A, Coupland G (2004) Photoreceptor regulation of CONSTANS protein in photoperiodic flowering. Science 303: 1003–1006
Vierstra RD (2003) The ubiquitin/26S proteasome pathway, the complex last chapter in the life of many plant proteins. Trends Plant Sci 8: 135–142
Yanovsky MJ, Kay SA (2003) Living by the calendar: how plants know when to flower. Nat Rev Mol Cell Biol 4: 265–275

Chapter 39

Comparative Molecular Biology in Photoperiodic Flowering Between the Short-Day Plant Rice and the Long-Day Plant *Arabidopsis*

Takeshi Izawa

Introduction

Flowering plants are largely categorized into short-day and long-day plants. In some species, no floral response to photoperiods is observed, which creates another group, day-neutral plants. In 1920, Garner and Allard reported that many flowering plants recognize the day-length to determine flowering-time and set seeds at appropriate seasons (see reviews by Mouradov et al 2002, Simpson and Dean 2002). Recent studies in a short-day plant, rice, and a long-day plant, *Arabidopsis thaliana*, revealed that plants utilize an evolutionarily conserved flowering pathway to establish opposite photoperiodic responses (see reviews by Izawa et al 2003, Yanovsky and Kay 2003). Here I summarize recent progress on molecular mechanisms of photoperiodic flowering to overlook genetic players at molecular levels to confer both short-day and long-day responses in photoperiodic flowering.

Photoperiodic Photoreceptors

The first molecular genetic evidence on photoperiodic photoreceptors was reported using *fha* mutants in *Arabidopsis* (Guo et al 1998). In this report, it was shown that the *CRYPTOCHROME2* (*CRY2*) gene plays an important role to confer floral promotion under LD conditions in *Arabidopsis*. Although many physiological studies have suggested that phytochromes are photoperiodic photoreceptors in higher plants, phytochromes are thought to have supporting roles in photoperiodic flowering of *Arabidopsis*. The reason is because the long-day promotion pathway regulated by *CONSTANS (CO)* does not require *PHY-*

National Institute of Agrobiological Sciences, Plant Molecular Genetics Group, Tsukuba, Ibaraki 305-8602, Japan

TOCHROME A and *B* (*PHY A* and *B*) genes at least under white-light conditions. Recent studies indicate that cry2 modifies CO protein to be an activator of *FLOWERING LOCUS-T (FT)* gene expression (Kobayashi et al 1999, Kardailsky et al 1999, Samach et al 2000, Saurez-Lopez et al 2001, Yanovsky and Kay 2002). Under far-red light rich conditions, phyA functions similarly in *Arabidopsis*. Note that the long-day floral promotion by cry2 requires a red-light signaling, suggesting a role of phytochromes in the long-day floral promotion pathway of *Arabidopsis*. Recently, *Arabidopsis FLAVIN-BINDING, KELCH-REPEAT, F-BOX PRTEIN 1* (*FKF1*) gene was shown to be a blue light receptor involved in photoperiodic flowering (Imaizumi et al 2003).

In a short-day plant, rice, phytochromes mainly function as photoperiodic photoreceptors. A phytochrome-deficient mutant of rice, photoperiodic sensitivity 5 (se5), exhibits drastic early flowering regardless of photoperiods and completely loses the photoperiodic response, indicating that long-day inhibition of flowering requires phytochromes in rice (Izawa et al 2000). In addition, Heading date 1 (Hd1) gene, the rice ortholog of *CO*, promotes and inhibits flowering under SD and LD conditions, respectively (Yano et al 2000). Therefore, in rice, phytochromes can make *Hd1* a repressor of *FT-like* genes to inhibit flowering under LD (Izawa et al 2002, Kojima et al 2002).

Light Signals Interacted with the Circadian Clock in Photoperiodic Flowering

Many physiological data have supported a model, "the external coincidence model," in which the coincidence between acute light signals and photoinducible phases set by the circadian clock determines the photoperiodic responses (see reviews by Izawa et al 2003, Yanovsky and Kay 2003). The first evidence in molecular genetics to support this model comes from the work in rice. Izawa et al. (2002) demonstrated that the phytochrome-deficient mutant of rice, *se5*, which exhibits no response to photoperiods, does not show any differences in phase setting of the circadian clock under LD and SD, and free-running rhythms in LL and DD. Meanwhile *FT-like* genes are up-regulated in *se5* although the *Hd1* mRNA pattern does not change. Here, *Hd1* mRNA expression is regulated by the circadian clock. Therefore, it is likely that phytochromes modify circadian clock-regulated *Hd1* activity. The expression patterns of *FT-like* genes suggest that phytochromes may make *Hd1* a repressor at subjective night and this repression may be released around dawn under SD due to a decrease of Pfr phytochromes during darkness. Therefore, it is likely that interaction between Pfr and Hd1 protein makes a strong repressor for *FT-like* expression under LD. Based on *FT-like* gene expression in *se5* under LD and SD, it is likely that molecular mechanisms of *FT-like* gene expression and the involvement of phytochromes differ between LD and SD (Izawa et al unpublished). Under SD, there might be phase-specific gene activation of *FT-like* genes at two time points, before and after dawn, and antagonizing gene repression by phytochrome signals. In contrast, under LD, there might be a flat gene expression of *FT-like* genes observed, which is repressed by phytochrome signals. This suggests that no phase-dependent activation system may work under LD. It is possible that a strong repressor of *FT-like* genes is up-regulated by phytochromes under LD in rice. The molecular nature of this photoperiodic gene expression is still unclear.

The data to support the external coincidence model was subsequently reported in *Arabidopsis* (Yanovsky and Kay 2002, Imaizumi et al 2003). *CO* mRNA is basically expressed at the subjective night by the circadian clock (Suarez-Lopez et al 2001). The duration of *CO* mRNA expression is a little bit longer under LD than SD. Recent study indicates that this photoperiodic response of *CO* mRNA expression pattern requires *FKF1* photoreceptor gene (Imaizumi et al 2003). FKF1 protein is expressed at dusk by the circadian clock. Under LD, *FKF1* can mediate the external blue light signals to express *CO* mRNA at dusk, since there is a coincidence between FKF1 protein and the external light signals only at dusk under LD. This is one of the major reasons why CO protein is expressed at dusk under LD. It is unknown how the FKF1 protein, which contains an F-box, is involved in *CO* mRNA expression. CO protein is further activated at dusk under LD with light signals mediated by cry2 and phyA to induce *FT* mRNA (Yanovsky and Kay 2002). This is the second coincidence between the circadian clock output and the external light. These results indicate that there are multiple points to integrate information of phases of the circadian clock with the external light conditions into *FT* mRNA expression to recognize the day-length in *Arabidopsis* (Figure 1). Recently, it was reported that CO protein activation is regulated

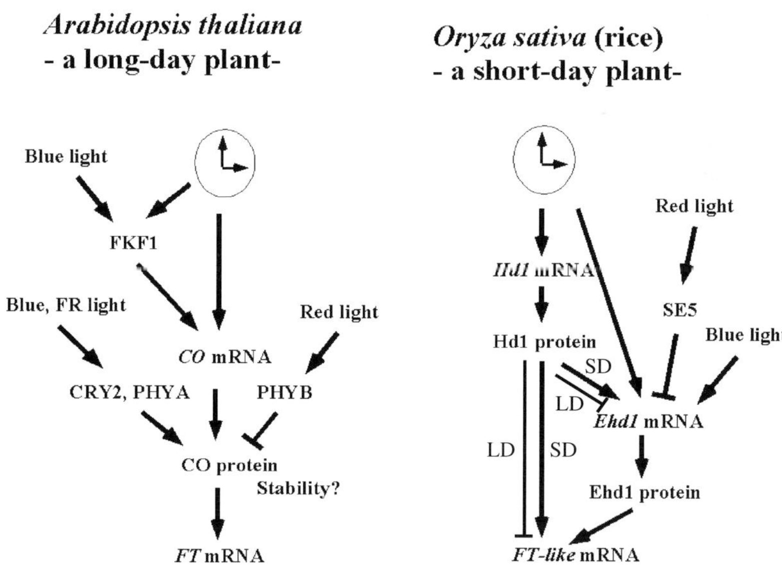

FIG. 1. Schematic models of photoperiodic pathway in *Arabidopsis* and rice. *Left*, A model in *Arabidopsis*. *Right*, A model in rice. *Hd1* and *FT-like* genes in rice are orthologous to *CO* and *FT* in *Arabidopsis*, respectively. The rice ortholog of *FKF1* exists in rice genome, but there is no report on it yet. In contrast, The *Arabidopsis* ortholog of *Ehd1* does not exist in the genome. *LD*, under long-day conditions; *SD*, under short-day conditions

by ubiquitin-mediated proteolysis, which is regulated by light/dark signals through phyB, phyA, and cry2 photoreceptors (Valverde et al 2004). Here, phyB is involved in degradation of CO protein in the nothing under SD and LD (Figure 1).

A Novel Phototransduction Pathway in Photoperiodic Flowering in Rice

Recently we have identified a novel gene, termed *Early heading date 1 (Ehd1)*, which plays a key role in the photoperiodic flowering of rice (Doi et al 2004). *Ehd1* encodes a B-type response regulator whose ortholog may not exist in *Arabidopsis*. In addition, in the steps of QTL cloning of *Ehd1*, we happened to demonstrate that *Ehd1* can function without the functional *Hd1* gene because Taichung 65, one of the parent cultivars, was identified to be deficient in both *Hd1* and *Ehd1*. This suggests that the *Ehd1* pathway can be independent from the conserved *Hd1/CO* photoperiodic flowering pathway. Furthermore, we have shown that *Ehd1* controls some *FT-like* and MADS-box gene expression, indicating that the *Ehd1* pathway is also integrated into the conserved *FT* (or its orthologs) gene expression. Since *Ehd1* may be controlled by the circadian clock, unidentified factors other than *Hd1* may mediate the circadian signals to control *Ehd1* mRNA.

We further demonstrated that the functional *Hd1* is able to increase the *Ehd1* expression under SD conditions using another cultivar with a functional *Hd1* allele and its nearly isogenic line which contains a non-functional *Hd1* allele. Therefore, *Hd1* also can mediate the phase information of the circadian clock into *Ehd1* gene expression. We next examined the *Ehd1* expression in *se5* and found that *Ehd1* is largely up-regulated by the *se5* mutation both under LD and SD. The *Ehd1* expression pattern characteristically differed in *se5* under LD compared with SD. Comparison of *FT-like* gene expression with *Ehd1* in *se5* suggests that the non-photoperiodic early flowering phenotypes observed in *se5* were largely given by the derepression of *Ehd1* mRNA expression. In addition, we have shown that *Ehd1* mRNA is induced only under blue light conditions and the early flowering phenotypes of *se5* require blue light. Taken together, both the several external light signals and the circadian clock-mediated signals are integrated into *Ehd1* expression, especially under SD, in rice (Izawa et al unpublished, Figure 1).

Summary

Integration between phase information of the circadian clock and the acute light signals into gene expression of floral inducer genes is the molecular basis of day-length recognition. The coincidence of these signals specific to certain photoperiods is a key for photoperiodic responses. Comparative biology in photoperiodic flowering between rice and *Arabidopsis* revealed both conserved and diverse floral pathways in molecular genetic webs. Understanding how opposite photoperiodic responses are constructed at the molecular levels is now blooming.

References

Doi K, Izawa T, Fuse T, Yamanouchi U, Kubo T, Shimatani Z, Yano M, Yoshimura A (2004) Ehd1, a B-type response regulator in rice, confers short-day promotion of flowering and controls FT-like gene expression independently of Hd1. Genes Dev 18: 926–936

Guo H, Yang H, Mockler TC, Lin C (1998) Regulation of flowering time by Arabidopsis photoreceptors. Science 279: 1360–1363

Imaizumi T, Tran HG, Swartz TE, Briggs WR, Kay SA (2003) FKF1 is essential for photoperiodic-specific light signalling in Arabidopsis. Nature 426: 302–306

Izawa T, Oikawa T, Tokutomi S, Okuno K, Shimamoto K (2000) Phytochromes confer the photoperiodic control of flowering in rice (a short-day plant). Plant J 22: 391–399

Izawa T, Oikawa T, Sugiyama N, Tanisaka T, Yano M, Shimamoto K (2002) Phytochrome mediates the external light signal to repress FT orthologs in photoperiodic flowering of rice. Genes Dev 16: 2006–2020

Izawa T, Takahashi Y, Yano M (2003) Comparative biology comes into bloom: genomic and genetic comparison of flowering pathways in rice and Arabidopsis. Curr Opin Plant Biol 6: 113–120

Kardailsky I, Shukla VK, Ahn JH, Dagenais N, Christensen SK, Nguyen JT, Chory J, Harrison MJ, Weigel D (1999) Activation tagging of the floral inducer FT. Science 286: 1962–1965

Kobayashi Y, Kaya H, Goto K, Iwabuchi M, Araki T (1999) A pair of related genes with antagonistic roles in mediating flowering signals. Science 286: 1960–1962

Kojima S, Takahashi Y, Kobayashi Y, Monna L, Sasaki T, Araki T, Yano M (2002) Hd3a, a rice ortholog of the Arabidopsis FT gene, promotes transition to flowering downstream of Hd1 under short-day conditions. Plant Cell Physiol 43: 1096–1105

Mouradov A, Cremer F, Coupland G (2002) Control of flowering time: interacting pathways as a basis for diversity. Plant Cell 14 suppl:S111–S130

Samach A, Onouchi H, Gold SE, Ditta GS, Schwarz-Sommer Z, Yanofsky MF, Coupland G (2000) Distinct roles of CONSTANS target genes in reproductive development of Arabidopsis. Science 288: 1613–1616

Simpson GG, Dean C (2002) Arabidopsis, the Rosetta stone of flowering time? Science 296: 285–289

Suarez-Lopez P, Wheatley K, Robson F, Onouchi H, Valverde F, Coupland G (2001) CONSTANS mediates between the circadian clock and the control of flowering in Arabidopsis. Nature 410: 1116–1120

Valverde F, Mouradov A, Soppe W, Ravenscroft D, Samach A, Coupland G (2004) Photoreceptor regulation of CONSTANS protein in photoperiodic flowering. Science 303: 1003–1006

Yano M, Katayose Y, Ashikari M, Yamanouchi U, Monna L, Fuse T, Baba T, Yamamoto K, Umehara Y, Nagamura Y, Sasaki T (2000) Hd1, a major photoperiod sensitivity quantitative trait locus in rice, is closely related to the Arabidopsis flowering time gene CONSTANS. Plant Cell 12: 2473–2484

Yanovsky MJ, Kay SA (2002) Molecular basis of seasonal time measurement in Arabidopsis. Nature 419: 308–312

Yanovsky MJ, Kay SA (2003) Living by the calendar: how plants know when to flower. Nat Rev Mol Cell Biol 4: 265–275

CHAPTER 40

The Photoperiodic Control of Flowering in Rice, a Short-Day Plant

KO SHIMAMOTO and SHUJI YOKOI

Introduction

Regulation of flowering is one of the most important processes of plants since it is closely related to the success of reproduction. Flowering time is controlled by a number of environmental factors such as day length, temperature, and water supply. Among them, the photoperiod is a key regulator of flowering and has been studied for many years (Figure 1). However, molecular genetic study of the photoperiodic regulation of flowering began only about 10 years ago. *Arabidopsis thaliana*, a long-day plant, has been extensively used to study the photoperiodic regulation of flowering, and a large number of genes involved in flowering time determination have been isolated and characterized. As a result we now understand the genes involved in day length control of flowering in a long-day plant relatively well. However, our knowledge on genes involved in flowering in short-day plants has been limited mainly because many model short-day plants have not been amenable to molecular genetic analysis until recently.

Rice is a short-day plant, and is becoming an increasingly important model monocot plant for molecular biological study because of the advance in the genome sequencing and its easiness in production of transgenic plants. Moreover, flowering time is one of the most important agricultural traits of rice, and rice breeders all over the world have been developing rice varieties whose flowering times are most suitable for the area where they are cultivated. Because of its importance as a breeding character, extensive genetic studies have been done on its photoperiodic flowering for a long time. However, molecular genetic study on the photoperiodic control of flowering began only recently. In this review we focus on recent progress in the identification of genes involved in flowering time control in rice.

Laboratory of Plant Molecular Genetics, Nara Institute of Science and Technology, 8916-5 Takayama, Ikoma 630-0101, Japan

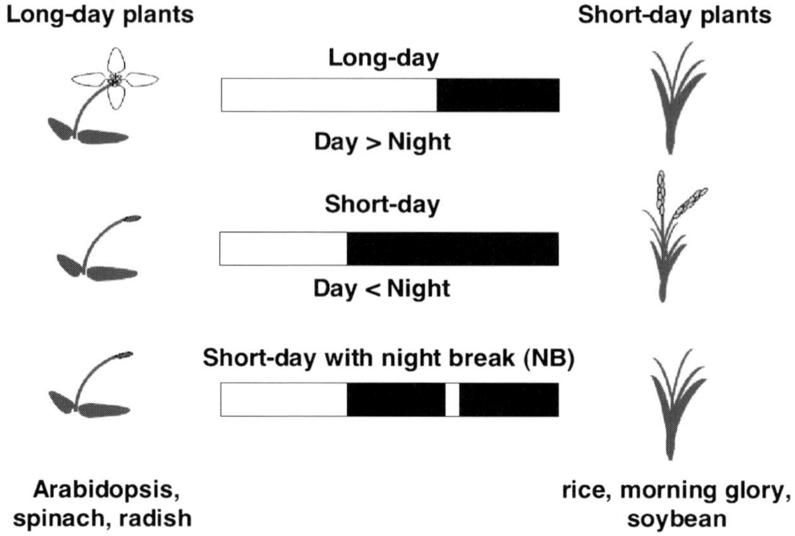

FIG. 1. Regulation of flowering in long-day plants (*LDPs*) and short-day plants (*SDPs*) and night break (*NB*)

Genes Involved in the Major Pathway of the Photoperiodic Regulation of Flowering in Rice

Three genes which constitute a major genetic pathway in the photoperiodic regulation of flowering in rice have recently been isolated (Figure 2). *OsGI*, an ortholog of *Arabidopsis GI*, *Hd1*, an ortholog of *Arabidopsis CO*, and *Hd3a*, an ortholog of *Arabidopsis FT* are shown to form the main pathway for the photoperiodic regulation of flowering in rice. *OsGI* was isolated as a gene whose mRNA was suppressed in the *se5* mutant which is insensitive to the photoperiod by a differential display method (Hayama et al 2002). *OsGI* has high homology with *GI* and contains a nuclear localization signal. *Hd1* was isolated by map-based cloning after the identification by quantitative trait locus (QTL) analysis (Yano et al 2000) and is closely related to *Arabidopsis CO*, which plays a central role in the photoperiodic control of flowering (Putterill et al 1995, Suarez-Lopez et al 2001). Two motifs, zinc finger motif and CCT motif, are highly conserved in *Hd1* and *CO*. *Hd3a* was similarly isolated by map-based cloning after the identification by QTL analysis (Kojima et al 2002). *Hd3a* and *FT* are highly conserved and both function as activators of flowering under inductive conditions (Kardailsky et al 1999, Kobayashi et al 1999).

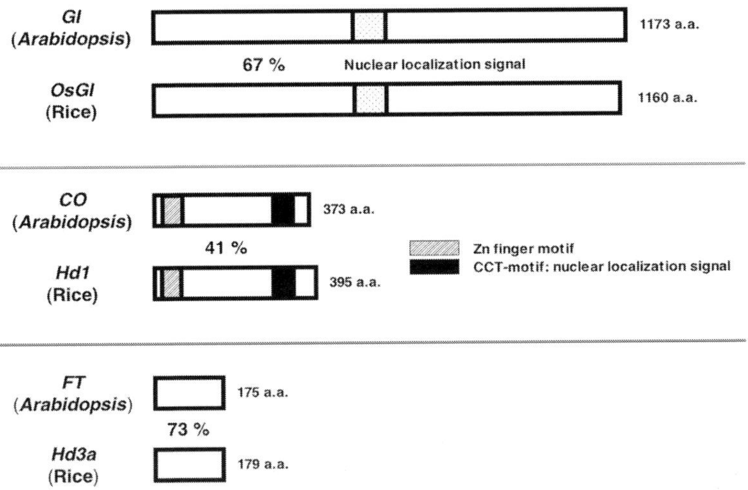

FIG 2. Structures and conservation of key genes involved in the photoperiodic control of flowering in rice and *Arabidopsis*

The *OsGI-Hd1-Hd3a* Pathway as a Main Pathway in the Regulation of Flowering in Rice

Genetic studies on flowering time in rice have generated a large body of data on genes involved in rice (Yano et al 2001). *Hd1* has been shown to be a positive regulator of flowering under short-day (SD) conditions (Yano et al 2000). Furthermore, *Hd3a* was also shown to be a positive regulator of flowering in SD (Kojima et al 2002). By the analysis of transgenic rice plants overexpressing *OsGI* or suppressing *OsGI* by RNAi, it was shown that *OsGI* acts as an activator of *Hd1* in both SD and long-day (LD) (Hayama et al 2003). However, *OsGI*, hence *Hd1* as well, function as negative regulators of flowering in LD by suppressing *Hd3a* (Hayama et al 2003). These results indicate that in rice, a major genetic pathway for the photoperiodic regulation of flowering is *OsGI–Hd1–Hd3a* and this pathway is conserved in rice and *Arabidopsis* (Figure 3). Therefore, these results provide an important hypothesis that the difference between rice, a short-day plant, and *Arabidopsis*, a long-day plant, resides in regulation of *Hd3a/FT* by *Hd1/CO*: *CO* activates *FT*, but *Hd1* suppresses *Hd3a* in LD. As described above, *Hd1* also activates *Hd3a* in SD. Thus, *Hd1* has a dual role in regulating *Hd3a* depending on the photoperiod.

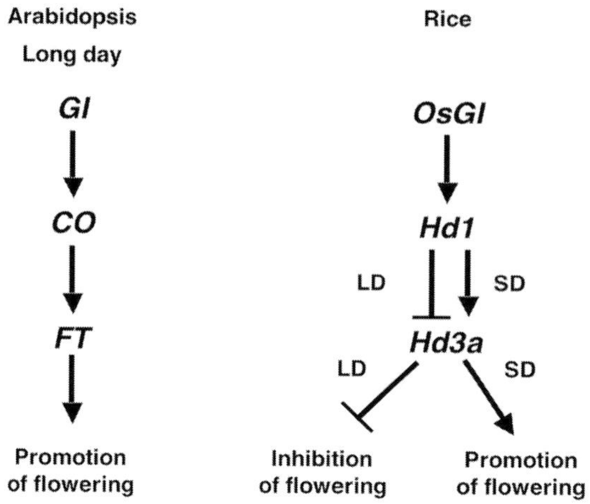

FIG. 3. The photoperiodic pathways for flowering in *Arabidopsis* and rice

Other Genes Involved in the Photoperiod Pathway in Rice

Hd6 was identified by QTL analysis and shown to encode the α-subunit of protein kinase CK2 (Takahashi et al 2001). It causes late flowering in LD. The *Arabidopsis* CK2 was shown to be involved in the circadian rhythm (Sugano et al 1998), therefore it is also conserved in rice and *Arabidopsis*. Genes which have been isolated as key regulators of the photoperiodic regulation of flowering are well conserved between rice and *Arabidopsis*. Therefore, one question arises: are there any rice-specific genes involved in regulation of flowering? Recently *Ehd1* was isolated by map-based cloning and shown to encode a B-type response regulator (Doi et al 2004). *Ehd1* acts to promote flowering in SD by activating expression of *Hd3a* and its homologs. Its interesting feature is that it activates downstream genes independently of *Hd1*, suggesting the presence of a branched pathway for promotion of flowering in SD. Although there are other genes genetically identified by QTL analysis, they have not been molecularly isolated. There is a good possibility that genes which are unique to rice flowering will be discovered in the near future.

Genes Involved in Phytochrome Function

The *se5* mutation is completely insensitive to the photoperiod and flowers very early in LD. The *SE5* gene was shown to encode an enzyme for biosynthesis of chromophore required for phytochromes (Izawa et al 2000). For detailed discussion of this section please see Chapter 39 by Izawa.

Other Genes Involved in Flowering in Rice

Recently a putative ortholog of *Arabidopsis SOC1/AGL20*, which plays an important role in flowering time determination (Lee et al 2000, Samach et al 2000), was isolated in rice and characterized (Tadege et al 2003, Lee et al 2004). Its overexpression causes early flowering in *Arabidopsis* (Tadege et al 2003). Furthermore, T-DNA insertion mutant of *OsSOC1* caused late flowering in rice, suggesting that it is an activator of flowering in rice (Lee et al 2004). These results suggest that *OsSOC1* is involved in regulation of flowering time in rice its relationship with the main *OsGI-Hd1-Hd3a* pathway remains to be studied.

Arabidopsis LEAFY plays an important role in flowering (Weigel et al 1992). Rice ortholog of *LEAFY*, *RFL*, has been isolated and the analysis of transgenic *Arabidopsis* indicates that it has conserved as well as diverged function (Kyozuka et al 1998, Chujo et al 2003). However, whether *RFL* plays a role in control of flowering time remains to be studied.

Three Pathways for the Photoperiodic Regulation of Flowering in Rice

In rice, flowering is induced by SD and suppressed by LD. Thus, SD promotion pathway and LD suppression pathway exist in rice. However, some rice varieties are induced to flower in LD although it takes longer time than SD. Therefore, there is a third pathway in rice which is LD promotion pathway. Pathways for LD suppression and LD promotion can be distinguished by developmental stage of plants. Long-day suppression continues from early stage to the entire growth stage. Then the LD promotion pathway becomes activated at later stage and it is independent of the LD suppression pathway. Therefore in LD, both the promotion pathway and the suppression pathway function simultaneously in the late developmental stage. As described above, the *OsGI–Hd1–Hd3a* pathway constitutes a main pathway for the regulation of flowering in rice and it forms both SD promotion pathway and LD suppression pathway depending on the regulation of *Hd3a* by *Hd1*. Actions of these two pathways are also developmentally regulated. The SD promotion pathway becomes activated at early stage while the LD suppression pathway function throughout the growth stage. The molecular nature of the third pathway, namely the LD promotion pathway, has not been identified yet. It will be of great interest to identify components of this pathway.

Night Break as a Tool to Study Molecular Mechanism of Regulation of Photoperiodic Flowering in Rice

Night break (NB) is a phenomenon in which a short exposure of light in the night causes inhibition of flowering in short-day plants (Thomas and Vince-Prue 1997). This phenomenon was first discovered in the study of the photoperiodic control of flowering in a short-day plant, *Xanthium* (Hamner and Bonner 1938). The night break effect was shown to be controlled by phytochrome in the very early stage of phytochrome study (Borthwick et al 1952). Although NB is clear in many short-day plants, its effect on long-day plants is not always clear and when its effect is found it requires longer time of light exposure (Thomas and Vince-Prue 1997). Numerous studies on NB have been performed in various short-day plants such as soybean, morning glory, and chrysanthemum. These studies identified a number of important findings on the photoperiodic regulation of flowering. First, the NB study showed that the circadian clock may be involved in the photoperiodic flowering. For instance, when the effect of NB was examined with long night in certain conditions the second weak peak of the effect is observed ca. 24h from the first peak which is usually detected at 7–9h of the 12–16h night. Second, it has been shown that there are two distinct effects of light in the photoperiodic flowering; one is to entrain the circadian clock and the second is the direct inhibitory effect in short-day plants. Lumsden and Furuya (1986) showed in morning glory that light fluences required for the effect on the circadian clock as revealed by the phase shift of and those on direct inhibition of flowering were different. This and other similar studies provide support for the external coincidence model which explains the mechanism of the photoperiodic control of flowering (Bünning 1936, Pittendrigh and Minis 1964).

Although a number of important findings in the photoperiodic flowering have been obtained from NB experiments, only few studies have dealt with NB at the molecular level. The main reason for the lack of molecular studies on NB is that the molecular biological and genomic tools are not well developed in those species which have been extensively used for the NB studies. Morning glory is one such example. We recently initiated an NB study using rice and established conditions under which a clear NB effect on flowering is observed. Since the rice genome is almost completely sequenced and transformation is easy, rice may be an ideal plant to study the NB effect on flowering at the molecular level. Night break study in rice would give interesting findings on the photoperiodic regulation of flowering in the future.

Summary and Conclusion

Studies on the photoperiodic regulation of flowering in rice have recently advanced and rice is becoming a model short-day plant. A number of genes involved in flowering time determination in rice have been isolated by various methods. One striking conclusion from these molecular studies is a remarkable conservation of genes which play important roles in the regulation of flowering time in rice and *Arabidopsis*. The major difference between rice, a SD plant, and *Arabidopsis*, a LD plant, was shown to be the regulation of *Hd3a/FT* by *Hd1/CO*. Under LD conditions, this regulation is positive in *Arabidopsis* while negative in rice, thus making LD and SD plants. Recent studies indicate that there are three pathways for the day length control of flowering in rice: (1) the SD activation pathway, (2) the LD suppression pathway, and (3) the LD activation pathway. Some genes are used in multiple pathways and others are pathway specific. They are also differentially regulated depending on developmental stages. Rice may be an ideal plant to study the NB effect on flowering. In the near future we hope to better understand the regulation of flowering in rice at the molecular level.

References

Borthwick HA, Hendricks SB, Parker MW (1952) The reaction controlling floral initiation. Proc Natl Acad Sci USA 38: 929–934

Bünning E (1936) Die endogene Tagesrhythmik als Grundlage der photoperiodischen Reaktion. Ber Deut Bot Ges 54: 590–607

Chujo A, Zhang Z, Kishino H, Shimamoto K, Kyozuka J (2003) Partial conservation of *LFY* function between rice and *Arabidopsis*. Plant Cell Physiol 44: 1311–1319

Doi K, Izawa T, Fuse T, Yamanouchi U, Kubo T, Shimatani Z, Yano M, Yoshimura A (2004) *Ehd1*, a B-type response regulator in rice, confers short-day promotion of flowering and controls *FT*-like gene expression independently of *Hd1*. Genes Dev 18: 926–936

Hamner KC, Bonner J (1938) Photoperiodism in regulation to hormones as factors in floral initiation and development. Bot Gaz 100: 388–431

Hayama R, Izawa T, Shimamoto K (2002) Isolation of rice genes possibly involved in the photoperiodic control of flowering by a fluorescent differential display method. Plant Cell Physiol 43: 494–504

Hayama R, Yokoi S, Tamaki S, Yano M, Shimamoto K (2003) Adaptation of photoperiodic control pathways produces short-day flowering in rice. Nature 422: 719–722

Izawa T, Oikawa T, Tokutomi S, Okuno K, Shimamoto K (2000) Phytochromes confer the photoperiodic control of flowering in rice (a short-day plant). Plant J 22: 391–399

Kardailsky I, Shukla VK, Ahn JH, Dagenais N, Christensen SK, Nguyen JT, Chory J, Harrison MJ, Weigel D (1999) Activation tagging of the floral inducer *FT*. Science 286: 1962–1965

Kobayashi Y, Kaya H, Goto K, Iwabuchi M, Araki T (1999) A pair of related genes with antagonistic roles in mediating flowering signals. Science 286: 1960–1962

Kojima S, Takahashi Y, Kobayashi Y, Monna L, Sasaki T, Araki T, Yano M (2002) *Hd3a*, a rice ortholog of the Arabidopsis *FT* gene, promotes transition to flowering downstream of *Hd1* under short-day conditions. Plant Cell Physiol 43: 1096–1105

Kyozuka J, Konishi S, Nemoto K, Izawa T, Shimamoto K (1998) Down-regulation of RFL, the FLO/LFY homolog of rice, accompanied with panicle branch initiation. Proc Natl Acad Sci USA 95: 1979–1982

Lee H, Suh SS, Park E, Cho E, Ahn JH, Kim SG, Lee JS, Kwon YM, Lee I (2000) The AGAMOUS-LIKE 20 MADS domain protein integrates floral inductive pathways in Arabidopsis. Genes Dev 14: 2366–2376

Lee S, Kim J, Han JJ, Han MJ, An G (2004) Functional analyses of the flowering time gene OsMADS50, the putative SUPPRESSOR OF OVEREXPRESSION OF CO 1/AGAMOUS-LIKE 20 (SOC1/AGL20) ortholog in rice. Plant J 38: 754–764

Lumsden PJ, Furuya M (1986) Evidence for two actions of light in the photoperiodic induction of flowering in *Pharbitis nil*. Plant Cell Physiol 27: 1541–1551

Pittendrigh CS, Minis DH (1964) The entrainment of circadian oscillations by light and their role as photoperiodic clocks. Am Nat 108: 261–295

Putterill J, Robson F, Lee K, Simon R, Coupland G (1995) The *CONSTANS* gene of *Arabidopsis* promotes flowering and encodes a protein showing similarities to zinc finger transcription factors. Cell 80: 847–857

Samach A, Onouchi H, Gold SE, Ditta GS, Schwarz-Sommer Z, Yanofsky MF, Coupland G (2000) Distinct roles of CONSTANS target genes in reproductive development of Arabidopsis. Science 288: 1613–1616

Suarez-Lopez P, Wheatley K, Robson F, Onouchi H, Valverde F, Coupland G (2001) *CONSTANS* mediates between the circadian clock and the control of flowering in *Arabidopsis*. Nature 410: 1116–1120

Sugano S, Andronis C, Green RM, Wang ZY, Tobin EM (1998) Protein kinase CK2 interacts with and phosphorylates the Arabidopsis circadian clock-associated 1 protein. Proc Natl Acad Sci USA 95: 11020–11025

Tadege M, Sheldon CC, Helliwell CA, Upadhyaya NM, Dennis ES, Peacock WJ (2003) Reciprocal control of flowering time by *OsSOC1* in transgenic *Arabidopsis* and by *FLC* in transgenic rice. Plant Biotech J 1: 361–369

Takahashi Y, Shomura A, Sasaki T, Yano M (2001) Hd6, a rice quantitative trait locus involved in photoperiod sensitivity, encodes the alpha subunit of protein kinase CK2. Proc Natl Acad Sci USA 98: 7922–7927

Thomas B, Vince-Prue D (1997) Photoperiodism in plants. Academic, London

Weigel D, Alvarez J, Smyth DR, Yanofsky MF, Meyerowitz EM (1992) LEAFY controls floral meristem identity in Arabidopsis. Cell 69: 843–859

Yano M, Katayose Y, Ashikari M, Yamanouchi U, Monna L, Fuse T, Baba T, Yamamoto K, Umehara Y, Nagamura Y, Sasaki T (2000) *Hd1*, a major photoperiod sensitivity quantitative trait locus in rice, is closely related to the Arabidopsis flowering time gene *CONSTANS*. Plant Cell 12: 2473–2484

Yano M, Kojima S, Takahashi Y, Lin H, Sasaki T (2001) Genetic control of flowering time in rice, a short-day plant. Plant Physiol 127: 1425–1429

Chapter 41

ZEITLUPE and the Control of Circadian Timing

David E. Somers

Introduction

The 24-h timing system known as the circadian clock regulates a wide range of physiological and developmental processes in plants. For this system to be relevant to the plant a close connection to the environment is essential. The entrainment of the central circadian pacemaker requires that the circadian system be responsive to the photocycles and thermocycles that the plant experiences each day with the rising and setting of the sun. This information is conveyed to the mechanism of a central oscillator, which sets the pace of the circadian system. One or more signaling pathways lead outward from the oscillator to control the period and phase of the various clock-controlled processes. At a molecular level the oscillator(s) is an autoregulatory transcription/translation feedback system, with one or more interlocked molecular loops (Young and Kay 2001).

Good progress has been made in recent years in identifying the photoreceptors that convey the daily changes in light intensity and quality to the circadian clock (Devlin 2002, Somers et al 1998). The phytochromes and cryptochromes are clearly important in controlling circadian period and phase, and now the *ZEITLUPE* (*ZTL*) gene family has emerged as a novel set of proteins that may fuse photoperception with protein degradation.

ZTL Gene Family

The founding member of the *ZTL* gene family was first identified in a screen for period length mutants in Arabidopsis (Somers et al 2000). *ztl-1* has an endogenous free-running period of 27h period, 3h longer than wild type. All circadian outputs thus far tested are lengthened by this mutation, though some rhythms

Department of Plant Cellular and Molecular Biology/Plant Biotechnology Center, Ohio State University, Columbus, OH 43210, USA

are more strongly affected than others (Somers et al 2000). A null mutation (*ztl-3*) shows the same phenotype, confirming that long period is the loss-of-function phenotype of *ZTL* (Jarillo et al 2001, Somers et al 2004). A graded series of enhanced expression lines of ZTL show increasingly shorter periods correlating with increasingly higher levels of ZTL, terminating in arrhythmicity at extreme levels of expression (Somers et al 2004).

The two additional family members, *LKP2* and *FKF1*, are closely related at the amino acid level (Somers 2001). LKP2 causes arrhythmicity when over-expressed, suggesting a role similar to ZTL (Schultz et al 2001). FKF1 loss-of-function mutants have no effect on circadian period, but cause a loss of photoperiod sensitivity and late flowering (Imaizumi et al 2003, Nelson et al 2000).

Structure: Protein Domains

The three proteins are comprised of three distinct domains, each of which has been well characterized in other protein contexts. Near the amino-terminus is a LOV/PAS domain. This region shares a large number of highly conserved amino acids with the LOV domains of phototropin (Arabidopsis) and WHITE-COLLAR 1 (WC-1; Neurospora). The phototropins are a small family of blue-light photoreceptors that mediate a number of light-induced growth and movement responses, most notably phototropism (Briggs and Christie 2002). The phototropin LOV domain binds flavin mononucleotide (FMN), which confers blue-light dependent kinase activity to the molecule. The homologous domain in WC-1 also binds flavin (FMN) and facilitates blue-light dependent DNA binding (Froehlich et al 2002, He et al 2002). Recently, the bacterially expressed LOV domain of FKF1 has been shown to bind FMN in vitro (Cheng et al 2003, Imaizumi et al 2003). Additionally, the FKF1 LOV domain can, in part, functionally substitute for the LOV domain in WC-1 (Cheng et al 2003), and can undergo light-induced changes in absorbance in vitro (Imaizumi et al 2003). Similar absorbance results were obtained with the isolated ZTL and LKP2 LOV domains, providing evidence that this gene family may constitute a novel type of blue-light photoreceptor.

The F-box and six kelch domains lie carboxy-terminal to the LOV motif, and are likely to act together as a biochemical unit in a manner well described for other F-box proteins. The Skp1-Cullin-F-box protein (SCF) complex is a type of E3 ubiquitin ligase that promotes proteasome-dependent protein degradation (Deshaies 1999). The role of the F-box protein is to specifically interact with a target protein and facilitate its physical positioning within the larger complex, thus allowing the transfer of the ubiquitin moiety to the target, marking it for destruction. One protein–protein interaction domain of F-box proteins lies C-terminal to the F-box and is often a WD40 or leucine-rich repeat (Willems et al 1999). This region confers the target protein specificity, and in the ZTL family the kelch repeats play this role. They are predicted to fold into a six-bladed β-propeller, with each of the 6 repeats forming one blade of the final structure (Adams et al 2000). Functional evidence for this role comes from the finding that the *ztl-1* mutation is an aspartate to asparagine replacement in the second kelch repeat, which eliminates target protein binding (Mas et al 2003, Somers et al 2000).

In yeast, the F-box motif interacts with the Skp1 protein to connect the target to the larger SCF ligase complex, and the Arabidopsis Skp-like proteins 1 and 2 (ASK1 and ASK2) can complex with ZTL in planta (Han et al 2004). This association remains if the LOV domain is deleted (F-kelch protein), indicating that this motif is not required for SCF participation. Consistent with this is our finding that overexpression of F-kelch protein causes arrhythmicity, similar to the effect of high expression of full length ZTL (Somers et al 2004). Deletion of the kelch domain (LOV-F protein) eliminates ASK1 in co-immunoprecipitations, but cullin is still detected, suggesting that other ASK associate with this truncated form of ZTL (Han et al 2004). Co-immunoprecipitation of ZTL and ASK1 is abrogated when missense mutations are introduced into the F-box motif in the context of the full-length protein, confirming the necessity of this motif. Taken together, all current evidence indicates that ZTL forms a functional SCF complex (SCF^{ZTL}) in planta.

Regulation

Often components of the central oscillator are rhythmically expressed, either at the level of mRNA or protein abundance, or both (Young and Kay 2001). *ZTL* mRNA is constitutively expressed in light/dark cycles and under constant conditions (Kim et al 2003, Somers et al 2000). However, strong oscillations of ZTL protein occur in 12 h light/12 h dark cycles, with minimum levels near Zeitgeber time 1 (ZT 1; 1 h after lights-on) and maximum levels at ZT 13–14 (1–2 h after lights-off). This 3- to 4-fold oscillation of ZTL damps to constitutively high levels in constant light (LL) and to constitutively low levels in constant darkness (DD) (Kim et al 2003). Further testing of ZTL protein half-life showed that degradation rates are highest at times of low ZTL abundance (i.e., ZT 1) and lowest when levels are at maximum (ZT 13). This oscillation in ZTL degradation rate persists in constant light. This demonstrates a clock-regulated phase-dependant proteolysis of ZTL, which itself is part of the proteolytic degradation pathway of the circadian system. Additionally, ZTL degradation is mediated by the proteasome, implicating some form of E3 ubiquitin ligase in its own proteolysis (Kim et al 2003). It is possible that ZTL is degraded along with its target, within the context of the SCF complex, by moving as larger complex to the proteasome. However, strong overexpression of a known target substrate (TOC1) does not detectably alter the diurnal rhythmic changes of ZTL abundance (Han et al 2004). ZTL also may be regulated by a separate, currently unknown, E3 ligase.

Localization

A common underlying theme in the transcription-translation feedback loop paradigm of the clock is a temporal alternation in the nuclear and cytoplasmic position of clock components (Roenneberg and Merrow 2001). Components of the proteasome and SCF E3 ligases can be found in the cytoplasm and nucleus, and in some cases nuclear localization/exclusion of F-box proteins can serve to regulate development (Blondel et al 2000). *ZTL(LKP1)* promoter–GUS fusions show widespread expression of ZTL throughout the plant body, including roots, with only minor expression in dry seed, hypocotyls and petioles (Kiyosue and Wada 2000). These results fit with data indicating functional autonomy of the circadian system in individual plant cells (Kim et al 2003, Thain et al 2000). Intracellular localization of ZTL using GFP–ZTL fusion proteins shows ZTL present in both the cytoplasm and the nucleus (Mason and Somers unpublished). Immunoblots of cell-fractionated whole plant extracts also show ZTL present in both the cytoplasmic and nuclear fractions (Kim and Somers, unpublished). To date we have not been able to detect a circadian-phase dependence on the intracellular localization of ZTL. These results allow that ZTL may be active in both cellular compartments, with different substrates or cofactors partnering with it in the two regions.

Function

The intriguing combination of a LOV domain prefixed to an F-box protein immediately raises the novel possibility of a light-regulated ubiquitin-dependent proteolysis of one or more oscillator components (Somers et al 2000, Somers 2001). Two obvious candidates are the myb transcription factors, CCA1 and LHY. Message and protein levels of both genes cycle robustly and peak levels are in the early subjective day, anti-phasic to ZTL. Overexpression of either causes arrhythmicity, and the single mutants shorten period. The *cca1 lhy* double mutant causes very rapid damping of circadian cycling, suggesting some degree of functional redundancy between the two (Mizoguchi et al 2002). However, yeast two-hybrid assays and attempts to co-immunoprecipitate ZTL and CCA1/LHY have been unsuccessful in demonstrating a molecular interaction between the two (Kim and Somers, unpublished). In addition, CCA1 levels are greatly reduced in *ztl* mutants, in contrast to the expectation of higher levels if ZTL facilitated CCA1 degradation (Somers et al 2004). Therefore, all available evidence suggests that CCA1/LHY levels are not directly regulated by ZTL.

As with *CCA1* and *LHY*, mutations in *TOC1* shorten circadian period. *CCA1/LHY* negatively regulate *TOC1* expression, and TOC1 positively controls *CCA1/LHY* message levels. These data have led to a very simple model of feedback regulation, whereby the central oscillator is comprised, in part, of CCA1/LHY binding to the *TOC1* promoter to inhibit *TOC1* expression, and of TOC1 positively upregulating *CCA1/LHY* gene expression through a currently unknown mechanism (Alabadi et al 2001, Young and Kay 2001). *TOC1* mRNA and protein levels are clock-controlled and peak near the middle of the subjective night, anti-phasic to CCA1/LHY, and consistent with the negative feedback model. As the TOC1 expression pattern is very similar to ZTL, it was interesting to find that their loss-of-function mutant phenotypes were opposite: *ztl* mutations lengthen circadian period and *toc1* mutations shorten period. Under these circumstances a direct interaction between the two would only be likely if one negatively regulated the activity of the other. Yeast two-hybrid assays do demonstrate a strong interaction between TOC1 and ZTL, although FKF1 and LKP2 can also activate transcription with TOC1 as bait in these assays. More compelling is the ability to co-immunoprecipitate TOC1 and ZTL from plant extracts, and the demonstration that TOC1 levels rise to constitutively high levels at all circadian times in the *ztl* null backgrounds (Mas et al 2003). Additionally, the *ztl-1* mutation in the kelch domain abrogates TOC1 interaction, consistent with the postulated role of the kelch region as the target interaction domain. When considered with the evidence that ZTL can interact with ASK and pull down other core components of the SCF complex (e.g., cullin and Rbx1), these results identify TOC1 as a substrate of the SCF^{ZTL} E3 ubiquitin ligase. However, still lacking is a direct demonstration of TOC1 phosphorylation or ubiquitination.

The mechanistic details of how the SCF^{ZTL} E3 ligase controls TOC1 levels are still unclear. High levels of ZTL should correlate with low TOC1 accumulation if SCF^{ZTL} E3 ligase activity is only dependent on ZTL abundance. The co-ordinate peak expression of the two proteins suggests that ZTL activity towards TOC1 is modulated by other factors. Some insight may come from the fact that in extended LL TOC1 and ZTL cycling damps to high levels, and in DD both proteins decrease in abundance (Kim et al 2003, Mas et al 2003). This suggests that light acts to stabilize both proteins, and conversely, darkness accelerates their degradation. However, the co-ordinate regulation of both proteins may not be causal. *ztl-1* protein continues to oscillate in LD, despite the absence of TOC1 interaction (Somers et al 2004), and conversely, ZTL levels in the wild type also oscillate normally in LD when TOC1 is strongly constitutively overexpressed (Kim and Somers unpublished). These results suggest that ZTL abundance is not dependent on TOC1, as has been suggested for other F-box protein/target substrate interactions (Li et al 2004).

ZTL may be constitutively active in extended darkness, leading to a drop in TOC1 abundance. In support of this notion, TOC1 levels remain high in DD in the *ztl-1* mutant. In turn, light may inactivate ZTL or its ability to interact with TOC1, allowing the constitutive rise in TOC1 levels in constant light. However, in LD cycles where the light and dark periods are only 12 h, a different relationship between the two proteins and illumination exists, as both are low during the first part of the photoperiod and rise to peak levels in the dark. Only during the second half of the night do TOC1 levels drop. If ZTL is dark-activated, some light-dependent aspect of ZTL function near the end of the light period is able to delay its targeting of TOC1 for the proteasome until some hours into the night. Alternatively, TOC1 may be recalcitrant to ZTL action at certain circadian times, possibly through phase-dependent phosphorylation. Other known F-box protein substrates require phosphorylation of the target proteins for successful interaction (Deshaies 1999). Overexpression of CKB3, the regulatory subunit of casein kinase 2 (CK2), shortens period, consistent with *TOC1* loss of function. However, the same phenotype is associated with *cca1* and *lhy* mutations, and these proteins can be phosphorylated by CK2 in vitro (Sugano et al 1998, Sugano et al 1999). There is no evidence for TOC1 as a substrate for CK2.

At present it is not clear what delays the drop in TOC1 levels while ZTL levels are at their peak. High resolution tracking of temporal changes in the ZTL/TOC1 interaction over the circadian cycle in different light environments may be one approach to this question.

Remaining Questions

ZTL has emerged as a key intermediary in the process of phototransduction to the clock. The presence of a flavin-binding LOV domain now suggests that it may be included along with the phytochromes and cryptochromes as a photoreceptive molecule that can alter the pace of the central oscillator. Full resolution of this possibility requires the development of a light-dependent in vitro assay for ZTL activity. Many other questions still remain.

Does ZTL act in planta as a molecular partner with the crys or phys, as some in vitro data suggest (Jarillo et al 2001), or does it act independently and parallel or further downstream of those receptors? These questions are being addressed genetically through *ztl* double mutant combinations with the *phy* and *cry* photoreceptors.

Are one or more ASK proteins dedicated to the SCF^{ZTL} complex? The proven ZTL interactions with ASK1 and ASK2 are not reflected by any changes in period in the *ask1* or *ask2* mutants (Han et al 2004). This suggests some functional redundancy among the more than 20 ASK proteins, and that other, unidentified, ASKs may be playing a more central role in the SCF^{ZTL} complex.

Is TOC1 the sole ubiquitination target of ZTL? *TOC1* belongs to a five-member gene family of pseudo-response regulators (*PRR*s) that are rhythmically expressed in sequentially staggered phases over a 12h time period (Somers 2001). One or more additional members of the TOC1 family may be targeted by SCF^{ZTL}, in addition to the possibility of other, unknown, clock components. Ongoing characterization of double *ztl toc1* mutants, and screens for suppressors of *ztl-1* are beginning to address these questions.

References

Adams J, Kelso R, Cooley L (2000) The kelch repeat superfamily of proteins: propellers of cell function. Trends Cell Biol 10: 17–24

Alabadi D, Oyama T, Yanovsky MJ, Harmon FG, Mas P, Kay SA (2001) Reciprocal regulation between TOC1 and LHY/CCA1 within the Arabidopsis circadian clock. Science 293: 880–883

Blondel M, Galan JM, Chi Y, Lafourcade C, Longaretti C, Deshaies RJ, Peter M (2000) Nuclear-specific degradation of Far1 is controlled by the localization of the F-box protein Cdc4. EMBO J 19: 6085–6097

Briggs WR, Christie JM (2002) Phototropins 1 and 2: versatile plant blue-light receptors. Trends Plant Sci 7: 204–210

Cheng P, He Q, Yang Y, Wang L, Liu Y (2003) Functional conservation of light, oxygen, or voltage domains in light sensing. Proc Natl Acad Sci USA 100: 5938–5943

Deshaies RJ (1999) SCF and Cullin/Ring H2-based ubiquitin ligases. Annu Rev Cell Dev Biol 15: 435–467

Devlin PF (2002) Signs of the time: environmental input to the circadian clock. J Exp Bot 53: 1535–1550

Froehlich AC, Liu Y, Loros JJ, Dunlap JC (2002) White Collar-1, a circadian blue light photoreceptor, binding to the frequency promoter. Science 297: 815–819

Han L, Mason M, Risseeuw EP, Crosby WL, Somers DE (2004) Formation of an SCF complex is required for proper regulation of circadian timing. Plant J 40: 291–301

He Q, Cheng P, Yang Y, Wang L, Gardner KH, Liu Y (2002) White Collar-1, a DNA binding transcription factor and a light sensor. Science 297: 840–843

Imaizumi T, Tran HG, Swartz TE, Briggs WR, Kay SA (2003) FKF1 is essential for photoperiodic-specific light signalling in Arabidopsis. Nature 426: 302–306

Jarillo JA, Capel J, Tang RH, Yang HQ, Alonso JM, Ecker JR, Cashmore AR (2001) An Arabidopsis circadian clock component interacts with both CRY1 and phyB. Nature 410: 487–490

Kim WY, Geng R, Somers DE (2003) Circadian phase-specific degradation of the F-box protein ZTL is mediated by the proteasome. Proc Natl Acad Sci USA 100: 4933–4938

Kiyosue T, Wada M (2000) LKP1 (LOV kelch protein 1): a factor involved in the regulation of flowering time in Arabidopsis. Plant J 23: 807–815

Li Y, Gazdoiu S, Pan ZQ, Fuchs SY (2004) Stability of homologue of Slimb F-box protein is regulated by availability of its substrate. J Biol Chem 279: 11074–11080

Mas P, Kim WY, Somers DE, Kay SA (2003) Targeted degradation of TOC1 by ZTL modulates circadian function in *Arabidopsis thaliana*. Nature 426: 567–570

Mizoguchi T, Wheatley K, Hanzawa Y, Wright L, Mizoguchi M, Song HR, Carre IA, Coupland G (2002) LHY and CCA1 are partially redundant genes required to maintain circadian rhythms in Arabidopsis. Dev Cell 2: 629–641

Nelson DC, Lasswell J, Rogg LE, Cohen MA, Bartel B (2000) FKF1, a clock-controlled gene that regulates the transition to flowering in Arabidopsis. Cell 101: 331–340

Roenneberg T, Merrow M (2001) Circadian systems: different levels of complexity. Philos Trans R Soc Lond B Biol Sci 356: 1687–1696

Schultz TF, Kiyosue T, Yanovsky M, Wada M, Kay SA (2001) A role for LKP2 in the circadian clock of Arabidopsis. Plant Cell 13: 2659–2670

Somers DE (2001) Clock-associated genes in Arabidopsis: a family affair. Philos Trans R Soc Lond B Biol Sci 356: 1745–1753

Somers DE, Devlin PF, Kay SA (1998) Phytochromes and cryptochromes in the entrainment of the Arabidopsis circadian clock. Science 282: 1488–1490

Somers DE, Schultz TF, Milnamow M, Kay SA (2000) *ZEITLUPE* encodes a novel clock-associated PAS protein from *Arabidopsis*. Cell 101: 319–329

Somers DE, Kim WY, Geng R (2004) The F-box protein ZEITLUPE confers dosage-dependent control on the circadian clock, photomorphogenesis, and flowering time. Plant Cell 16: 769–782

Sugano S, Andronis C, Green RM, Wang ZY, Tobin EM (1998) Protein kinase CK2 interacts with and phosphorylates the Arabidopsis circadian clock-associated 1 protein. Proc Natl Acad Sci USA 95: 11020–11025

Sugano S, Andronis C, Ong MS, Green RM, Tobin EM (1999) The protein kinase CK2 is involved in regulation of circadian rhythms in *Arabidopsis*. Proc Natl Acad Sci USA 96: 12362–12366

Thain SC, Hall A, Millar AJ (2000) Functional independence of circadian clocks that regulate plant gene expression. Curr Biol 10: 951–956

Willems AR, Goh T, Taylor L, Chernushevich I, Shevchenko A, Tyers M (1999) SCF ubiquitin protein ligases and phosphorylation-dependent proteolysis. Philos Trans R Soc Lond B Biol Sci 354: 1533–1550

Young MW, Kay SA (2001) Time zones: a comparative genetics of circadian clocks. Nat Rev Genet 2: 702–715

Part VIII
Epilogue

Chapter 42

Epilogue: Eighteen Years of Progress in Photomorphogenesis

Winslow R. Briggs

Science has a way of progressing in giant steps. Between these steps one may find extended periods of consolidation and exploitation of the findings and a rush to incorporate them into the main body of knowledge. The excitement generated by the discoveries often drives an intense competition between laboratories that consolidates them and lays the groundwork for the next giant steps.

The field of photomorphogenesis is no different from any other fields of science. After almost a century and a half of photophysiology, culminating in some careful quantitative action spectroscopy, the first plant photoreceptor, eventually designated phytochrome, was isolated in 1959 (Butler et al 1959), clearly a giant step. However, it took another 23 years before phytochrome A was purified in an unambiguously undegraded form (Vierstra and Quail 1983, Litts et al 1983). Meanwhile we learned a great deal about phytochrome structure and properties, and two years thereafter, we obtained the first phytochrome amino acid sequence (Hershey et al 1985). The years between 1959 and 1985 were mostly devoted to expanding our knowledge of the biochemical properties of this unique photoreceptor.

At the end of this period, the Tobin laboratory had just demonstrated conclusively by run-on transcription that transformation of Pr to Pfr led to changes in gene transcription (see Tobin and Silverthorne 1985), a second giant step. Photomorphogenesis was entering into the molecular era and another explosion of knowledge was about to take place.

The 16th Yamada Conference that Professor Masaki Furuya organized in 1986 occurred just after the second of these steps, but much too early to reflect any of the consolidation and exploitation that was to take place shortly thereafter. The conference also occurred before most workers in the field had discovered the extraordinary power of molecular genetics, and especially before they discovered the model plant, *Arabidopsis thaliana*. The contrast between what was current and exciting in 1986 and what was current and exciting in 2004 is truly dramatic. Let's compare the status of just a few of the areas that received intensive cover-

Department of Plant Biology, Carnegie Institution of Washington, 260 Panama Street, Stanford, CA 94305, USA

age in 2004, as reported at the 58th Yamada conference organized by Professor Masamitsu Wada, with their status at the time of the Yamada Conference in 1986.

At the 1986 Yamada Symposium, both Lee Pratt and Marie Michèle Cordonnier and Akira Nagatani and his colleagues presented clear evidence, based on monoclonal antibody studies, that there must be two biochemically distinct phytochromes present in oats and pea, respectively, something also shown by Tokuhisa et al (1985). Until that time, the word "phytochrome" referred to a single biochemical entity. For a while thereafter, "phytochrome" was used as a collective noun referring to a two-member family of chromoproteins. Clear differences in amino-acid sequence between "green" and "etiolated" phytochrome were already elucidated (Abe et al 1985), and the existence two biochemically distinct phytochromes was clearly established. By the time of the 2004 Yamada Conference, the existence of five phytochrome genes in *Arabidopsis* encoding polypeptides differing significantly in amino acid sequence, and of multiple phytochrome genes in other species was old news, a great deal was known about their properties and their role in photomorphogenesis, and Robert Sharrock was even able to present evidence of heterodimer formation between some of the *Arabidopsis* phytochromes.

Even more dramatic, however, is the progress that has been made through molecular genetics and the use of mutants. It seems hard to believe today but in the 1986 Yamada Conference proceedings the word *Arabidopsis* is nowhere to be found in the index (nor could this author find it anywhere in the text). Peter Quail and co-workers (16th Yamada Conference, Chapter II, 2) made the only mention of any kind of higher-plant mutant, the *aurea* mutant of pea, lacking spectrally and immunologically detectable phytochrome. They reported that the mutant contained normal levels of phytochrome mRNA, and attributed the low level of the chromoprotein to a mutation that produces a highly unstable phytochrome, subject to rapid turnover. By 2004, mutants with lesions in all five *Arabidopsis* phytochromes were available. Single, double, triple, and even quadruple phytochrome mutants had been characterized, and a great deal was known about the independent, antagonistic, and/or overlapping functions of the five phytochromes in photomorphogenesis. By 2004, it was also well established that phyA mediated both the incredibly photosensitive very low-fluence response to red light and the high-irradiance response to far red light whereas phyB was the major player in classic red–far-red-reversible responses and in the high-irradiance response to red light.

The focus of the 1986 Yamada Conference was almost entirely on phytochrome, and many of the speakers devoted their reports to its structure and biochemical properties. As mentioned above, the role of phytochrome in the regulation of gene expression, long proposed from physiological studies, had just been demonstrated for a very few genes, first in the Tobin laboratory (16th Yamada Conference, Chapter II, 3) and then by subsequent workers. These studies were only the first hints of the explosion of information emanating over the next eighteen years from the skillful use of molecular genetics. By 2004, gene-chip technology had identified hundreds of genes that were up- or down-regulated by phytochrome phototransformation to Pfr (Tepperman et al 2001). Furthermore, groups of genes could be clustered according the kinetics for the light-induced changes in their transcription, the extent of transcriptional change induced, and the biochemical or regulatory pathways in which many of their products were involved.

In 1986, virtually nothing could be said about the nature of the signal-transduction pathways activated by phytochrome phototransformation to Pfr, much less the nature of the components of these pathways at that time. Only a few laboratories were considering genetic approaches, plant molecular biology was still in its infancy, and workers in the field had yet to discover *Arabidopsis*. Contrast this situation with what we heard in 2004: Peter Quail was able to discuss mutants in downstream components of the various phytochrome (and cryptochrome) signal-transduction pathways that had allowed the identification and characterization of over 30 different *Arabidopsis* genes involved in phytochrome-mediated photomorphogenesis, and permitted a dramatic increase in our understanding of complex regulatory pathways involved, pathways undreamed of in 1986. The application of molecular genetics had produced another giant step forward.

At the time of the 16th Yamada Conference, phytochrome was regarded a soluble cytoplasmic protein, although there was physiological evidence that at least for some systems it might be membrane associated. However, by 2004, Eberhard Schäfer could report in detail on the dynamics of the redistribution of phytochrome-GFP from cytoplasm to nucleus in response to light signals and the remarkable formation of "speckles" within the nucleus, including rapid association of Pfr with the transcription factor PIF3. However, a report on the properties and functions of membrane-associated phytochrome must await the next Yamada Conference.

About the only known function of proteases in photomorphogenesis in 1986 was their role in phytochrome A Pfr degradation. Indeed a major difference between "etiolated plant phytochrome" and "green plant phytochrome" was that the former (phyA) was rapidly degraded as Pfr whereas the latter (phyB) was not. It was also an imperative to include proteolysis inhibitors during phytochrome purification as the earlier literature was littered with papers reporting the purification of what turned out to be proteolytically degraded phytochrome. The notion that proteolysis might play a role in any regulatory pathway, whether induced by light, hormone, stress, or insect or pathogen attack, was simply not considered. Yet by 2004 it was clear, as described by Xing-Wang Deng, that proteolysis, in the form of the targeting specific regulatory proteins for degradation, was of major consequence in photomorphogenesis. Likewise, David Somers was able to shed some insight into the role of so-called F-box proteins specifically involved in targeting other proteins for degradation in the regulation of circadian rhythms.

In 1986, it had been long established that circadian rhythms played an important role in flowering responses to day length (16th Yamada Conference, chapter IV, 7 by Daphne Vince-Prue and Atsushi Takimoto), and that phytochrome played an important role in setting the phase of the circadian rhythm in plants. However, the nature of the central oscillator was a complete mystery, and the erroneous notion prevailed that circadian rhythms were the exclusive property of eukaryotic cells. By 2004, however, Steve Kay was able to provide a detailed overview of the central oscillators as they occurred in higher plants, cyanobacteria, the ascomycete *Neurospora crassa*, the fly *Drosophila melanogaster*, and mammals. An external signal (for example light, acting through a photoreceptor) induces the transcription of a transcription factor that activates transcription of a second transcription factor that then represses transcription of the first factor. Interaction of these factors with upstream and downstream elements provides the mechanism by which circadian oscillations in countless processes are gated. Although the proteins in the various systems are completely different, the basic mechanism is identical and a great deal is known both about the various upstream and downstream elements, as David Somers reported for *Arabidopsis* and Takao Kondo reported for cyanobacteria. Again it was the creative use of molecular genetics that permitted this enormous progress, without question another major advance.

At the time of the 1986 Yamada Conference, the existence of blue-light receptors in plants had been known for over a century and a half. As early as 1817, Poggioli reported that light in the blue region of the visible spectrum was by far the most effective in inducing directional leaflet movements in the leguminous shrub *Mimosa pudica* (what we now call solar tracking) and oriented leaf growth in *Raphanus rusticanus*. Over the years an enormous literature on blue-light responses developed based on action spectroscopy and a wide range of physiological techniques both with higher and lower plants and fungi (see Senger and Briggs 1981, for example). However, the nature of what was frequently called **THE** blue light receptor was still unknown and arguments raged as to whether the chromophore was a flavin, a carotenoid, a pterin, or even a retinylidene homologue (Briggs and Iino 1983). With the phytochrome photoreceptor finally in hand and biochemically accessible (if not yet genetically accessible) it should come as no surprise that the 1986 conference focused almost entirely on phytochrome. However, there was one chapter by Edward Lipson (16th Yamada Conference, Chapter V, 4) devoted to the blue-light responses of the phycomycete *Phycomyces blakesleeanus*, long a favorite subject for photobiological research. It is interesting to note that Lipson was able to make more solid conclusions from the use of mutants in *Phycomyces* than Peter Quail could about the *aurea* mutant in pea! Although there was a second chapter on blue-light receptors, it didn't even evolve from a formal presentation, but rather was the summary of two somewhat disjointed round-table discussions (Horst Senger and Edward Lipson are to be congratulated in their "rejointing" these discussions in their chapter!).

By 2004, the situation with respect to blue-light receptors had changed dramatically. *Arabidopsis* mutants had enabled Ahmad and Cashmore to isolate and characterize the first of three *Arabidopsis* cryptochromes in 1993, and by 2004, the cyanobacterial homolog of one of them (cry3 or cry-DASH for ***Drosophila***,

Arabidopsis, ***Synecchocystis***, and ***Homo sapiens***) had been characterized by X-ray crystallography (Brudler et al 2003).

Anthony Cashmore was able to describe the interactions between the C-terminal and N-terminal domains of a cryptochrome in repressing cryptochrome responses and reviewed major work from his laboratory showing that light released this repression. Both Chentao Lin and Min Ni could elucidate some of the complexities of the signal transduction pathways both at the photoreceptor level (Lin) and downstream from the photoreceptor in cryptochrome-regulated responses (Lin and Ni), and Margaret Ahmad provided some insights into the molecular mechanism by which light might induce a biochemical change in cryptochrome to initiate cryptochrome-mediated responses.

Although directional leaf movements and growth in response to the direction of incident blue light were first noted by Poggioli in 1817, as mentioned above, it took over a century and three quarters to identify and characterize one of the two responsible photoreceptors phototropin 1 (Christie et al 1998). However, by 2004, it was already well known that there were two phototropins in *Arabidopsis*, phot1 and phot2, and that in addition to serving as photoreceptors for phototropism, they also mediated blue light-activated leaf expansion (Sakamoto and Briggs 2002), stomatal opening as discussed by Ken-ichiro Shimizaki, chloroplast movements as discussed by Masamitsu Wada, and the transient and rapid inhibition of stem elongation of etiolated seedlings (Folta and Spalding 2001). Wolfhart Rüdiger was able to describe in detail blue light-activated hierarchical autophosphorylation of phot1, highlighting the complex nature of this early biochemical response to blue light, and four reports (those Trevor Swartz, Satoro Tokutomi, Roberto Bogomolni, and Joachim Heberle) provided a detailed account of the early biophysical and biochemical consequences of phototropin photoactivation, a unique photoreaction involving formation of a cysteinyl adduct between phototropin cysteines and the C(4a) carbon of the flavin mononucleotide chromophores. More than one and three quarter centuries after Poggioli's initial observations, blue-light photobiology accomplished its own major steps and presented the field with two new families of photoreceptors—the cryptochromes and the phototropins.

This account of some of the changes occurring between the two Yamada Conferences in the field of photomorphogenesis is hardly complete as numerous workers making presentations at the 2004 conference workers reported great progress in areas not mentioned above. However, it does highlight some of the dramatic advances that occurred between the conferences over a remarkably short time, and indicates the strength, vigor, and promise of a rapidly advancing field. It must be mentioned here that some things don't change: Professor Masaki Furuya gave inspiring lectures at both Yamada Conferences.

References

Abe H, Yamamoto KT, Nagatani A, Furuya M (1985) Characterization of green tissue-specific phytochrome isolated immunochemically from pea seedlings. Plant Cell Physiol 26: 1387–1399

Ahmad M, Cashmore AR (1993) *HY4* gene of *A. thaliana* encodes a protein with the characteristics of a blue-light photoreceptor. Nature 366: 162–166
Butler WL, Norris KH, Siegelman HW, Hendricks SB (1959) Detection, assay, and preliminary purification of the pigment controlling photoresponsive development of plants. Proc Natl Acad Sci USA 45: 1703–1708
Briggs WR, Iino M (1983) Blue-light-absorbing photoreceptors in plants. Philos Trans R Soc Lond B 303: 347–359
Brudler R, Hitomi K, Daiyasu H, Toh H, Kucho K, Ishiura M, Kanehisa M, Roberts VA, Todo T, Tainer JA, Getzoff ED (2003) Identification of a new cryptochrome class: structure, function and evolution. Mol Cell 11: 59–67
Christie JM, Reymond P, Powell GK, Bernasconi P, Raibekas AA, Liscum E, Briggs WR (1998) *Arabidopsis* NPH1: A flavoprotein with the properties of a photoreceptor for phototropism. Science 282: 1698–1701
Folta KM, Spalding EP (2001) Unexpected roles for cryptochrome 2 and phototropin revealed by high-resolution analysis of blue light-mediated hypocotyl growth inhibition. Plant J 26: 471–478
Hershey HP, Barker RF, Idler KB, Lissemore JL, Quail PH (1985) Analysis of cloned cDNA and genomic sequences for phytochrome: complete amino acid sequences for two gene products expressed in etiolated *Avena*. Nucleic Acids Res 13: 8543–8559
Litts JC, Kelly JM, Lagarias JC (1983) Structure–function studies on phytochrome. Preliminary characterization of highly purified phytochrome from *Avena sativa* enriched in the 124 kilodalton species. J Biol Chem 258: 11025–11031
Poggioli S (1817) Della influenza che ha il raggio magnetico sulla vegetatione delle piante. Bologna—Coi Tipi di Annesio Nobili Opusc Scientif Fasc I, pp 9–23
Sakamoto K, Briggs WR (2002) Cellular and subcellular localization of phototropin 1. Plant Cell 14: 1723–1735
Senger H, Briggs WR (1981) The blue light receptor(s): primary reactions and subsequent metabolic changes. In: Smith KC (ed) Photochemical and photobiological reviews vol. 6, pp 1–38
Tepperman JM, Zhu T, Chang HS, Wang X, Quail PH (2001) Multiple transcription-factor genes are early targets of phytochrome A signaling. Proc Natl Acad Sci USA 98: 9437–9442
Tobin EM, Sliverthorne J (1985) Light regulation of gene expression in higher plants. Annu Rev Plant Physiol 36: 569–593
Tokuhisa JG, Daniels SM, Quail PH (1985) Phytochrome in green tissue: spectral and immunochemical evidence for two distinct molecular species of phytochrome in light-grown *Avena sativa* L. Planta 164: 321–332
Vierstra RD, Quail PH (1983) Purification and initial characterization of 124-kilodalton phytochrome. Biochemistry 22: 2498–2505

Subject Index